Physics of Thin Films

Advances in Research and Development

VOLUME 19

OPTICAL CHARACTERIZATION OF REAL SURFACES AND FILMS

Serial Editors

MAURICE H. FRANCOMBE
Department of Physics
The University of Pittsburgh
Pittsburgh, Pennsylvania

JOHN L. VOSSEN
John Vossen Associates
Technical and Scientific Consulting
Bridgewater, New Jersey

Contributors to This Volume

DAVID L. ALLARA

ILSIN AN

P. CHINDAUDOM

ROBERT W. COLLINS

B. DRÉVILLON

YOUMING LI

YIWEI LU

R. E. NEWNHAM

HIEN V. NGUYEN

ATUL N. PARIKH

S. TROLIER-MCKINSTRY

K. VEDAM

V. YAKOVLEV

Physics of Thin Films

Advances in Research and Development

OPTICAL CHARACTERIZATION OF REAL SURFACES AND FILMS

Guest Volume Editor

K. Vedam

*The Pennsylvania State University
Department of Physics and Materials Research Laboratory
University Park, Pennsylvania*

VOLUME 19

Academic Press
San Diego New York Boston
London Sydney Tokyo Toronto

This book is printed on acid-free paper. ⊚

COPYRIGHT © 1994 BY ACADEMIC PRESS, INC.

ALL RIGHTS RESERVED.
NO PART OF THIS PUBLICATION MAY BE REPRODUCED OR
TRANSMITTED IN ANY FORM OR BY ANY MEANS, ELECTRONIC
OR MECHANICAL, INCLUDING PHOTOCOPY, RECORDING, OR
ANY INFORMATION STORAGE AND RETRIEVAL SYSTEM, WITHOUT
PERMISSION IN WRITING FROM THE PUBLISHER.

ACADEMIC PRESS, INC.
A Division of Harcourt Brace & Company
525 B Street, Suite 1900, San Diego, California 92101

United Kingdom Edition published by
ACADEMIC PRESS LIMITED
24-28 OVAL ROAD, LONDON NW1 7DX

International Standard Serial Number: 0079-1970

International Standard Book Number: 0-12-533019-7

PRINTED IN THE UNITED STATES OF AMERICA
94 95 96 97 98 99   EB   9 8 7 6 5 4 3 2 1

Contents

Contributors ix
Preface xi

In Situ Studies of Crystalline Semiconductor Surfaces by Reflectance Anisotropy

B. Drévillon and V. Yakovlev

I. Introduction 2
II. The Reflectance Anisotropy Technique 3
   A. Reflectance Anisotropy and the Other Characterizations Based on Polarized Light 3
   B. RA Configuration Based on the Use of a Photoelastic Modulator 6
   C. Comparison With Other RA Configurations 9
   D. Nature of the RA Under Crystal Growth Conditions 10
III. Applications of RA to the Growth of III-V Semiconductors 11
   A. The Various Deposition Techniques 11
   B. Reflectance Anisotropy of GaAs and AlAs under MBE Conditions 12
   C. Applications of RA to Migration-Enhanced Atomic Layer Epitaxy (MEALE) 19
   D. Reflectance Anisotropy of GaAs Under Chemical Beam and Vacuum Chemical Epitaxy Growth Conditions 22
   E. Growth of Lattice-Matched Multilayered Structures by LP-MOCVD 29
   F. Reflectance Anisotropy of GaAs Under Atmospheric Pressure (AP) MOCVD Growth Environments 32
IV. Growth of Lattice-Mismatched III-V and II-VI Structures 38
V. Summary and Conclusions 43
   References 45

Real-Time Spectroscopic Ellipsometry Studies of the Nucleation, Growth, and Optical Functions of Thin Films, Part I: Tetrahedrally Bonded Materials

Robert W. Collins, Ilsin An, Hien V. Nguyen, Youming Li, and Yiwei Lu

I. Introduction	50
A. Real-Time Ellipsometry at a Single Photon Energy	50
B. Real-Time Spectroscopic Ellipsometry	52
II. Techniques of Real-Time Spectroscopic Ellipsometry	54
A. Instrumentation	55
B. Data Collection and Interpretation	56
III. Studies of Tetrahedrally Bonded Thin Films	61
A. Hydrogenated Amorphous Silicon	61
B. Hydrogenated Amorphous Silicon-Carbon Alloys	85
C. Microcrystalline Silicon	93
D. Diamond	105
IV. Summary	121
Acknowledgments	121
References	122

Real-Time Spectroscopic Ellipsometry Studies of the Nucleation, Growth, and Optical Functions of Thin Films, Part II: Aluminum

Hien V. Nguyen, Ilsin An, and Robert Collins

I. Introduction	127
II. Theoretical Background	130
A. The Optical Functions of Bulk Aluminum	131
B. Size Effects of the Optical Functions	133
C. The Generalized Effective Medium Theory	139
III. Experimental Apparatus for Real-Time Monitoring	147
IV. Results and Discussion	149
A. Overview of the Optical Functions of Aluminum Thin Films in the Nucleation and Growth Stages	150
B. Analysis of the Optical Functions of Continuous Films	162
C. Analysis of the Optical Functions of Particle Films	168
V. Summary	183
Acknowledgments	187
References	187

Optical Characterization of Inhomogeneous Transparent Films on Transparent Substrates by Spectroscopic Ellipsometry

P. Chindaudom and K. Vedam

I. Introduction	191
II. Inhomogeneity in Thin Films	193
III. Experimental	196
A. Spectroscopic Ellipsometry	196
B. RAE with a Compensator	198
C. Detector Non-Linearity	202
D. Data Analysis	205
E. Measurements on Vitreous Silica as a Test of Accuracy of our Modified SE System	207
IV. Experimental Results on Transparent Films	215
A. Particulars of Films Studied	215
B. Preliminary Results on MgO and LaF_3 Films	216
C. Results on Transparent Films On Vitreous Silica Substrates	220
V. Discussion	231
A. General	231
B. Effect of Surface Roughness	233
C. Optical Functions of the Dielectric Film Materials	234
IV. Summary and Conclusion	243
References	244

Characterization of Ferroelectric Films by Spectroscopic Ellipsometry

S. Trolier-McKinstry, P. Chindaudom, K. Vedam, and R. E. Newnham

I. Introduction	249
II. Description of the SE System	254
III. Experimental Procedure	256
IV. Results and Discussion	258
A. Epitaxial and Oriented Films on Single Crystal Substrates	258
B. *In-situ* Annealing Studies on Ferroelectric Films	264
IV. Relation Between Film Microstructure and Electrical Properties	272
VI. Conclusions	275
Acknowledgments	276
References	276

Effects of Optical Anisotropy on Spectro-Ellipsometric Data for Thin Films and Surfaces

Atul N. Parikh and David Allara

I. Introduction	279
II. The Generalized Approach	283
A. Definition of the Fundamental Electromagnetic Problem	283
B. Implementation of the Yeh 4 × 4 Transfer Matrix Method	290
III. Comparison with Available Data and Previous Methods	291
IV. Simulations of Anisotropic Effects	295
A. Construction of a Model Anisotropic Medium from an Ensemble of Oriented Oscillators	295
B. Single Interface Ambient-Substrate Systems	295
C. Multiple Interface, Ambient–Film–Substrate Systems	309
Acknowledgment	312
References	313
Author Index	315
Subject Index	323

Contributors

Numbers in parenthesis indicate the pages on which the authors' contribution begins.

DAVID L. ALLARA (279), Department of Materials Science and Engineering and Department of Chemistry, The Pennsylvania State University, University Park, Pennsylvania 16802

ILSIN AN (49, 127), Department of Physics and Materials Research Laboratory, The Pennsylvania State University, University Park, Pennsylvania 16802

P. CHINDAUDOM (191, 249), NECTECH, Ministry of Science Technology Environment, Rama IV, Bangkok 10400, Thailand

ROBERT W. COLLINS (49, 127), Department of Physical and Materials Research Laboratory, The Pennsylvania State University, University Park, Pennsylvania 16802

B. DRÉVILLON (1), Laboratoire de Physique des Interfaces et des Couches Minces, Ecole Polytechnique, 91128 Palaiseau, France

YOUMING LI (49), Department of Physics and Materials Research Laboratory, The Pennsylvania State University, University Park, Pennsylvania 16802

YIWEI LU (49), Department of Physics and Materials Research Laboratory, The Pennsylvania State University, University Park, Pennsylvania 16802

R. E. NEWNHAM (249), Materials Research Laboratory, The Pennsylvania State University, University Park, Pennsylvania 16802

HIEN V. NGUYEN (49, 127), Department of Physics and Materials Research Laboratory, The Pennsylvania State University, University Park, Pennsylvania 16802

ATUL N. PARIKH (279), Department of Materials Science and Engineering, The Pennsylvania State University, University Park, Pennsylvania 16802

S. TROLIER-MCKINSTRY (249), Materials Research Laboratory, The Pennsylvania State University, University Park, Pennsylvania 16802

K. VEDAM (191, 249), Department of Physics and Materials Research Laboratory, The Pennsylvania State University, University Park, Pennsylvania 16802

V. YAKOVLEV (1), Instruments SA Inc., Edison, New Jersey 08820

Preface

It is well known that *real* thin films of metals, semiconductors and insulators are in general far from homogeneous due to their internal microstructure which is strongly dependent on the preparatory conditions. Consequently the optical properties of the films were seldom reproducible until recently. To cite just one example, Professor Abeles in his extensive review article in this Series on the "Optical Properties of Metallic Films" [*Physics of Thin Films 6*, 151–204 (1971)] confined his attention only to the theoretical aspects and did not discuss the experimental results since "they are more subject to revision and modification in the near future" [*ibid* p. 152]. Further, the assumptions invoked in the interpretation of the experimental data, as for example whether the film is homogeneous or not, etc., also play a dominant role in the final results. This is brought out very clearly in the results of the Round Robin Experiments conducted by the Optical Society of America [Appl. Opt. *23*, 3571–3596 (1984)] in which seven laboratories from around the world were asked to determine the optical constants of Rh metal and Sc_2O_3 using nominally same thin films. It is interesting to observe from Tables 17 and 18 of this article that, in the case of Rh film all the workers assumed the Rh film to be homogeneous, whereas in the case of Sc_2O_3 film only three of the seven groups assumed it to be inhomogeneous. Further even in those three cases, they had to invoke an arbitrary assumption that the inhomogeneity in the film varies linearly with thickness. It is now evident that the assumptions in both the cases are not justified, since (i) the microroughness of the surface was totally ignored and (ii) microstructure and hence the void distribution in the film is seldom linearly dependent on the thickness.

Almost similar statements can be made to the case of surfaces as well. *Real* surfaces are in fact quite different from the ideally perfect surface which corresponds to the plane surface terminating an ideally perfect semi-infinite solid. Even if we ignore the cleavage steps, tear lines, dislocations and damaged regions or the imperfections introduced during the preparatory stage as well as the contaminant overlayer if any on the surface, the abrupt termination of the solid and the presence of the so-called dangling bonds on the outermost layer can cause the symmetry as well as the physical and chemical properties of this layer to be different from those of the bulk specimen.

During the last five to ten years two major breakthroughs have been made in the development of two optical techniques to characterize in real time and *in situ* the films and surfaces respectively. With the first one-Real Time Spectroscopic Ellipsometry (RTSE), we can now nondestructively and noninvasively determine the spectra of the relative changes in both amplitude and the very sensitive "phase" of the reflected light as a function of photon energy in a few milliseconds and thus measure and store such a library of spectra collected during entire film growth in real-time as well. With such a wealth of data we are now able to perform detailed regression analysis of these spectral data first at successive intervals of time and later also globally with the entire data, to finally obtain statistically meaningful results. In other words we are now for the first time able to obtain meaningful and reproducible results on (i) the morphological and/or the microstructural features of the inhomogeneities in the film and (ii) the true optical functions of film-materials. The second optical technique Reflectance Anisotropy (RA) can also nondestructively and noninvasively probe the surface of the growing crystal, to follow (a) the minute variations in the crystallographic symmetry of the growing surface layer during epitaxial film growth and thus (b) in its optical properties. Hence it is now appropriate for us to collect a review of these new optical techniques and the summary of the results obtained thus far, so that the scientists and engineers at large can benefit from this collection of reviews from the pioneers who developed these techniques. At the same time it will also enable us to (i) assess whether the full potentialities of these techniques have been realized or not, (ii) determine their limitations and deficiencies of these techniques and (iii) also point out the areas that need further work and/or development. The present volume in this series on the Physics of Thin Films aims to address these issues.

The first article by Drévillon and Yakovlev provides a critical evaluation of the extensive literature on "reflectance anisotropy" (RA) technique, a field in which Drévillon is one of the pioneers. This is a normal incidence optical probe that uses the reduced symmetry of the surface layer to enhance the typically low sensitivity of reflectance measurements to surface phenomena. Unlike the various electron beam based surface analytical techniques, RA is not limited to ultra high vacuum (uhv) environment and hence has been used successfully to study the growth processes of III-V semiconductors in various deposition conditions ranging from Molecular Beam Epitaxy (MBE) requiring uhv conditions, to metalorganic chemical vapor deposition (MOCVD) under atmospheric pressure H_2 environment. It is shown that in all these cases RA spectroscopy can follow the changes in the various surface reconstructions and also obtain real-time control of the chemical cycles of atomic layer epitaxy (ALE). Further in the case of low pressure MOCVD, RA transient measurements were utilized to control the deposition parameters to optimize the quality of the heterojunctions in GaInP/GaAs superlattice structures.

The second article authored by Collins and his students provides an excellent review of the experimental details, the physics involved and a summary of the results obtained on the growth of technologically important tetrahedrally-bonded films [such as hydrogenated amorphous silicon (a-Si:H), and silicon carbon alloys (a-$Si_{1-x}C_x$:H)] using the Real Time Spectroscopic Ellipsometry (RTSE), a technique pioneered by Collins. It is evident that single wavelength ellipsometry and/or reflectance and absorptance spectroscopy, even with real-time measurements cannot untangle the numerous processes that take part simultaneously during the film growth, such as (i) evolving microstructure (i.e., void volume fraction), (ii) changes in the chemical composition (particularly in the case of alloys), (iii) changes in the crystal structure (particularly in the case of thin metal films) and (iv) surface roughening (or smoothening) effect. On the other hand RTSE is shown to untangle all this and thus provides insights into the monolayer scale surface processes that control the ultimate properties of the material. As a direct result of these studies by RTSE with on line control of the deposition parameters it was possible to control the surface smoothening and obtain improved photoresponse and electronic performance of both these amorphous materials. Similar RTSE studies on the nucleation and growth of diamond films by heated-filament (1950°C) assisted CVD, (i.e., even under the most adverse conditions for any real-time optical studies), have enabled them to identify and overcome numerous problems such as tungsten contamination from the filament at the diamond/substrate interface etc., and grow excellent diamond films by on-line monitoring and control.

The third article by Hien, An and Collins deals with an important problem both from the physics and technological points of view, namely the optical properties of thin metallic films, a satisfactory understanding of which has eluded physicists for well over a century. The development of RTSE has enabled Collins and his group to follow the evolution of the optical function of Aluminum film as a continuous function of thickness throughout the nucleation, coalescence and bulk film regimes. Detailed analyses of their extensive data reveal that at the very early stages of the film growth, the film is composed of partially coalesced disordered particles with a constant electron mean free path λ of 7.5 ± 2 Å even when the film thickness is over the percolation threshold of 45–50 Å. The above value of

λ is found to be the same irrespective of the method of film preparation, or rate of film deposition or the size of the particle. As the film thickness increases to 55–60 Å, there is an abrupt transition when λ increases by an order of magnitude indicating conversion of the defective particles into high quality single-crystalline grains extending through the thickness of the film. For thicknesses above the transition, λ is found to increase gradually with thickness as would be expected for electron scattering at grain boundaries. All the published optical and electrical data in the literature on aluminum, can now be explained for the first time with this picture.

The fourth article in this volume by Chindaudom and Vedam deals with the nondestructive characterization of inhomogeneous transparent optical coatings on transparent substrates with the help of Spectroscopic Ellipsometry (SE). It is shown that the spectral variation of Δ, the relative change in "phase" in the reflected light, contains information on the inhomogeneities in the thin film, while the corresponding spectral variation of Ψ contains information on the optical function of the film material. Hence circularly polarized light was used as the incident beam in a rotating analyzer type SE system, which in turn made it possible to measure the spectral variation of Δ to a high degree of accuracy, even though Δ itself was close to 0°. Examination of a number of different fluoride and oxide optical coatings (deposited by electron-beam-evaporation) on vitreous silica substrate reveal that *all* these optical coatings were inhomogeneous due to the presence of microrough surface layer and/or a voided interface between substrate and film or inhomogeneous distribution of voids throughout the bulk of the film. Besides depth-profiling the film, SE characterization studies yield also the optical function i.e., the refractive index and its dispersion with wavelength, of the film material. Such data were not available in the literature for some of these materials.

The fifth article in this volume authored by Trolier-McKinstry and her colleagues discusses the characterization of transparent ferroelectric thin films by SE, as well as *in situ* annealing the as-deposited films. All ferroelectric thin films (lead based perovskite films deposited by rf magnetron sputtering, multi-ion-beam sputtering and sol-gel spin-on techniques) studied, display some level of inhomogeneity in the form of low density regions distributed through the thickness of the film and/or surface roughness. It is shown that many of the apparent size effects reported for ferroelectric thin films are probably associated with either poor crystallinity or defective microstructure, rather than intrinsic changes in the ferroelectric properties with film thickness. Results of SE studies on *in situ* annealing of deposited ferroelectric films are also discussed.

The sixth and final article by Parikh and Allara deals with a long outstanding problem in SE that has not been addressed satisfactorily until now, i.e., spectro-ellipsometry of anisotropic materials in its most generalized approach. This includes the material under consideration as optically biaxial, and optically absorbing; and it can be in the form of thin film or act as substrate with lossy overlayers which may or may not be anisotropic. The experimental variables considered are variable angles of incidence, wavelength range varying from 350 nm to 850 nm. Such a problem may appear too esoteric; but it is not far from the case of uniaxial Langmuir-Blodgett films on optically biaxial SbSI substrate. In fact Parikh and Allara have alluded to many such examples in which the effects of anisotropy do arise and play dominant role in ellipsometric measurements. Parikh and Allara have constructed the algorithms for the generalized approach mentioned above, tested the validity of the calculations with the few selected experimental observations and then consider some hypothetical models to generate the spectra of the ellipsometric parameters (Δ, Ψ) in three dimensional [wavelength, angle of incidence, and Δ (or Ψ)] space to gain physical insight into the variety of possible anisotropic effects.

In Situ Studies of Crystalline Semiconductor Surfaces by Reflectance Anisotropy

B. DRÉVILLON

*Laboratoire de Physique des Interfaces et des Couches Minces,
Ecole Polytechnique, Palaiseau, France*

AND

V. YAKOVLEV

*Instruments SA Inc.
Edison, New Jersey*

I. Introduction	2
II. The Reflectance Anisotropy Technique	3
A. Reflectance Anisotropy and the Other Characterizations Based on Polarized Light	3
B. RA Configuration Based on the Use of a Photoelastic Modulator	6
C. Comparison with Other RA Configurations	9
D. Nature of the RA under Crystal Growth Conditions	10
III. Applications of RA to the Growth of III–V Semiconductors	11
A. The Various Deposition Techniques	11
B. Reflectance Anisotropy of GaAs and AlAs under MBE Conditions	12
C. Applications of RA to Migration-Enhanced Atomic Layer Epitaxy (MEALE)	19
D. Reflectance Anisotropy of GaAs under Chemical Beam and Vacuum Chemical Epitaxy Growth Conditions	22
1. Surface Transformation between As- and Ga-Stabilized Surfaces	22
2. Growth Oscillations under CBE Conditions	26
3. *In Situ* Control of Surface V/III Ratio	27
E. Growth of Lattice-Matched Multilayered Structures by LP-MOCVD	29
F. Reflectance Anisotrophy of GaAs under Atmospheric Pressure (AP) MOCVD Growth Environments	32
1. Surface Reconstructions	32
2. AP Atomic Layer Epitaxy	35
IV. Growth of Lattice-Mismatched III–V and II–VI Structures	38
V. Summary and Conclusions	43
References	45

Copyright © 1994 by Academic Press, Inc.
All rights of reproduction in any form reserved.
ISBN 0-12-533019-7

I. Introduction

The measurement of surface anisotropy of semiconductors under normal incidence has recently emerged as a very attractive technique. The method has been termed reflectance difference spectroscopy (RDS) (*1–3*) or reflectance anisotropy (RA) (*4*) and reflectance anisotropy spectroscopy (RAS) (*5*). Both latter terms refer directly to the physical origin of the signal and will be preferred in this review. More generally, RA can be considered as a polarization-modulation technique, like ellipsometry and dichroïsm. RA is the determination of the difference between normal-incidence reflectances for light polarized along the two principal axes of the surface. However, as compared to the other polarized-light techniques, RA measurements can only be sensitive to the surface of a material, even though the light penetrates deep into the bulk. As a matter of fact, because of its lower symmetry, the surfaces of nominally isotropic crystals can display a measurable anisotropy (*1*). As a consequence, this diagnostic was successfully applied to real-time crystal growth investigations of III–V semiconductors using high-vacuum epitaxial techniques (*3,6–9*). In contrast with most of the other surface characterizations based on electron beams, RA is not restricted to a high-vacuum environment. Thus, this optical technique appears particularly attractive for a variety of new applications, including the *in situ* study of III–V crystal growth by higher pressure vapor phase techniques (*10–14*).

Besides applications of RA to crystal growth, this optical probe has extensively been used to study crystal surface processes of crystalline semiconductors such as chemical treatment of Si(110) (*2,16*), GaAs(110) and InP surfaces (*17*), the Ge(001) surface (*18*), oxidation of Si(110) (*19*), passivation of GaAs(001) by sulfur (*20*) and sodium sulfide solutions (*21*), photoelectrochemical deposition of gold on *p*-type GaAs(100) electrodes (*22*), contactless electrical characterization of *p*-type ZnSe (*23*), hydrogen adsorption of the Si(110) surface (*24*), and *in situ* determination of free-carrier concentrations in GaAs under organometallic chemical vapor epitaxy growth conditions (*25*).

This review is organized as follows. The RA technique is described in Section II. The RA probe is first compared to the various spectroscopic characterizations based on the use of polarized light. The relation between the RA signal and the surface anisotropy is discussed. The various RA configurations are then compared. Because RA configurations based on the use of a photoelastic modulator are extensively used for real time investigations, this experimental technique is presented in more detail. Then, the more representative applications of RA to the growth of III–V semiconductors are described in Section III. In particular, *in situ* studies of the growth of GaAs using different techniques such as molecular beam epitaxy (MBE), chemical beam and vacuum chemical epitaxy (CBE and VCE) are compared. Growth of III–V compounds using metalorganic chemical vapour deposition (MOCVD) in various pressure regimes is also addressed. Finally, RA studies of lattice-mismatched III–V and II–VI structures are described in Section IV.

II. The Reflectance Anisotropy Technique

A. Reflectance Anisotropy and the Other Characterizations Based on Polarized Light

The optical response of a sample to the excitation provided by a light beam is given by its refractive index n. In the case of an anisotropic medium, n may vary with the light polarization. The dependence upon polarization of the real part of n is generally referred to as birefringence, while the variation of the imaginary part is usually called dichroism. Another important mechanism of polarization modification occurs at an interface between two media. Under oblique incidence, it is well known that the polarizations of the reflected and the transmitted waves generally differ from the incident polarization. Thus, one generally distinguishes two types of diagnostics of polarized light: on the one hand, techniques investigating optical anisotropy under normal incidence; on the other hand, techniques operating under oblique incidence, which allows the analysis of isotropic media. The instruments corresponding to the latter case are termed ellipsometers. The two techniques can be used in reflection or in transmission configurations, as shown in Fig. 1. However, the ellipsometric technique is generally involved in the study of nontransparent materials. As a consequence, ellipsometry in the reflection configuration is much more widespread than in the transmission configuration. Dichographs are more generally used in chemistry or biology. They are called linear or circular, according to the eigenpolarization of the medium. Thus, simply speaking, a RA instrument can be considered as a normal incidence ellipsometer or as a dichroism setup operating in reflection.

Investigations using polarized light under normal incidence provide useful probes for samples that display an anisotropy in the plane perpendicular to the light beam. In principle, isotropic materials are not adapted for such investigations. Nevertheless, it was recently pointed out that, because of

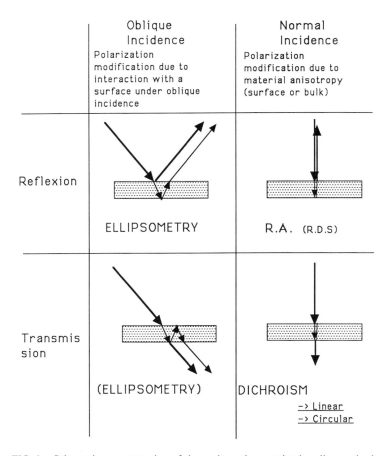

FIG. 1. Schematic representation of the various characterization diagnostics based on the use of polarized light (4).

their lower symmetry, surfaces (or interfaces) of nominally isotropic crystals can display a measurable anisotropy in reflection coefficients (2). Thus, RA appears very attractive for probing the growing surface, whatever the photon energy, whereas other reflection techniques are sensitive to the bulk material, as a function of the penetration depth d_λ of the light at a given wavelength λ. For instance, in the case of the application of ellipsometry to the growth of semiconductors, UV light must be in order to be more sensitive to the surface region. Nevertheless, even in the latter case the probed depth extends at least afew nanometers.

Following the analogy with ellipsometry, the RA formalism can be described by considering the two optical eigenaxes of the sample, x and y. In general, the ellipsometric technique (in reflection configuration) measures the ratio ρ between the two corresponding complex reflectances r_x and r_y (26):

$$\rho = r_x/r_y = \tan\Psi \exp(i\Delta). \tag{1}$$

Considering oblique incidence, the optical eigenaxes are the directions respectively parallel (p) and perpendicular (s) to the plane of incidence. Thus, ellipsometry is the measurement of r_p/r_s. In the case of a crystal surface, the eigenaxes can be deduced from symmetry considerations. For instance, [011] and [01$\bar{1}$] are the eigenaxes of the (100) surfaces of zincblende-type semiconductors, and r_{011} and $r_{01\bar{1}}$ are the reflectances associated with these directions. RA consists in the measurement of the relative difference:

$$r_a = \frac{(r_{011} - r_{01\bar{1}})}{r_{011}}. \tag{2}$$

Then one can define, instead of (1):

$$\rho = r_{01\bar{1}}/r_{011} = \tan\Psi \exp(i\Delta'). \tag{3}$$

Equations (1) and (3) are equivalent if $\Delta' = \Delta - \pi$.

The presentation of RA measurements in terms of Ψ and Δ can be very useful for emphasizing the similarities between RA and ellipsometry. Moreover, in the various ellipsometric techniques, the equations relation Ψ and Δ to the detected signals are generally well known (26). It is important to notice that the same equations can be used under normal incidence configuration (RA).

In the case of small anisotropies,

$$r = r_{011} \sim r_{01\bar{1}} \sim \frac{(r_{011} + r_{01\bar{1}})}{2}. \tag{4}$$

Then, Eq. (4) can be approximated by

$$r_a = 2\Psi' + i\Delta', \tag{5}$$

where $\Psi = \Psi' - \pi/4$ and $\Delta' = \Delta - \pi$.

In this case the small quantities Ψ' and Δ' can be considered as the most useful variables. In particular, Eqs. (4) and (5) show that the various definitions used in the literature are equivalent:

$$2\Psi' = \text{Re}(r_a) = \frac{1}{2}\frac{(R_{011} - R_{01\bar{1}})}{R},$$

$$\Delta' = \text{Im}(r_a) = (\Delta_{011} - \Delta_{01\bar{1}}),$$

with

$$R = |r|^2.$$

When large anisotropies are considered, Eq. (1) becomes

$$r_a = \tan\Psi \exp(i\Delta') - 1.$$

B. RA Configuration Based on the Use of a Photoelastic Modulator

At this point, let us come back to the typical intensities of the RA signals. As already mentioned, the bulk material is isotropic. Thus, the surface anisotropy generally induces small RA contributions, relative surface anisotropy ranging from 10^{-4} to 10^{-3} being typically reported. This can be a strong limitation when dealing with real-time measurements because of the presence of noises induced by the experimental environment (rotating pumps, vibrations...). Phase-modulated ellipsometry (PME) is based on the use of a photoelastic modulator, generating a periodic phase shift $\delta(t)$, at a frequency $\omega = 20-80\,\text{kHz}$, between orthogonal amplitude components of the transmitted beam. Thus, as compared to other techniques, PME takes advantage of the high frequency modulation. As a consequence, the most successful *in situ* RA applications are performed using a photoelastic modulator (27). At the first order (28–30):

$$\delta(t) = A_\lambda \sin\omega t, \qquad (6)$$

where A_λ is the modulation amplitude. Finally, the detected intensity takes the general form

$$I(t) = I[I_0 + I_s \sin\delta(t) + I_c \cos\delta(t)], \qquad (7)$$

where I_0, I_s and I_c are trigonometric functions of Ψ, Δ and the orientation of the optical components. The modulation amplitude is generally chosen such as $J_0(A_\lambda) = 0$, where $J_n(A_\lambda)$ are the Bessel functions. Then, it can be easily shown from (6) and (7) that the quantities I_0, I_s and I_c are related to the harmonics of the detected signal S_0, S_ω and $S_{2\omega}$ by

$$I_s = \frac{1}{2J_1(A)T_1} \frac{S_\omega}{S_0} \tag{8}$$

and

$$I_c = \frac{1}{2J_2(A)T_2} \frac{S_{2\omega}}{S_0},$$

where T_1 and T_2 are attenuation coefficients induced by the detection system.

Particular configurations of the optical elements allow simple determinations of Ψ and Δ. In the case of small anisotropies, the best sensitivity is obtained by using the configuration shown in Fig. 2. In latter case, one obtains in the small anisotropies approximation

$$I_0 = 1,$$
$$I_s \approx -(\pm)_P(\pm)_A \Delta', \tag{9}$$
$$I_c \approx -(\pm)_P(\pm)_M 2\Psi'.$$

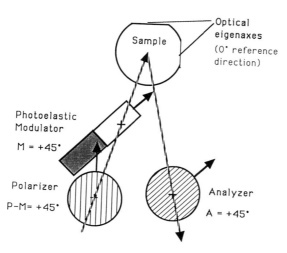

FIG. 2. Orientation of the optical devices corresponding to the measurement configuration of the RA signal (4).

Equations (5) and (9) show that I_s and I_c are related to the imaginary and real part, respectively, of the RA signal. More generally, the optical configuration corresponding to Fig. 2 is generally used in RA measurements.

An example of the optical setup of an RA spectrometer is displayed in Fig. 3(*4*). It can be considered as the normal incidence configuration of the spectroscopic PME. The light source can be either a 75 W xenon lamp or a low-power laser. A mechanical shutter is included between the light source and the optical fiber; it allows the evaluation of the dc background. In order to increase the compactness of the spectrometer, optical fibers can be used in both arms. Let us recall that since the beam goes through a fixed polarizer before being modulated and through an analyzer before being detected, PME is insensitive to any polarization effect due to the optical fibers (*30–32*). The angle between the modulator and the polarizer is fixed to 45°. The modulator and the analyzer can be automatically rotated by means of stepping motors. The spot size on the sample can be estimated to less than 1 mm². The energy of the light is analyzed by a grating monochromator. In the simplest setup, a photomultiplier is used as detector. In this case, the available wavelength range is 230 to 830 mm.

The data acquisition system of the RA spectrometer is described in detail elsewhere (*4,31,32*). Briefly speaking, the digital readout system is based on a fast ADC (12-bit, 1 MHz) combined with a digital signal processor (DSP, Motorola 56001) specially dedicated to the fast Fourier transform computation. Finally, the dc component together with the first harmonics of the detected signal $(S_0, S_\omega, S_{2\omega}, \ldots)$ are continuously transmitted to the personal

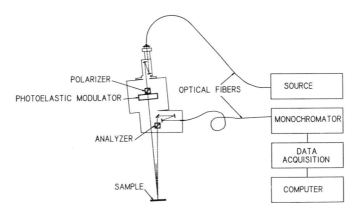

FIG. 3. Schematic description of the reflectance anisotropy spectrometer (*4*).

computer, the maximum data acquisition rate being fixed to 1 kHz. The data acquisition system allows further external connections, as previously described (*4,31,32*). In particular, the present system can easily deal with a second low-frequency modulation. This facility can also be used to record RA measurements performed with rotating samples, as recently suggested (*33*).

A complete description of the calibration and operation of the RA spectrometer is out of the scope of this review. The latter topics together with a detailed investigation of the various parasitic contributions can be found elsewhere (*4,27,34*). Let us mention only the experimental problem induced by the optical windows. In any *in situ* application of RA, the reactor windows can be submitted to inhomogeneous strains, inducing birefringence effects that perturb the RA measurements (*27*). The adaptation of low-strain windows (*35*) cannot be extended to the MOCVD reactors based on the use of a cylindrical quartz tube. It can be shown that the tube birefringence affects the imaginary part of the RA signal (*4,27*). Therefore, in the latter case, the imaginary part of the RA signal can be measured *in situ* only when large anisotropies are investigated, as in the case of the growth of lattice-mismatched semiconductors (*13*).

C. Comparison with Other RA Configurations

The RA system just presented is based upon the use of a photoelastic modulator. It can be compared to the other RA setups based upon modulation techniques. The first RA measurements were performed using rotating samples and a lock-in amplifier (*1,36*). This configuration only allows the measurement of real quantities. Thus, this technique is not sensitive to the anisotropy of the phase shift at the reflection on the sample as defined by Eqs. (2) and (5). Moreover, this technique is incompatible with growth processes involving fixed samples. In contrast, as already mentioned, the RA spectrometer just described above can be used with (and without) a rotating sample. Then, an adaptation of a rotating analyzer ellipsometer to RA was proposed for *in situ* applications with a fixed sample (*3*). It allows the determination of the real part of the RA signal. In the case of real-time studies, the main limitation comes from the relatively low frequency of the mechanical rotation of the analyzer (50–100 Hz). Nevertheless, *in situ* measurement of anisotropies in the 10^{-3} range was reported. However, RA setups based on the phase modulation technique can achieve one order of magnitude improvement in detected sensitivity.

Another RA configuration based on modulation techniques was recently proposed. In this case, a Pockels cell single-wavelength configuration was applied to study photoelectrochemical deposition of gold on p-type GaAs(100) electrodes (22) and MBE of GaAs (37). Introduction of a polarization modulation by the Pockels cell, which allows a rapid switching of the polarization of the laser beam by electro-optic modulation, gives the capability for a time resolution similar to the performance of photoelastic modulation (milliseconds) (37). However, this configuration provides only the real part of the RA signal.

On the other hand, an RA setup based on the use of two detectors was successfully used (38). The sample is illuminated with a polarized light tilted to 45° as referred to the optical eigenaxes. Then the two ortogonal polarizatons are separated and detected, allowing the measurement of the real part of the RA. *In situ* measurements were performed. Nevertheless, the absence of modulation can be a limitation for more general real-time investigations.

Other normal-incidence RA techniques were also presented. They consist of strictly normal-incidence techniques, using semitransparent plates (39,40). The signal processing allows the determination of $\text{Re}\, r_a$ and $\text{Im}\, r_a^2$. Thus, they do not appear to be well adapted to measurements of small anisotropies ($\text{Im}\, r_a^2 \approx 10^{-7}$–$10^{-8}$). Therefore, these last techniques do not provide a crucial improvement as compared to the rotating sample setup.

D. Nature of the RA under Crystal Growth Conditions

As already pointed out, RA appears as a very attractive surface probe of semiconductor crystals, whereas other optical diagnostics are generally dominated by bulk effects because of the penetration depth of the light into the material.

However, a possible contribution from the bulk material to the RA signal may be discussed. The volume of cubic semiconductors is known to be isotropic. However, the spatial dispersion of the dielectric tensor $\varepsilon(\omega, \mathbf{k})$ has to be considered. More precisely, ε depends not only on the direction of the electric field, but also on the direction of the wavevector (41). As a consequence, for some directions of light propagation, the bulk crystal may display a slight birefringence. This behavior induces corrective terms in the expressions ε of the order of $(a/\lambda)^2$, where a is the lattice parameter. These corrective terms are small but measurable (42,43). In particular, they can contribute to the measured anisotropy of some samples (2).

Considering now the surface contributions, it is reasonable to distinguish intrinsic and extrinsic effects. The dominant intrinsic contribution arises from many-body screening effects, which can induce a large optical anisotropy of the (110) surfaces of cubic crystals (44). Extrinsic surface contributions result from the presence of physo- and chemisorbed molecules (or radicals) at the crystal surface. For instance, chemisorption can possibly induce changes in the electronic polarizability of the outermost atomic plane, e.g. electronic transitions between energy levels of the local atomic structures. The extrinsic surface effects are the main objective of the present presentation since they are directly related with the growth processes. Thus, the RA signal associated with the latter effects gives unique information about surface chemistry, surface structure (at the atomic scale), and anisotropic microroughness.

As a consequence, the measurement of the complex RA signal allows the determination of the anisotropic contribution to the surface dielectric function of a sample under investigation, and then the performance of a quantitative analysis of the recorded spectra. Therefore, the results are obtained in a form that can be compared directly to optical models. Considering the (100) surface, the surface dielectric anisotropy $\Delta(\varepsilon d)$ is determined as (1)

$$\Delta(\varepsilon d) = (\varepsilon_{011} - \varepsilon_{01\bar{1}})d = \frac{\lambda}{4\pi}(\varepsilon_s - 1)r_a, \qquad (10)$$

where ε_{011} and $\varepsilon_{01\bar{1}}$ are surface dielectric functions corresponding to the [011] and [01$\bar{1}$] directions, respectively; d represents the effective surface-layer thickness; ε_s is the average dielectric function of the bulk material; and λ is the wavelength of light. The surface dielectric function has the dimensions of length, as it describes polarization per unit area instead of polarization per unit volume as in the bulk. The ability to obtain spectral dependencies $\Delta(\varepsilon d) = \Delta(\varepsilon_1 d) - i\Delta(\varepsilon_2 d)$ is a significant advance not only because they are the functions that describe the fundamental response of the surface, but also because $\Delta(\varepsilon_2 d)$ is a local in energy and thus yields directly the anisotropy of the spectrum of elementary excitations that cause the change in the surface dielectric response (23).

III. Applications of RA to the Growth of III–V Semiconductors

A. The Various Deposition Techniques

One of the main differences among the various growth techniques of III–V semiconductors such as molecular beam epitaxy (MBE), chemical beam epitaxy (CBE), vacuum chemical epitaxy (VCE) and metalorganic chemical vapour deposition (MOCVD) is the pressure in the deposition chamber: UHV conditions for MBE, $10^{-6}-10^{-4}$ mbar for CBE, $10^{-4}-0.3$ mbar for VCE, and from a few millibars to around atmospheric pressure for MOCVD. Only MBE and CBE vacuum conditions allow RHEED measurements. Thus, the RA technique appears most attractive for the control and monitoring of MOCVD, as the pressure conditions in this case do not allow applications of *in situ* high-vacuum electron beam surface probes.

Furthermore, real-time application of RA to pulsed MBE or VPE gives an opportunity to precisely control monolayer thickness and to achieve atomic layer epitaxy (ALE) or migration enhanced epitaxy (MEE). In the pulsed VPE, compound semiconductors are produced by the separate and alternate exposure of the substrate to reactants containing elements of the compound. In the ALE technique, a single atomic layer of a crystal is deposited on a substrate before the modification of the process, the next atomic layer then being grown. However, the self-limiting ALE suffers from carbon contamination, and the extent to which the growth conditions occur depends critically on the reactor design, the reactants used, their surface reaction kinetics and growth temperature (*45–47*). Another realization of pulsed VPE, metalorganic migration enhanced epitaxy (MOMEE), at elevated temperatures is much less subject to impurity capture due to surface decomposition of organometals, and the appearance of the metallic surface, but it requires a precise monitoring of one monolayer per cycle of metallic layer deposition (*48*).

RA is also a valuable complementary probe for molecular beam epitaxy (MBE), providing a quantitative information about metallic Ga island formation and the relative surface concentrations of Ga–Ga and As–As dimer bonds at GaAs homoepitaxy (*3*). The surface chemical information deduced from RA can be correlated with the structural information available from reflection high-energy electron diffraction (RHEED).

B. REFLECTANCE ANISOTROPY OF GaAs AND AlAs UNDER MBE CONDITIONS

RA was first applied to study GaAs and AlAs growth in an MBE chamber equipped with a standard RHEED system and dealt with abrupt alternate changes in As and Ga (or Al) fluencies during homoepitaxy on (001) surfaces (*3*). During these sequences, the surface structures successively change from As-stabilized (2×4) to Ga-stabilized (4×2) (see Fig. 4).

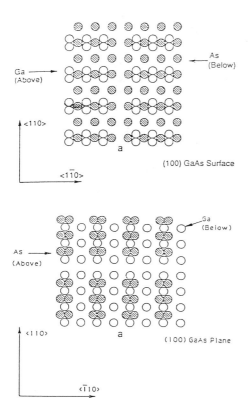

FIG. 4. Schematic representation of (a) 2 × 4 As-stabilized and (b) 4 × 2 Ga-stabilized surfaces.

Spectral dependencies of the differences in optical anisotropies between the (2 × 4) and (4 × 2) reconstructions showed that the maximum differences occurs at 2.0 –2.5 and 3.5 eV for GaAs and AlAs, respectively (3). The results suggest that the observed optical anisotropies arise from optical absorption associated with the As–As and Ga–Ga (or Al–Al) dimer bonds, respectively. The strong dependence of this anisotropy on photon energy then allows one to spectrally distinguish between Ga–Ga and Al–Al surface dimer bonds. In particular, the measurement of the relative surface concentration of the dimer bonds has been performed in the case of $Al_{0.5}Ga_{0.5}As$ growth (27).

Real-time measurements at a photon energy of 3.54 eV, i.e., away from the 2.0–2.5 eV anisotropy maximum, reveal a direct correlation between RA and RHEED responses (see Fig. 5) (*34*). The obvious similarity between the RA signal at 3.54 eV and the simultaneously measured RHEED transient, which is induced by a structural evolution, reveals that the RA signal at this photon energy is predominantly determined by surface structure, and not surface chemistry. On the contrary, the essentially linear time dependence of the RA signal with Ga deposition up to 1 ML of Ga (Fig. 5) indicates that at 2.48 eV the RA signal is determined by the relative surface concentrations of Ga–Ga and As–As dimer bonds (*27*). Thus, the RA signals can probe either surface chemistry or surface structure, depending on the photon energy. Measurements at structure sensitive energies make RA essentially equivalent to RHEED and give a unique opportunity to apply RA as a

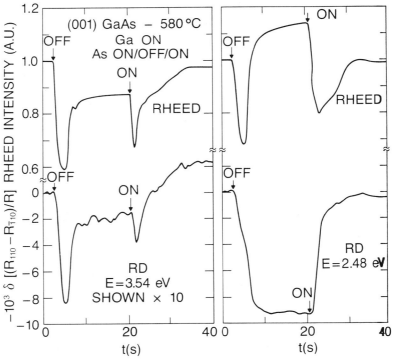

FIG. 5. RHEED (upper) and RA (lower) transients obtained by interrupting and resuming the As flux during otherwise normal growth of (001) GaAs, at 1 (001) atomic bilayer per 4.8 s (*27*).

structural probe of non–ultrahigh-vacuum growth techniques. Moreover, Fig. 5 reveals the complimentarity of RHEED and RA: Some features are revealed by both surface probes, while other features can be detected by only one experimental technique.

A typical diagnostic, associated with RHEED measurements during crystal growth, consists of the "RHEED oscillations," namely, a cyclic variation in the intensity of the diffracted beam whose period corresponds to the time required to grow one (001) atomic bilayer (*49*). RHEED oscillations are very useful to monitor crystal growth and deposition rates and to measure absolute alloy compositions. Analogous oscillations of the RA signal were observed during the MBE growth of (001) GaAs and AlAs (*34,50*). In contrast to RHEED, which provides only surface structural information, RA oscillations are due to cyclic changes in either surface chemistry (i.e., relative coverage) or surface structure, depending on the photon energy of probing beam. The RA oscillations were detected both at surface chemistry-sensitive photon energy (around 3.5 eV) and structure-sensitive energy (*51*). In both cases the RA period is precisely the time required to grow a single (001) atomic bilayer as determined by RHEED (see in particular Fig. 6 for GaAs). However, in the case of the chemically sensitive probe there is a phase shift between RA and RHEED responses, reflecting a difference in the origins of the signal. At the same time, the oscillatory part of the structure-sensitive RA signal is in phase with RHEED, because both are being determined by surface structure under the same growth conditions. The structure-sensitive RA signal is determined by the difference between the depolarization or screening charge that is developed for the two orthogonal polarizations. Since local-field perturbations range over distances of the order of the structural anomalies, i.e., the step heights, the surface correlation length for the structure-sensitive RA signal will be extremely short, of the order of the step height itself (*6*).

It has previously been shown that a proper choice of the photon energy allows the attainment of a specific, chemical or structural, sensitivity. Nevertheless, the RA signal includes both chemical and structural components, as the surface reconstructions depend crucially on the surface dimer coverage (*35,52,53*). Thermal As_2 desorption experiments performed on GaAs(100) in an MBE growth chamber equipped with an RA spectrometer at variable temperature from 510 to 630°C reveal a series of reconstructions as indicated by RHEED (*37*). It was observed that the appearance of the various reconstructions, successively $c(4 \times 4)$–(2×4)–(3×1)–(4×2), is correlated with steps in the RA signal. As each reconstruction corresponds to a specific range of the "stable" coverage, it appears that a systematic

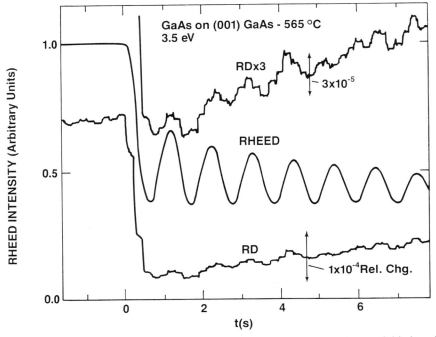

FIG. 6. Averages of 33 RHEED (middle) and RA (upper, lower) signals upon initiation of GaAs growth at 1.1 s per 1 Ga ML on an As-stabilized (2 × 4) GaAs surface. Magnitude scales are shown for the RA data, which were taken well away from the chemical peak of GaAs. The upper and lower traces are the same RA data shown with different scaling factors (51).

relationship exists between RA signal evolutions and the approximate coverage of As or Ga expected for the reconstructions. Transitions to lower or higher coverages that are stable within a definite temperature interval indicate a change in the structure of the surface. The reconstructed surface necessarily reveals a chemically distinct stable overlayer of As and Ga dimers having well-defined coverage limits. Figure 7 illustrates the sensitivity of RA spectroscopy to surface reconstructions (54). Surface dielectric anisotropy of (001) GaAs under steady-state conditions, which was determined by the RA spectrometer in the case of As-stabilized surface, reveals that the temperature-dependent surface phase boundaries are not sharp, with the possible exception of the limit between the c(4 × 4) and the (2 × 4) reconstructions (23). A strong increase in anisotropy during the thermal desorption process shows that oxide removal induced by the increase of

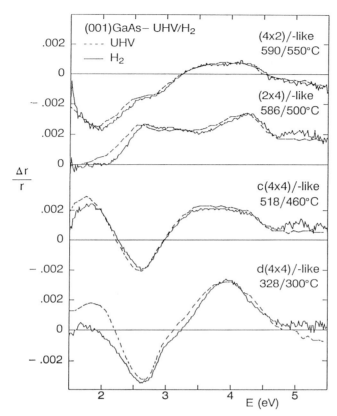

FIG. 7. RA spectra of the primary reconstructions on (001) GaAs in UHV and APH$_2$. Sample temperatures for UHV/H$_2$ ambients are indicated. The surfaces in UHV were prepared under an As$_4$ beam equivalent pressure of 5×10^{-9} Torr for the (4 × 2) reconstruction and 5.7×10^{-5} Torr for the rest (54).

temperature is complex, involving morphological modifications (roughening) of the interface.

Determinations of surface dielectric anisotropy, which can be deduced from the measurement of the complex RA signal, reveal features related to surface reconstructions and change of relative surface dimers coverage (see Fig. 8) (33). The measurements were performed by interrupting of As flux both for singular (001) GaAs surface S and for two vicinal GaAs surfaces A and B cut 6° off (001) toward (111)a and (111)B, respectively. The photon

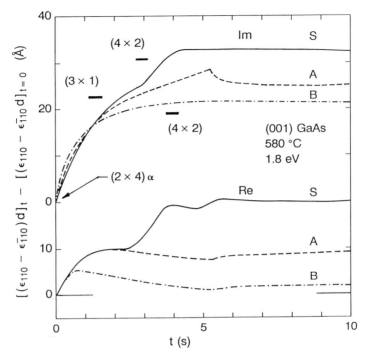

FIG. 8. Changes in the real and imaginary parts of $\Delta(\varepsilon d)$ that occur for a singular (001)GaAs surface S and for two vicinal GaAs surfaces A and B cut 6° off (001) toward (111)A and (111)B, respectively. The onsets of various RHEED reconstruction patterns are also shown (33).

energy 1.8 eV is used because $\Delta(\varepsilon_1 d)$ and $\Delta(\varepsilon_2 d)$ correspond to (2 × 4) and (4 × 2) reconstructions and then are sensitive to surface coverage by As–As and Ga–Ga dimers, respectively. This energy corresponds to a maximum sensitivity of $\Delta(\varepsilon_2 d)$ to Ga–Ga dimers, together with a weak sensitivity to the (2 × 4) reconstructions. Therefore, the latter photon energy allows one to discriminate between surface-structure and dimer-coverage contributions to the RA signal. The data of Fig. 8 (recorded at 1.8 eV) clearly evidence the evolution of the previous surfaces through various phases as they vary from As-stabilized to Ga-stabilized conditions. The more pronounced nonlinear kinetic variations $\Delta(\varepsilon_2 d)$ in Fig. 8, by comparison to the evolution of the real part, show that Ga enrichment occurs not only by Ga deposition, but also by As evaporation (23). The kinetic curves reveal that the evolution of the different surface phases is determined by the relative concentrations of Ga and As. On all three surfaces, the most rapid process after As interruption is the nearly exponential transient with a time constant of 1 s or less displayed only by $\Delta(\varepsilon_1 d)$. This time dependence, together with the quasi-absence of this feature in $\Delta(\varepsilon_2 d)$ variations, identifies this process as the evaporative loss of As–As dimers that are maintained in dynamic equilibrium with the surface when the As flux is interrupted (35). The slower and smoother time dependencies observed in the $\Delta(\varepsilon_2 d)$ variations reflect the correlative accumulation of Ga–Ga dimers. While the $\Delta(\varepsilon_2 d)$ curve displays a final equilibrium near 4.3 s for surface S, the $\Delta(\varepsilon_1 d)$ response indicates that this surface continues to evolve to 5.5 s. The $\Delta(\varepsilon d)$ data for A show a remarkable behaviour, breaking sharply near 5.3 s. This feature is interpreted as a supersaturation phase transition involving the formation of metallic Ga droplets, which nucleate to provide an alternate sink for Ga when a critical excess surface concentration has been reached (33).

The latter results clearly demonstrate that the possibility of determining the surface dielectric anisotropy by measuring both real and imaginary parts of the RA signal extends drastically the physical meaning of the RA technique and allows real-time surface structural and chemical information to be obtained from kinetic measurements at optimum photon energy.

C. Applications of RA to Migration-Enhanced Atomic Layer Epitaxy (MEALE)

Migration-enhanced atomic layer epitaxy (MEALE) is a very promising technique for growth improvement of MBE and CVD (47). A basic mechanism responsible for growth improvement is the enhancement of the surface mobility or migration of absorbed atomic metallic species of group III when separately supplied from group V species. To realize this approach, MBE is modified by using alternating or pulsed molecular beams synchronized with the monolayer-by-monolayer growth sequence characteristic of MBE. A stationary MEE growth process requires a precise monitoring of one monolayer per cycle of metallic layer deposition. Thus, RA appears, to be a very appropriate probe to establish the monolayer deposition. However, as pointed out earlier, the steps or saturations of the RA transients that could indicate a complete metallic monolayer can be related to the change of surface coverage or reconstructions. Therefore, in order to determine unambiguously the monolayer coverage, one should establish the relation between the RA signal signature and the surface coverage (or structure) at constant growth conditions.

A study of the surface chemistry–sensitive RA transients (at 2.0 eV photon energy of probing beam) during Ga deposition at constant temperature in the range $T_s = 350$–$560°C$ and precise monitoring of Ga coverage reveals the specific relations between the RA signature and surface structure or coverage (see Fig. 9) (*38*). It can be observed that the maximum of the RA peak is achieved for a constant Ga dose that corresponds to 0.75 ML, as referred to the initial negative peak (Fig. 9) for $T_s \leqslant 450°C$ or from the origin for $T_s = 500°C$. This behaviour indicates that the initial (2 × 4) As-stabilized surface already contains a substantial concentration of Ga because of thermal desorption of As. Ga doses corresponding to an excess of 1 ML lead to a nearly constant RA signal for all substrate temperatures. That excess

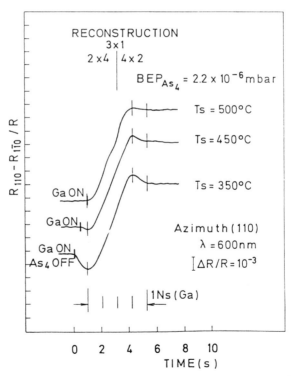

FIG. 9. RA signal transients for three different substrate temperatures, T_s. Thin vertical lines refer to a common scale of Ga doses included at the bottom. Vertical bars mark actual instants for As_4-OFF/Ga-ON. Observed surface reconstructions are indicated at the top (*38*).

Ga accumulates on the surface in the form of droplets, as evidenced by light scattering.

In order to get further insights into the RA signal evolution at Ga coverage up to 1 ML and to relate these features with the surface structure and stoichiometry, the following light absorption mechanism was postulated (*38*). According to the model proposed by Chadi (*55*), the structure of the As-stabilized (001) GaAs(2 × 4) reconstructed surface represents 4* periodicity, which arises from an ordered array of three As dimers and one missing dimer per unit cell. The As surface coverage for perfect (2 × 4) surface is 0.75 ML. However, the (2 × 4) surface reconstruction remains within a certain range ($\pm$ 0.25 ML) of As coverage around its ideal value of 0.75 ML. The ability of the As-stabilized surface to accommodate impinging Ga at bonding surface sites depends on its initial defect structure or its step anisotropic distribution. The one monolayer Ga deposition period can be divided into four stages.

The first stage is coverage up to 0.25 ML Ga. An initial decrease of the RA signal at low T_s (see Fig. 9) for increasing Ga doses is explained by Ga atoms being preferentially bonded at As-rich steps along the [110] direction. Rows of Ga atoms, nucleated at these steps, might absorb [1$\bar{1}$0] polarized light in a similar way to the Ga–Ga dimers nucleated on the flat (001) surface. In the latter case, [110] polarized light is preferentially absorbed because the Ga–Ga dimers run along the [110] direction (see Fig. 4). Accordingly, the Ga dose needed to reach the RA initial minimum by saturation of the steps is a measure of the initial density of the step sites along the [1$\bar{1}$0] direction on the static surface. This density depends on substrate temperature, As flux and the previous history of the sample, as well as the polishing angle for vicinal surfaces. For low substrate temperatures (350°C), this dose achieves a maximum of about 0.25 ML (Ga).

The second stage is Ga coverage up to 0.5 ML. Under this stage, 2-D nucleation of Ga–Ga dimers on still (2 × 4) surface takes place. According to the model of Farrel *et al.* (*56*), no more Ga can be incorporated after achievement of the 0.5 ML coverage without the simultaneous supply of arsenic or a change in reconstruction. Under pulsed MBE conditions, Ga is supplied alone and surface structure changes precisely for 0.5 ML (Ga), through an intermediate (3 × 1), into the Ga-stabilized (4 × 2) phase.

Under the third stage (Ga coverage up to 0.75 ML), a monotonic growth of the RA signal originates primarily from an increase of polarized light absorption because of an accumulation of surface Ga–Ga dimers. Further Ga deposition up to 1 ML coverage (fourth stage) reveals a decrease of the RA signal, especially pronounced at T_s higher than 520°C, which might originate from anisotropic light absorption by Ga accumulated along

[1$\bar{1}$0]-oriented steps on the (4 × 2) reconstructed surface (*38*).

Analysis of the RA signal during pulsed MBE carefully monitored to deposit 1 ML (GA) per cycle on the ideal (001) plane (6.25 10^{14} sites/cm^2) in the temperature range 150–550°C gives the possibility of optimizing the temperature for atomic layer MEE (*38*). In the lowest temperature range (150–250°C), the signal amplitude does not reach its maximum value, indicating that the amount of Ga is not sufficient to reach 1 ML Ga coverage at the end of each period. As a result, the surface is never Ga-stabilized during the cycle and no migration enhancement takes place. Consequently, crystal quality is degraded. In the temperature range allowing maximum and constant RA peak-to-peak amplitude (400–500°C), the surface achieves full 1 ML Ga coverage for a short time during each cycle immediately before the As$_4$ flux is switched *on* and the growth cycle is repeated. Under these conditions, growth proceeds by alternate full coverage of Ga and As at the surface, then allowing atomic layer-by-layer growth. At higher temperatures (>550°C), a clear signal saturation at the end of the Ga supply sequence and a corresponding reduction in the peak-to-peak signal amplitude are observed (*38*). The latter behaviour shows that the surface never reaches complete As coverage during each cycle, preventing the achievement of ALE conditions. Therefore, RA allows the identification of migration-enhanced ALE growth conditions and the monitoring of consecutive cycles of full surface coverage of both Ga and As. By interrupting Ga and As$_4$ fluxes immediately before the RA signal maximum or minimum, respectively, one can realize migration MEALE and thus optimize the crystal quality. This conclusion appears very important in case of CVD techniques where RHEED can not be used to determine surface coverage.

D. Reflectance Anisotropy of GaAs under Chemical Beam and Vacuum Chemical Epitaxy Growth Conditions

1. Surface Transformation between As- and Ga-Stabilized Surfaces

CBE and VCE techniques cover the pressure range between MBE and MOCVD. Arsine (AsH$_3$) or tertiarybutylarsine (TBA, CAsH$_2$(CH$_3$)$_3$) are generally used as group-V precursors; triethylgallium [TEG, Ga(C$_2$H$_5$)$_3$], trimethylgallium [TMG, Ga(CH$_3$)$_3$] are used as group-III precursors. CBE uses metalorganic precursors at typical pressures from 10^{-6} to 10^{-4} mbar, which allow RHEED measurements. Thus, CBE allows correlations to be established between RA and RHEED responses for metalorganic chemistry. VCE at pressures from 10^{-4} to 0.3 mbar connects UHV conditions of CBE to low-pressure MOCVD. VCE can, as an intermediate technique, bridge the gap between growth controlled by molecular flow and growth controlled by diffusion.

The RA trajectory at the photon energy $E = 19\,\text{eV}$ (near the transition energy characteristic of Ga dimers (*33*)) for a Ga saturation experiment on the (001) GaAs surface in a CBE system is shown in Fig. 10b, TEG being used as the precursor (*57*). The signal evolution reveals a transition from the (2×4) to the (3×1) reconstructions. A similar feature is exhibited by RA signals recorded in MBE conditions (Figs. 8 and 9). This feature is transformed into a specific peak under VCE conditions (Fig. 10a). There-

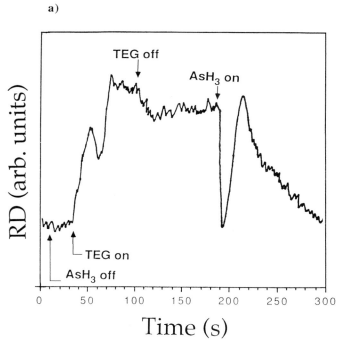

FIG. 10. RA responses recorded when an As-stabilized surface is exposed to TEG (a) in the CBE system and (b) in the VCE system. The surface reconstructions determined by RHEED in the CBE system are indicated in the lower part of (a). The RA transeints due to supply of As precursors are also shown (*57*).

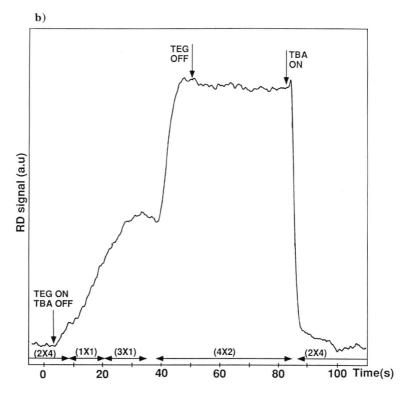

FIG. 10 — (continued)

fore, one could conclude that it reflects a structural surface rearrangement and should not be crucially dependent on the nature of the precursor. However, another CBE experiment performed on the (001) surface of GaAs at the same photon energy (1.9 eV) with TMG as a precursor displays a more complicated behaviour. Figure 11 shows that the latter feature disappears when the TMG dose decreases (8). RHEED indicates that an initial structure (2 × 4) is transformed into (4 × 6). In all cases during the initial stage, formation of Ga dimers (as a result of the catalytic decomposition of the organometals on reactive As sites) dominates the other processes, as revealed by an increase in the RA signal. Although a steady increase in RA signal is observed, the RA signal level reflects in fact only

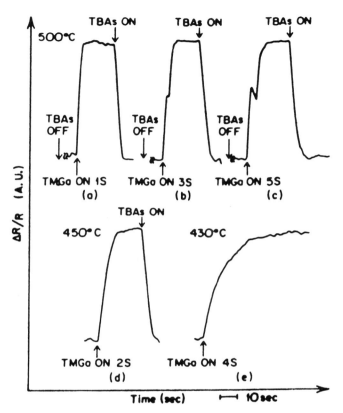

FIG. 11. RA transients on (001) GaAs for initiation of TMG and TBA exposures. (a) 1 s, (b) 2 s and (c) 5 s TMG exposure at 500°C. (d) 2 s TMG exposure at 450°C. (e) 4 s TMG exposure at 430°C (8).

those Ga atoms displaying a dimer bonding configuration. Those deposited Ga atoms that are still bonded to CH_3 methyl radicals cannot be detected by RA. It was suggested that extended TMG exposure on a Ga-rich surface brings about the adsorption of surplus CH_3 radicals that are attached to nearb allium dimers and break Ga–Ga dimer bonds (8). A decrease in the RA signal is observed through this decrease of the Ga–Ga dimer groups. These observations show that signal features corresponding to structural and chemical modifications can be attributed to metalorganic chemical reactions.

2. Growth Oscillations under CBE Conditions

Growth oscillations, which have initially been observed under MBE conditions (*51*), can also be detected in the pressure range from VCE conditions (10^{-4} mbar) up to conditions resembling low-pressure MOCVD (LP-MOCVD) (0.3 mbar) in the growth temperature range 430–570°C (*7, 58–61*). Depending on growth conditions, damped oscillations in the RA signal can be obtained by injecting TEG after AsH_3 surface stabilization (*7, 58*). As many as 30 oscillations are observed before signal stabilization at a level characteristic of the V/III ratio used. Gradual transition of growth conditions from VCE into those of LP-MOCVD by adding an increasing amount of hydrogen allows one to follow how this modification affects the oscillations (*58*). The occurrence of the oscillations depends on the V/III ratio. However, although the range of V/III ratios over which clear oscillations can be obtained becomes narrower at higher pressures (see Fig. 12), similar oscillation amplitudes could still be obtained for optimal LP-MOCVD and VCE conditions (*61*). Besides, studies over a wide range of hydrogen partial pressures (up to 0.3 mbar) have shown that the occurrence of RA oscillations is dependent neither on the pressure nor on the

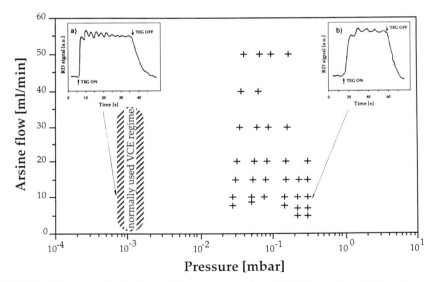

FIG. 12. Mapping of arsine flow and pressure range for which RA growth oscillations have been found (+ symbols). Inserts show examples for oscillations at the extremes of the experiments (*61*).

introduction of hydrogen. Furthermore, a comparison of the RA signal with and without hydrogen shows that hydrogen acts as a carrier gas and that its possible adsorption on the surface does not prevent the monolayer-by-monolayer growth mode. The oscillation period exactly corresponds to the growth of one bilayer of GaAs (2.8 Å) (58), the nature of the oscillations being interpreted in terms of alternate surface coverage of Ga and As single layers. However, a recent study of I. Kamiya et al. attributes the oscillations to a cyclic variation between one and two layers of As (62).

In conclusion, as in the case of RHEED, RA oscillations can be used to determine the growth rate. However, the correlation between both oscillations is not completely systematic. In particular, no resolvable RA oscillations have been obtained in CBE equipment in the pressure range 10^{-6}–10^{-4} mbar, although very good RHEED patterns as well as oscillations were achieved (57).

3. In Situ *Control of Surface V/III Ratio*

RA is a suitable technique for *in situ* control of surface V/III ratio. The material quality depends strongly on the V/III ratio, which controls morphology, impurities, carrier concentrations, optical properties, etc. The routinely used determination of the volume V/III ratio from mass flow controllers gives an error of approximately 5–10% in the flow ratio. Effects of the precursors cracking in the reactor cell increase the error in the effective surface ratio. Changing the ratio from 6.5 to 8.5 in MOCVD results in a transition from 3×10^{14} cm^{-3} *n*-type to 3×10^{14} cm^{-3} *p*-type material. As a consequence, acquiring a precise knowledge of the V/III ratio appears to be an important task (63).

The steady-state RA signal (after the decay of growth oscillations) has been recorded at $E = 1.9$ eV (Ga dimer–sensitive region) during VCE growth while the V/III ratio was decreased by lowering the arsine flow (57). The observed signal increase (see Fig. 13) toward a maximum can be correlated with raising the Ga dimer concentration when gradually decreasing the AsH$_3$ flow. A lower As flow leads to a surface with a higher concentration of Ga dimers; as a consequence, the Ra signal moves toward a level corresponding to a Ga-terminated surface. Further reduction of the arsine flow leads to a decrease of the RA signal. Figure 13 shows that the RA signal is very sensitive to the V/III ratio. As soon as the flow rate is decreased below a critical limit, the RA signal reveals a drastic change in the surface structure. A clear degradation of morphology can be observed when growth is carried out below this critical ratio (9). Performing the same

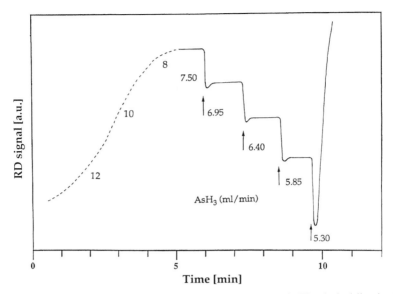

FIG. 13. The steady-state RA level during continuous growth. The dashed line in the left part of the figure shows schematically the behaviour of the RA signal while the arsine flow is decreased from a high V/III ratio. The right part shows a more carefully controlled experiment at lower V/III ratios where arsine flow was further decreased stepwise. Arrows indicate points at which the arsine flow was changed, and the numbers indicate the new flow rate. The TEG flow was kept constant during the experiment. The growth rate was 1 μm/h at the temperature 520°C (57).

experiment in the CBE equipment and studying the surface by RHEED reveals a change of surface reconstruction from (3 × 1) to the Ga-terminated (4 × 2) when passing through the critical ratio (8). Morphology and photoluminescence behaviours of the layers grown at a flow rate of about 7.5 mL/min, which corresponds to a maximum of the RA signal, are found to be optimized (9). The decrease of the RA signal for both higher and lower flow rates (Fig. 13) indicates that at least two processes are competing (9). At higher flow rates, the impinging rate of arsine becomes higher, which results in a shorter lifetime of the Ga dimers, and thereby a smaller average concentration. At lower flow rates, the impinging rate of arsine is so weak that excess TEG-related species can exist on the surface. They can then break some of the absorbing Ga-dimer bonds. At flow rates lower than the critical one, the TEG-related species cannot be incorporated as elemental

Ga at the same time as new TEG species stick to the surface. The latter behaviour leads to a rapid degradation of the surface.

Besides the monitoring of the growth rate from RA signal oscillations and surface V/III ratio from absolute level of the signal, a real-time control of cycles of chemical pulses and surface evolutions in an ALE-like mode can be achieved via the characteristic transients of the RA signal (*59,60*).

E. GROWTH OF LATTICE-MATCHED MULTILAYERED STRUCTURES BY LP-MOCVD

RA has also been used extensively to study the growth mechanisms of III–V semiconductors by low-pressure (100 mb) MOCVD (*12–14,64*). In the last study the RA spectrometer has been directly adapted to the LP.MOCVD reactor (*64*). Some results on heterojunction formation of lattice-matched compounds are summarized here (*14*).

An important goal of RA is the assessment of interface quality while growth is occuring. Ra can indeed be used for this purpose. Figure 14 shows the variations of the RA signal corresponding to the growth of two GaInP/GaAs structures. The RA signatures differ significantly. The electron Hall mobility at 77 K (30,000 and 90,000 cm^2 V^{-1} s^{-1}) reveals different qualities for the two interfaces. The two GaInP/GaAs samples, in fact, were grown using identical growth parameters, but the higher-quality film was grown after modifying the reactor geometry. A major difference in the two RA signatures is the time constant of the initial variation: 2 s for sample (a), and 0.2 s for sample (b). According to a more complete RA analysis, the initial variation of the RA signal is mainly due to the change in element V source on the surface (*14*). Here, the relatively long time constant observed in Fig. 14a suggests that the switching from arsine to phosphine was not abrupt enough. Therefore, the geometry of the group V section in the gas panel was modified; as a consequence, fast initial transients can be obtained (Fig. 14b). Conversely, the quality of the GaInP/GaAs films is significantly improved. The evolution of the RA signal during the seconds following the initial transient is not yet understood. However, it is very sensitive to film composition and growth rate. Any delay in the switching of the different gases altered its behaviour. For other materials too, such as GaInAs/InP, the RA signature was found to be sensitive to small deviations from optimal conditions (*14,64*).

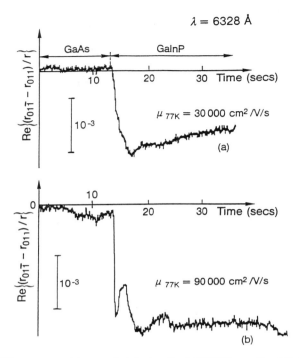

FIG. 14. RA measurements recorded during the MOCVD growth of GaInP/GaAs using two different reactor geometries (14).

The sensitivity of RA to small deviations from optimal conditions is extremely useful for monitoring complex structures. In the case of a quantum well or a superlattice, it is not possible to assess individually the quality of each interface using mobility measurements. The observation of the RA signature can be very useful for this purpose. The RA signature of the GaInP/GaAs interface has been observed during the growth of several GaInP/GaAs/GaInP quantum wells (Fig. 15). For small GaAs thickness (10 s or less), the signature clearly departs from that of the high-quality interface shown in Fig. 14b. This difference is greatly reduced by improving the switching sequence.

In Fig. 16, the RA record of a GaInP/GaAs superlattice exhibits features quite similar to those of Fig. 14b. It also shows that the interfaces are reproducible. It is then possible to compare the signatures of heterojunctions

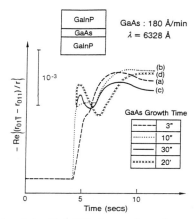

FIG. 15. RA records of the GaInP/GaAs interface of quantum wells with different well thicknesses *(14)*.

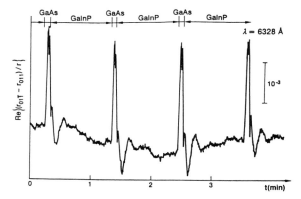

FIG. 16. RA measurements recorded during the growth of GaInP–GaAs superlattice by MOCVD *(14)*.

independently of the underlying structure only because RA is not sensitive to bulk effects. This is an extremely useful feature of RA as compared to other techniques such as ellipsometry. Ellipsometric records of superlattices do not allow direct comparison of the different periods *(66)*.

F. REFLECTANCE ANISOTROPY OF GaAs UNDER ATMOSPHERIC PRESSURE (AP) MOCVD GROWTH ENVIRONMENTS

1. Surface Reconstructions

Up to now, because of the lack of appropriate surface analytical techniques other than RA, little has been known about surface structures and reactions involved in atmospheric pressure MOCVD growth, in particular in the case of atomic layer epitaxy (ALE). Experimental results and interpretations of MOCVD processes are often controversial. Because surfaces in UHV are very reactive, it has been speculated that the (001) GaAs surface under AP could be quite different from surfaces in UHV, e.g., terminated by hydrogen in MOCVD (67). However, a comparison of the RA spectra taken in UHV and AP environments unambiguously indicates that the primary reconstructions (4 × 2) (no As layer and Ga dimers along [110]), (2 × 4) (one layer of As dimers along [110]), and c(4 × 4) (two layers of As dimers along [110]) that occur on the (001) GaAs in UHV also occur under AP H_2 (see Fig. 7) (54), He and N_2. For equivalent substrate temperatures and As fluxes, primary reconstructions, similar to (4 × 2), (2 × 4), c(4 × 4) and a high-As coverage disordered structure d(4 × 4) in UHV, are observed in AP. These reconstructions are terminated by Ga or As dimers that then specify the RA spectra (Fig. 7). The features of the spectra recorded under both UHV and AP conditions originate from electronic transitions between energy levels of the local atomic structures and can be uniquely related to specific surface dimers (68). The 1.9 eV spectral feature of the (4 × 2) surface is due to transitions between bonding Ga dimer orbitals and empty Ga lone-pair states. The 2.6 eV and the 4.2 eV features can uniquely be associated with the electronic transitions between filled As lone-pair states and free As antibonding dimer orbitals, and between bonding and antibonding As dimer orbitals, respectively (68). Thus, the presence of Ga and As dimers can be directly inferred from the latter RA spectral features. The 2.6 eV feature is positive for (2 × 4) and negative for c(4 × 4) because of a 90° difference in the dimer orientations (69). While the overall line shapes of the RA spectra are directly related to each primary reconstruction, details depend on the As supply rate and the substrate temperature since As coverage is determined by competition between adsorption and desorption.

The RA spectra obtained in an AP H_2 environment (AsH_3 partial pressure of 2.3 Torr) at a substrate temperature of 450°C, i.e., in typical growth conditions of self-limiting ALE (45,46), show a striking agreement with the UHV spectra (see Fig. 17) (62). These results show that formation

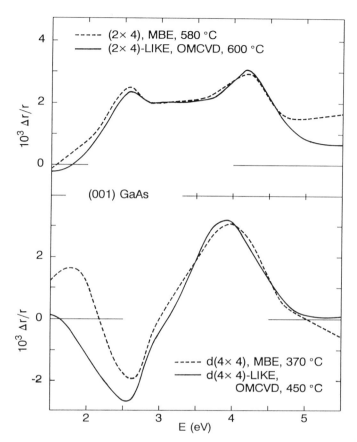

FIG. 17. Typical RA spectra of MBE-prepared (2 × 4) and "disordered" d(4 × 4) surfaces of (001) GaAs in UHV and AP-MOCVD (H_2) (62).

of the As dimers occurs even in the presence of AsH_3, i.e., under an environment where growth actually takes place. The c(4 × 4)/d(4 × 4) structure is found at higher temperatures during MOCVD because of the higher As partial pressure.

Results of the study of the variations of As-terminated structures as functions of sample temperature and As supply rate in both UHV and AP H_2 are summarized, in Arrhenius form, in Fig. 18 (62). This phase diagram is a further evidence of the similarity between (001) GaAs surfaces in UHV and in AP H_2. Moreover, Fig. 18 indicates that the difference in surface

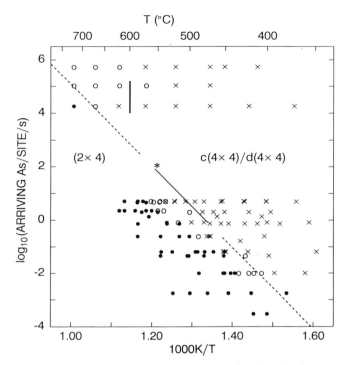

FIG. 18. (2×4)–$c(4 \times 4)/d(4 \times 4)$ phase diagram as a function of substrate temperature and incorporation rate of As atoms, determined from partial pressures of As-containing precursors (As_4 in MBE and AsH_3 in MOCVD). Dots, crosses and circles represent (2×4), $c(4 \times 4)/d(4 \times 4)$, and marginal structures, respectively, as established by RA. The asterisk (*) at 550°C corresponds to experimental conditions of Ref. 7 for the observation of the RA oscillations (62).

reactivity in the two environments is primarily due to the difference in As coverage, which is driven by strongly different partial pressures of As-containing precursors (As_4 in MBE and AsH_3 in MOCVD) (70). The phase diagram shows that growth in MBE occurs with the (2×4) reconstruction, whereas growth in MOCVD occurs under $c(4 \times 4)/d(4 \times 4)$-like stabilized conditions. The high–As-coverage disordered $d(4 \times 4)$ reconstruction, which can be distinguished from $c(4 \times 4)$ in RA spectra by the peak at 4.0 eV (Fig. 7), is a phase in which the long-range order of the $c(4 \times 4)$ reconstruction is partially broken by adsorption of excess As (62).

2. AP Atomic Layer Epitaxy

In the $TMG-AsH_3$ system, three models of self-limiting (SL) ALE have been suggested (*45,46,71,72*). In the selective adsorption model, it is assumed that TMG does not adhere to Ga, presumably the outer-layer species resulting from the exposure of (001) GaAs to TMG (*73*). Thus, when the surface is completely covered by Ga, its reaction with TMG disappears. In the adsorbate inhibition model, the same behaviour is obtained by assuming that each surface Ga is capped with a methyl radical, CH_3 (*74*). In the flux balance model, SL-ALE is a result of a dynamic equilibrium between Ga deposited from TMG and Ga desorbed as monomethylgallium, $GaCH_3$ (*75*). All these models assume that SL-ALE occurs at a substrate temperature of 400–550°C, the (001) GaAs surface alternating between single-layer coverage by Ga (possibly terminated with methyl radicals) and As. However, the phase diagram (see Fig. 18) shows that under SL-ALE conditions the surface is covered by more than one monolayer of As. The role of the As multilayers in the ALE or in MOCVD remains unclear (*70*).

ALE can unambiguously be addressed by permanently examining the state of the surface during an ALE cycle. Real-time RA spectroscopy has been applied to study ALE on (001) GaAs (*72*). Multitransient RA spectroscopy has been performed by recording kinetic RA measurements (RA transient) at photon energies from 1.5 to 5.5 eV at regular intervals ($\Delta E = 0.05$ eV) during each SL-ALE exposure cycle for a total number of 250 cycles (Figs. 19 and 20). Figure 19 shows the RA transients at 2.6 eV (As dimers transition) and 1.95 eV, near the transient energy of Ga dimers. Each cycle is divided into five regimes: the evolution of the AsH_3-stabilized surface during the H_2 purge (3–36); the purged surface during exposure to TMG (36–62); the TMG-saturated surface during excess exposure to TMG (62–81); the TMG-saturated surface during the H_2 purge (81–206); and the purged surface during further exposure to AsH_3 (206–248). Spectrum 36 shows that the 4 s H_2 purge removes only a small fraction of As, and that the surface immediately prior to TMG exposure is still covered by at least two layers of As (*72*). This observation is not consistent with any previously proposed simple model of SL-ALE. The comparison of primary reconstruction spectra (Fig. 7) with spectra 50 and 62 (Fig. 20) indicates that c/d(4 × 4) dimers are being replaced by As dimers oriented along [1$\bar{1}$0], i.e., the original two outer layers of As are being reduced to only one. Upon exposure to AsH_3 the H_2-purged surface returns directly to the AsH_3-saturated state through the accumulation of the first and second layers of As (206–248). The negative trend of the RA signal at 1.9 eV both upon

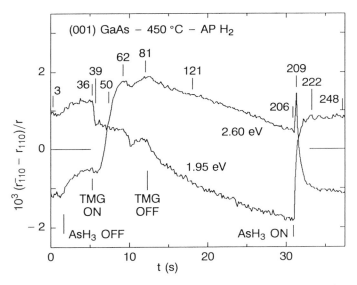

FIG. 19. Two of a set of 81 250-point transients illustrating the evolution of (001)GaAs over the ALE-like cycle of 4 s H_2, 7 s H_2 + TMG, 19 s H_2, and 10 s H_2 + AsH_3 (72).

exposure to TMG (36–81) and purging H_2 (81–206) indicates an increase in Ga dimers.

The TMG exposure is the most important part of the cycle because it is involved in the rate-limiting step when growth is limited by surface reactions and in the self-limiting process essential for the deposition of a single atomic layer. Figures 19 and 20 show that under TMG exposure (points 39–61), the surface is converting from c(4 × 4) to (2 × 4). During the H_2 purging (81–206), the TMG-saturated surface is transformed to a Ga-rich, probably (1 × 6)/(4 × 6)-like reconstruction (70). In contrast with the previous models of SL-ALE which suppose surface transformation from As-terminated to Ga-terminated during the TMG exposure, the RA studies reveal termination between double c(4 × 4) and single (2 × 4) layers of As (70, 72). Early investigations of the kinetic limitations of MOCVD, which reveal linear–exponential dependence in reaching monolayer coverage (Fig. 21) at the SL-ALE growth mode, can be attributed to study of the transition from c(4 × 4)/d(4 × 4) to (2 × 4) reconstructions (10, 11, 76). Separate thickness measurements show that the RA signal saturated value of the TMG exposure stage is reached when one monolayer of Ga has been deposited (72). Prolonging the exposure to TMG could lead to the formation of Ga

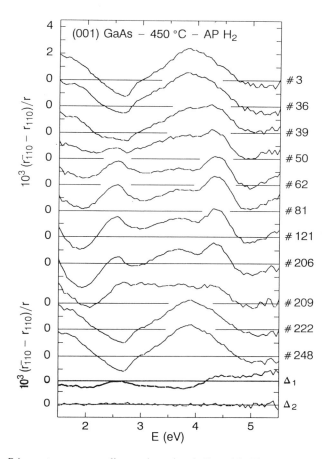

FIG. 20. RA spectra corresponding to the points indicated in Fig. 19 (72).

clusters that are not crystallographically oriented and thus do not contribute directly to the RA signal.

Hence, the full range of the RA transient under TMG exposure represents deposition of one Ga monolayer on the underlying As, although the surface remains As-terminated. The exposure time required for the formation of one monolayer may be obtained from these curves (77). As MOCVD is sensitive to process variations, the RA transients are also very sensitive to temperature and possible other time-varying effects. Hence, *in situ* RA monitoring of monolayer growth appears necessary.

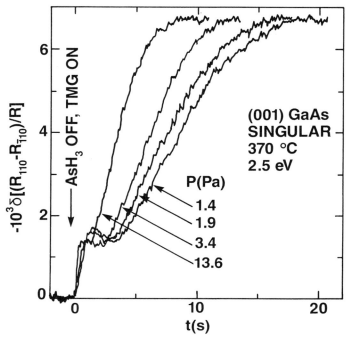

FIG. 21. Time dependence of RA signals at 2.5 eV for (001)GaAs in an AP MOCVD reactor upon initiation of TMG exposures at $t = 0$ for various TMG partial pressures as shown. For $t < 0$, the surface was stabilized in AsH_3 (10).

IV. Growth of Lattice-Mismatched III–V and II–VI Structures

The previous measurements have been performed on lattice-matched systems that are characterized by low values of the RA signal. In contrast, RA measurements related to lattice-mismatched structures can display a large RA signal (even > 0.1) (13, 64, 78).

Extensive RA studies of MOCVD growth of lattice-mismatched structures have been performed (InP/GaAs, InAs/InP, ZnTe/GaAs and CdTe/GaAs) (13, 64, 78). RA measurements display a general behaviour characterized by long-term damped oscillations after the initial transient stage (see Figs. 22 and 23). The damping is related to the penetration depth of a

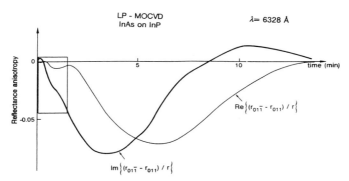

FIG. 22. RA record of the MOCVD growth of an InAs layer on an InP substrate, using a photon energy of 1.96 eV (*13*).

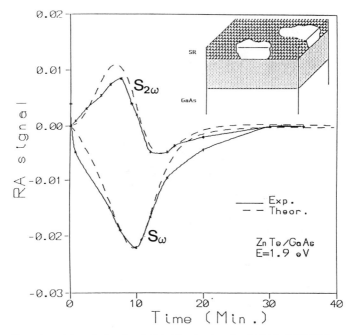

FIG. 23. Experimental and simulated RA kinetic signals of ZnTe MOCVD growth on GaAs at 365°C (*78*). $S_\omega = \text{Im}(\Delta r/r)$ and $S_{2\omega} = \text{Re}(\Delta t/r)$.

probing beam. In the region of material transparency, the damping factor is mainly determined by evolution of bulk or surface structures. In principle, the origin of the oscillations can be related to both bulk and surface anisotropy. Deposition of lattice-mismatched layers is correlated with the development of intrinsic strain. In the case of $In_xGa_{1-x}As$ on GaAs, the compressive strain leads to a tetragonal distortion of the unit cell below a certain critical thickness L_c (79). Above L_c, plastic relaxation occurs

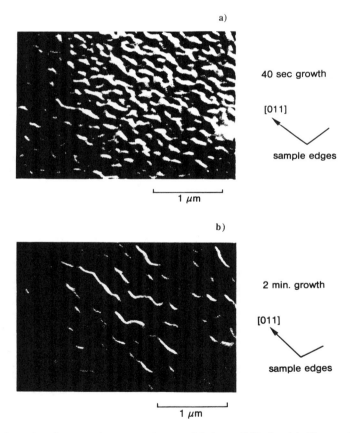

FIG. 24. Scanning electron microscope pictures of InAs on InP after (a) 40 s growth, (b) min growth, (c) 3 min 30 s growth and (d) 1 h growth (13).

increasingly by generation of misfit dislocations. The asymmetric strain relaxation leads to a change of crystal symmetry from tetragonal in unrelaxed layers to monoclinic in partially relaxed layers (79). Therefore, the observed damped oscillations of the RA signal might be considered as strain-relaxation effects in pseudomorphic growth. Nevertheless, taking into account scanning electron microscope (SEM) observation of surface morphology of deposited layers, the oscillations shown in Figs. 22 and 23 have been attributed to interference effects in the growing layer, their quasi-period directly corresponding to deposition rates (13). More precisely, the oscillation amplitude is supposed to be determined by the formation and the evolution of an anisotropic surface microroughness at a typical size of a tenth of a nanometer (80). However, RA monitoring of continuous lattice-mismatched MBE growth of InAs on GaAs at 2.4 eV does not reveal oscillations (81). The latter behaviour can probably be corelated to a fast smoothing of the anisotropic surface roughness within less than an interference quasi-period.

The surface morphology of InAs on InP has been observed by SEM (Fig. 24) (13). The lateral dimension of the roughness is found to increase with deposition time, its anisotropic shape appearing clearly. It consists mainly of rectangular holes of different sizes with all their edges parallel to [011] and [01$\bar{1}$]. This observation is consistent with the prediction that they should be the privileged symmetry directions. Scanning tunneling microscopy observation of 50 MBE-grown InAs monolayers on GaAs reveals the same kind of surface roughness (81). Surface morphology of CdTe (or ZnTe)/GaAs epilayers are also rough, exhibiting hillocks and pyramids (82). Based on these observations, the following optical optical model of an evolutionary anisotropic surface roughness can be suggested. The anisotropy of the surface microroughness leads to the appearance of the RA signal, its amplitude depending both on the effective thickness of the roughness and its anisotropy. Formation of the roughness results in an increase in the oscillation amplitude, whereas its smoothing leads to a decrease in the amplitude. The period of the oscillations is a direct measurement of the deposition rate. Considering the surface roughness as a single layer that is described by an effective thickness d_{SR} and a set of effective dielectric functions $\varepsilon_j (j = 1, 2$ and 3 or $x, y, z)$, determined by Bruggeman effective medium approximation (BEMA) (83). BEMA predicts the effective dielectric function ε_j for the electric field in the jth direction, provided that the corresponding depolarization factor q_j for the effective medium is given. This factor describes the screening of the electric field through the rough layer. The effective dielectric functions for the anisotropic surface roughness are calculated from

$$\frac{1-\varepsilon_j}{1+k_j\varepsilon_j}(f_v - 1) + \frac{\varepsilon_m - \varepsilon_j}{\varepsilon_m + k_j\varepsilon_j} = 0, \quad j = 1, 2, 3, \tag{11}$$

where ε_m is the dielectric function of the deposited material, f_v is the void volume fraction, and k_j is the screening parameter. The latter quantity accounts for the accumulation of charge at the boundaries of the layer and is related to the depolarization factor q_j, for the direction j along the electric field, according to the relation $k_j = (1 - q_j)/q_j$. The depolarization factor q_j incorporates the effect of the shape of the individual regions that make up the surface roughness. Anisotropic roughness morphologies, such as stripes or ellipsoids, lead to anisotropic depolarization factors. The main axes of this anisotropy are expected to be parallel to the symmetry directions of the surface as it is seen in Fig. 24. Thus, two different screening factors $q[011]$ and $q[01\bar{1}]$ for (100) surfaces (or $q[110]$ and $q[\bar{1}10]$ for (001) surfaces) are assumed for light polarized along the [011] and [01$\bar{1}$] ([110] and [$\bar{1}$10]) directions, respectively. As a consequence, two different effective dielectric functions $\varepsilon[011] = \varepsilon(\varepsilon_m, f_v, q[011])$ and $\varepsilon[01\bar{1}] = \varepsilon(\varepsilon_m, f_v, q[01\bar{1}])$ are obtained from Eq. (11). In the Wiener limits, for any composition and microstructure, the depolarization factor is varying within the limits $0 < q_j < 1$ (84). Let z be the axis normal to the growing surface. The depolarization factors ($q[011]$, $q[01\bar{1}]$, q_z) of the surface roughness layer along the three dimensions must satisfy the conditions $q[011] + q[01\bar{1}] + q_z = 1$. If the surface roughness consists of hemispherical isotropic hillocks, the depolarization factors for such a 2-D symmetric system $q[011] = q[01\bar{1}] = (1 - q_z)/2$, and then the RA signal should be zero. But if the hillocks have a trianglular shape or exhibit more complicated forms (stripes or ellipsoids), then $q[011] \neq q[01\bar{1}]$, and therefore $\varepsilon[011] \neq \varepsilon[01\bar{1}]$. As a consequence, a significant RA signal will appear. This RA signal is expected to depend on the surface morphology. In order to simulate the deposition kinetic with constant deposition rate v_d the following evolution of the surface roughness of thickness d_{SR} is assumed:

$$d_{SR} = at^b e^{-ct}, \tag{12}$$

where parameters a and b characterize mainly the initial stage of the evolution, c characterizing the relaxation (surface smoothing). Then the complex anisotropic reflectance of the three-layer system (substrate—dense layer of thickness $d = v_d t$—anisotropic surface roughness) is calculated using the conventional multilayer formalism (26). Figures 23 and 25 show a rather good agreement between experimental and simulated curves for both kinetic measurements of ZnTe AP-MOCVD growth on GaAs and spectroscopic

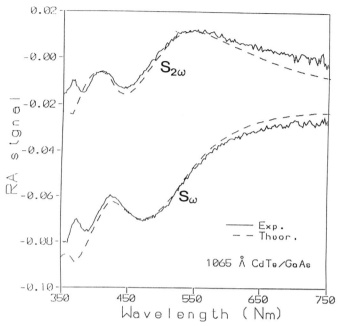

FIG. 25. Experimental and simulated RA spectroscopic measurements on 1,000 Å CdTe MOCVD-grown epilayer (78). $S_\omega = \text{Im}(\Delta r/r)$ and $S_{2\omega} = \text{Re}(\Delta r/r)$.

measurements of CdTe on GaAs (78). The kinetic simulation provides the following fitting parameters of ZnTe/GaAs deposition: $V_d = 0.8 \text{ Å s}^{-1}$, $a = 34 \text{ Å}$, $b = 1.6$ and $c = 0.26 \text{ s}^{-1}$. The simulation reveals that surface roughness reaches a maximum thickness around 125 Å after about 6 min of the deposition and then relaxes to a smooth and/or isotropic morphology. A good agreement is also observed in the case of the LP-MOCVD growth of lattice-mismatched III–V (see Fig. 26). Moreover, Fig. 26 displays a comparison between the two effective medium approximations Bruggeman and Maxwell–Garnett, a better agreement being obtained in the first case (26, 83, 84).

V. Summary and Conclusions

In this review, we have summarized recent *in situ* investigations of crystalline semiconductor surfaces in a growth environment by reflectance

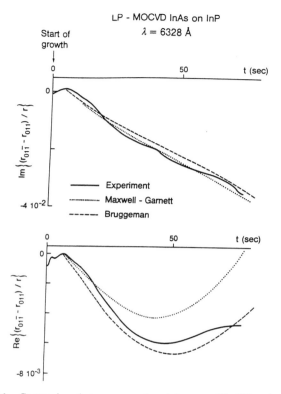

FIG. 26. Comparison between experiment (same as Fig. 22) and theory for the RA record at 1.96 eV of the first minute of growth of InAs on InP (13).

anisotropy (RA). RA is a normal-incidence optical probe that uses symmetry to enhance the typically low sensitivity of reflectance measurements to surface phenomena. As compared to the conventional electronic surface techniques, RA is non-invasive and can be used in any ambient.

Because of the recent great increase in published studies based on this technique, it can be inferred that RA is emerging as a very promising high-sensitivity surface probe. As a matter of fact, RA has been successfully applied to study the growth processes of III–V semiconductors in various deposition conditions from ultravacuum (MBE or CBE) up to atmospheric pressure (MOCVD). In particular, the RA responses of (001) surfaces of GaAs during crystal growth by MBE and MOCVD have been compared. The latter case illustrates how the determination of the complex surface dielectric anisotropy can extend the range of applications of the technique, providing surface-structural and chemical information from kinetic measurements at optimum photon energy. Taking advantage of its real-time capability, RA can be used in continuous or complex pulsed deposition systems such as migration-enhanced atomic layer epitaxy (MEALE). Besides monitoring the growth rate from RA signal oscillations and relative surface V/III ratio from the signal level, the real-time control of the chemical cycles of ALE has been achieved. In the case of low-pressure MOCVD, optimization of heterojunctions from RA transient measurements have also been presented. Growth of lattice-mismatched III–V and II–VI semiconductors induces large variations in the RA signals, which can be related to the development and the relaxation of an anisotropic microroughness through effective medium models.

Further improvements in experimental capabilities can also be anticipated from the analogy between RA and ellipsometry. As emphasized in the present review, the polarization modulation during RA measurements is generally provided by a photoelastic modulator. Therefore, RA can take advantage of the recent developments in similar ellipsometric techniques, such as real-time spectroscopic facility (32,85) and the extension toward infrared (32,86,87). As a consequence, RA will provide more information than can be expected from a surface analysis probe that can only operate in ultrahigh vacuum. Thus, RA will become extremely useful for providing the control necessary to realize future generations of semiconductor and optoelectronic devices.

References

1. D. E. Aspnes, *J. Vac. Sci. Technol.* **B3**, 1498 (1985).
2. D. E. Aspnes and A. A. Studna, *Phys. Rev. Lett.* **54**, 1956 (1985).
3. D. E. Aspnes, J. P. Harbison, A. A. Studna and L. T. Florez, *Phys. Rev. Lett.* **59**, 1687 (1987).
4. O. Acher and B. Drévillon, *Rev. Sci. Instrum.* **63**, 5332 (1992).
5. F. Manghi, R. Del Sole, A. Selloni and E. Molinary, *Phys. Rev. B* **41**, 9935 (1990).
6. D. E. Aspnes, J. P. Harbison, A. A. Studna, L. T. Florez and M. K. Kelly, *J. Vac. Sci. Technol.* **B6**, 1127 (1988).
7. G. Paulsson, K. Deppert, S. Jeppesen, J. Jönsson, L. Samuelson and P. Schmidt, *J. Crystal Growth* **105**, 312 (1990).
8. B. Y. Maa and P. D. Dapkus, *Appl. Phys. Lett.* **58**, 2261 (1991).
9. J. Jönsson, G. Paulsson and L. Samuelson, *J. Appl. Phys.* **70**, 1737 (1991).
10. D. E. Aspnes, E. Colas, A. A. Studna, R. Bhat, M. A. Koza and V. G. Keramidas, *Phys. Rev. Lett.* **61**, 2782 (1988).

11. E. Colas, D. E. Aspnes, R. Bhat, A. A. Studna, M. A. Koza and V. G. Keramidas, *J. Cryst. Growth* **94**, 613 (1989).
12. S. M. Koch, O. Acher, F. Omnes, M. Defour, M. Razeghi and B. Drévillon, *J. Appl. Phys.* **68**, 3364 (1990).
13. O. Acher, S. M. Koch, F. Omnes, M. Defour, M. Razeghi and B. Drévillon, *J. Appl. Phys.* **68**, 3564 (1990).
14. S. M. Koch, O. Acher, F. Omnes, M. Defour, B. Drévillon and M. Razeghi, *J. Appl. Phys.* **69**, 1389 (1991).
15. H. H. Farrell, M. C. Tamargo, T. J. Gmitter, A. L. Weaver and D. E. Aspnes, *J. Appl. Phys.* **70**, 1033 (1991).
16. V. L. Berkovits, V. A. Kisilev, and V. I. Safarov, *Surf. Sci.* **211/212**, 489 (1989).
17. D. E. Aspnes and A. A. Studna, *J. Vac. Sci. Technol.* **A5**, 546 (1987).
18. D. J. Wentik, H. Wormeester, P. de Boeij, C. Wijers and A. Van Silfhout, *Surf. Sci.* **274**, 270 (1992).
19. H. Wormeester, A. M. Molenbroek, C. M. J. Wijers and A. van Silfhout, *Surf. Sci.* **260**, 31 (1992).
20. V. L. Berkovits and D. Paget, *Appl. Phys. Lett.* **61**, 1835 (1992).
21. V. L. Berkovits, V. N. Besselov, T. N. L'vova, E. B. Novikov, V. I. Safarov, R. V. Khasieva and B. V. Tsarenkov, *J. Appl. Phys.* **70**, 3707 (1991).
22. S. R. Armstrong, A. G. Taylor and M. E. Pemble, *J. Phys. Condens. Matter.* **3**, S85 (1991).
23. D. E. Aspnes, A. A. Studna, L. T. Florez, Y. C. Chang, J. P. Harbison, M. K. Kelly and H. H. Farrell, *J. Vac. Sci. Technol.* **B7**, 901 (1989).
24. G. P. M. Poppe, H. Wormeester, A. Molenbroek, C. M. J. Wijers and A. van Silfhout, *Phys. Rev. B* **43**, 12122 (1991).
25. H. Tanaka, E. Colas, I. Kamiya, D. E. Aspnes, and R. Bhat, *Appl. Phys. Lett.* **59**, 3443 (1991).
26. R. M. Azzam and N. M. Bashara, "Ellipsometry and Polarized Light." North Holland, Amsterdam, 1977.
27. D. E. Aspnes, J. P. Harbison, A. A. Studna and L. T. Florez, *J. Vac. Sci. Technol.* **A6**, 1327 (1988).
28. S. N. Jasperson and S. E. Schnatterly, *Rev. Sci. Instrum.* **40**, 761 (1969).
29. B. Drévillon, J. Perrin, R. Marbot, A. Violet and J. L. Dalby, *Rev. Sci. Instrum.* **53**, 969 (1982).
30. O. Acher, E. Bigan and B. Drévillon, *Rev. Sci. Instrum.* **60**, 65 (1989).
31. B. Drévillon, J. Y. Parey, M. Stchakovsky, R. Benferhat, Y. Josserand and B. Schlayen, *SPIE Symp. Proc.* **1188**, 174 (1989).
32. B. Drevillon, *Progr. in Crystal Growth and Charact. of Mat.* **27**, 1 (1993).
33. D. E. Aspnes, Y. C. Chang, A. A. Studna, L. T. Florez, H. H. Farrell and J. P. Harbison, *Phys. Rev. Lett.* **64**, 192 (1990).
34. D. E. Aspnes, J. P. Harbison, A. A. Studna and L. T. Florez, *Appl. Phys. Lett.* **52**, 957 (1988).
35. A. A. Studna, D. E. Aspnes, L. T. Florez, B. J. Wilkens, J. P. Harbison and R. E. Ryan, *J. Vac. Sci. Technol.* **B7**, 901 (1989).
36. S. E. Acosta-Ortiz and A. Lastras-Martinez, *Solid State Comm.* **64**, 809 (1987).
37. S. R. Armstrong, M. E. Pemble, A. G. Taylor, B. A. Joyce, J. H. Neave, J. Zhang, and D. A. Klug, *Appl. Surf. Sci.* **54**, 493 (1992).
38. F. Briones and Y. Horikoshi, *Jpn. J. Appl. Phys.* **29**, 1014 (1990).
39. R. M. A. Azzam, *Opt. Comm.* **20**, 405 (1977).
40. R. M. A. Azzam, *J. Optics (Paris)* **12**, 317 (1981).

41. V. M. Agranovich and V. L. Ginzburg, "Spatial Dispersion in Crystal Optics and Theory of Excitons." Springer-Verlag, Berlin, 1984.
42. J. Pastrnak and K. Vedam, *Phys. Rev.* **B3**, 2567 (1971).
43. Y. P. Yu and M. Cardona, *Solid State Comm.* **9**, 1421 (1971).
44. W. L. Mochan and R. G. Barrera, *J. Phys. (Paris) Suppl.* **45**, 207 (1984).
45. S. P. Den Baars, P. D. Dapkus, C. A. Beyler, A. Hariz and K. M. Dzurko, *J. Cryst. Growth* **93**, 195 (1988).
46. P. D. Dapkus, B. Y. Maa, Q. Chen, W. G. Jeong and S. P. Den Baars, *J. Cryst. Growth* **107**, 73 (1991).
47. N. Kobayashi, T. Makimoto, Y. Yamauchi, and Y. Horikoshi, *J. Appl. Phys.* **66**, 640 (1989).
48. Y. Horikoshi, M. Kawashima, and H. Yamaguchi, *Jpn. J. Appl. Phys.* **27**, 169 (1988).
49. H. Sugiura, M. Kawashima, and Y. Horikoshi, *J. Cryst. Growth* **81**, 9 (1987).
50. B. A. Joyce, J. Zhang, T. Shirata, J. H. Neave, A. Taylor, S. Armstrong, M. E. Pemble, and C. T. Foxon, *J. Cryst. Growth* **115**, 338 (1991).
51. J. P. Harbison, D. E. Aspnes, A. A. Studna, L. T. Florez and M. K. Kelly, *Appl. Phys. Lett.* **52**, 2046 (1988).
52. D. E. Aspnes, L. T. Florez, A. A. Studna, and J. P. Harbison, *J. Vac. Sci. Technol.* **B8**, 936 (1990).
53. M. Wasssermeier, I. Kamiya, D. E. Aspnes, L. T. Florez, J. P. Harbison and P. M. Petroff, *J. Vac. Sci. Technol.* **B9**, 2263 (1991).
54. I. Kamiya, D. E. Aspnes, H. Tanaka, L. T. Florez, J. P. Harbison and R. Bhat, *Phys. Rev. Lett.* **68**, 627 (1992).
55. D. J. Chadi, *J. Vac. Sci. Technol.* **A5**, 834 (1987).
56. H. H. Farrell, J. P. Harbison, and L. D. Peterson, *J. Vac. Sci. Technol.* **B5**, 1482 (1987).
57. L. Samuelson, K. Deppert, B. Junno, J. Jonsson and G. Paulsson, *SPIE Symp. Proc.* **1678**, 268 (1992).
58. J. Jonsson, K. Deppert, S. Jeppersen, G. Paulsson, L. Samuelson and P. Schmidt, *Appl. Phys. Lett.* **56**, 2414 (1990).
59. L. Samuelson, K. Deppert, S. Jepperesen, J. Jonsson, G. Paulsson and P. Schmidt, *J. Cryst. Growth* **107**, 68 (1991).
60. G. Paulsson, K. Deppert, S. Jeppersen, J. Jonsson, L. Samuelson and P. Schmidt, *J. Cryst. Growth* **111**, 115 (1991).
61. K. Deppert, J. Jonsson, and L. Samuelson, *Appl. Phys. Lett.* **61**, 1588 (1992).
62. I. Kamiya, H. Tanaka, D. E. Aspnes, L. T. Florez, E. Colas, J. P. Harbison and R. Bhat, *Appl. Phys. Lett.* **60**, 1238 (1992).
63. T. Nakanisi, *J. Cryst. Growth* **68**, 282 (1984).
64. O. Acher, F. Omnes, M. Razeghi and B. Drévillon, *Mater. Sci. Eng.* **B5**, 223 (1990).
65. M. Razeghi, The MOCVD challenge, Adam-Hilger, Bristol, 1989.
66. F. Hottier, J. Hallais and F. Simondet, *J. Appl. Phys.* **51**, 1599 (1980).
67. M. E. Pemble, D. S. Buhaenko, S. M. Francis, P. A. Goulding and J. T. Allen, *J. Cryst. Growth* **107**, 37 (1991).
68. Y. C. Chang and D. E. Aspnes, *Phys. Rev.* **B41**, 12002 (1990).
69. M. Wassermeier, I. Kamiya, D. E. Aspnes, L. T. Florez, J. P. Harbison and P. M. Petroff, *J. Vac. Sci. Technol.* **B9**, 2263 (1991).
70. I. Kamiya, D. E. Aspnes, H. Tanaka, L. T. Florez, J. P. Harbison and R. Bhat, *J. Vac. Sci. Technol.* **B10**, 1716 (1992).
71. J. R. Creighton and B. A. Banse, *Mater. Res. Soc. Symp. Proc.* **222**, 15 (1991).
72. D. E. Aspnes, I. Kamiya, H. Tanaka and R. Bhat, *J. Vac. Sci. Technol.* **B10**, 1725 (1992).
73. M. Ozeki, K. Mochizuki, N. Ohtsuka and K. Kodama, *Appl. Phys. Lett.* **53**, 1509 (1988).

74. J. Nishizawa, T. Kurabayashi, H. Abe and N. Sakurai, *J. Vac. Sci. Technol.* **A5,** 1572 (1987).
75. M. L. Yu, N. I. Buchan, R. Souda and T. F. Keuch, *Mater. Res. Soc. Symp. Proc.* **222,** 81 (1991).
76. D. E. Aspnes, R. Bhat, E. Colas, V. G. Keramidas, M. A. Koza and A. A. Studna, *J. Vac. Sci. Technol.* **A7,** 711 (1988).
77. D. E. Aspnes, R. Bhat, E. G. Colas, L. T. Florez, J. P. Harbison and A. A. Studna, "Optical control of deposition of crystal monolayers," U.S. Patent No. 4,931,132, June 5, 1990.
78. V. Sallet, R. Druilhe, J. E. Bouree, O. Acher, V. Yakovlev, B. Drevillon, and R. Triboulet, *Mat. Sci. Eng.* B16, 118 (1993).
79. M. Grundmann, U. Lienert, D. Bimberg, A. Fischer-Dolbrie and J. N. Miller, *Appl. Phys. Lett.* **55,** 1765 (1989).
80. D. E. Aspnes, *Phys. Rev. B* **41,** 10334 (1990).
81. S. M. Scholz, A. B. Muller, W. Richter, D. R. T. Zahn, D. I. Westwood, D. A. Woolf and R. H. Williams, *J. Vac. Sci. Technol.* **B10,** 1710 (1992).
82. A. Raizman, M. Oron, G. Cinader and H. Shtrikman, *J. Appl. Phys.* **67,** 1554 (1990).
83. D. A. G. Bruggeman, *Ann. Phys. Leipzig* **24,** 636 (1935).
84. D. E. Aspnes, *Thin Solid Films* **89,** 249 (1982) and references therein.
85. W. M. Duncan and S. A. Henck, *Appl. Surf. Sci.* 63 (1993).
86. N. Blayo, B. Drévillon and R. Ossikovski, *SPIE Symp. Proc.* **1681,** 116 (1992).
87. A. Canillas, E. Pascual and B. Drévillon, *Rev. Sci. Instrum.* 64, 2153 (1993).

Real-Time Spectroscopic Ellipsometry Studies of the Nucleation, Growth, and Optical Functions of Thin Films, Part I: Tetrahedrally Bonded Materials

ROBERT W. COLLINS, ILSIN AN, HIEN V. NGUYEN, YOUMING LI, AND YIWEI LU

Department of Physics and Materials Research Laboratory,
The Pennsylvania State University, University Park, Pennsylvania

I. Introduction 50
   A. Real-Time Ellipsometry at a Single Photon Energy 50
   B. Real-Time Spectroscopic Ellipsometry 52
II. Techniques of Real-Time Spectroscopic Ellipsometry 54
   A. Instrumentation 55
   B. Data Collection and Interpretation 56
III. Studies of Tetrahedrally Bonded Thin Films 61
   A. Hydrogenated Amorphous Silicon 61
      1. Microstructural Analysis from Real-Time Observations 62
      2. Optical Functions from Real-Time Observations 70
      3. Process–Property Relationships in a-Si:H 75
   B. Hydrogenated Amorphous Silicon–Carbon Alloys 85
      1. Optical Gaps from Real-Time Observations 85
      2. Role of H_2-Dilution of Reactive Gases 88
   C. Microcrystalline Silicon 93
      1. Optical and Microstructural Analysis from Real-Time Observations . . 94
      2. Size Effects in Microcrystallites 99
   D. Diamond 104
      1. Substrate Treatment and Annealing 105
      2. Microstructural Analysis from Real-Time Observations 108
IV. Summary 116
   Acknowledgments 121
   References 122

I. Introduction

Because of the stringent demands on the performance specifications of thin films and the complexity of the processes used to fabricate them, techniques for real-time monitoring of thin-film and interface characteristics during preparation and processing have gained considerably in importance. If the data collected in real time can also be interpreted in real time, then it becomes possible to control materials characteristics through a closed-loop adjustment of process variables, such as the gas flow ratios that establish alloy composition. Even if real-time interpretation is not possible, then the information deduced in a post-process analysis of real-time measurements can be applied to a better understanding of the process. From a technological standpoint, one can assess reproducibility, identify problems, and arrive expeditiously at the appropriate process variables. In contrast, the fundamental mechanisms underlying processing–(materials performance) relationships may not be evident from post-process measurements.

Although the information desired from a real-time measurement is a function of the thin-film application, the following characteristics are of general interest: microstructure, including growth mode, density, and thickness; alloy composition; electronic properties, including band structure, band gap, doping and defect density; surface temperature; etc. Obtaining any such information at all is a challenge, considering that the probe must be non-invasive, non-perturbing, and hence contactless. Traditionally, diffraction techniques have been used most widely to provide feedback on the growth of materials under ultrahigh-vacuum conditions by molecular beam epitaxy (MBE). Surface probes involving either electrons and ions, however, cannot be used in the high-pressure, reactive environments associated with the plasma-enhanced chemical vapor deposition (PECVD) of thin films. In contrast, real-time probes of film growth based on photons, if sufficiently powerful, will permit a wider range of applications from MBE to PECVD.

A. Real-Time Ellipsometry at a Single Photon Energy

At first glance, it may appear that any optical probe will not be powerful enough to extract the film characteristics of general interest just listed. However, steady progress in real-time ellipsometry techniques over the last 15 years has been demonstrated through research on a wide range of thin-film applications ($1-9$). The power of ellipsometry derives from its ability to provide the relative amplitude ratio and the phase shift between the p and s electric field components of a fully polarized light wave, and thus, the changes incurred in these values when such a light wave is reflected from a surface under study (10). Here p and s refer to the directions parallel and perpendicular to the plane of incidence that contains the wave vector of the incident and reflected light waves. The amplitude and phase change parameters (ψ, Δ) obtained in ellipsometry can also be expressed in terms of the changes upon reflection in the angular orientation and ellipticity of the polarization ellipse associated with the incident wave. In contrast to ellipsometry, a reflectance measurement provides only one parameter, the ratio of the reflected to incident irradiances. If one performs ellipsometry on an ideal interface between an ambient medium and an atomically flat, opaque solid, then the index of refraction and the extinction coefficient [or both real and imaginary parts of the dielectric function, $(\varepsilon_1, \varepsilon_2)$] of the solid at a given photon energy $h\nu$ can be extracted from the (ψ, Δ) measurement at $h\nu$, as long as the angle of incidence is known. A reflectance measurement at $h\nu$ does not provide this capability.

By using rapid, automatic polarization state modulation or detection devices, real-time ellipsometry measurements of (ψ, Δ) at a single photon energy during thin-film growth can be performed with acquisition/repetition times from $4\,\mu s$ (11) to tens of seconds (9). The instrument design and data acquisition/accumulation routines are generally tailored to match the deposition rate of the process under study, keeping in mind the sensitivity of the ellipsometry angles to submonolayer changes in the film thickness. Such sensitivity arises from the high-accuracy phase measurement. Typically, however, one ensures that the acquisition time, t_a, for a single (ψ, Δ) corresponds to the accumulation of less than one monolayer (i.e., $t_a(s) < (2\,\text{Å})/[r(\text{Å}/s)]$, where r is the instantaneous deposition rate) in order to avoid thickness-averaging errors in (ψ, Δ). The repetition time for measurement is then chosen on the basis of the thickness scale for the processes of interest. For the homogeneous, layer-by-layer growth of a perfectly uniform transparent or absorbing film on an ideal, fully characterized substrate, real-time ellipsometry at a single photon energy $h\nu$ can provide $\{\varepsilon_1(h\nu), \varepsilon_2(h\nu)\}$ of the film and its thickness versus time (6, 12). Information such as the band gap, composition, and microstructure of the film are not readily accessible from the optical functions at the single photon energy, at least under most thin-film growth conditions.

Among the exceptions to this limitation is in the recent case of the epitaxial growth of well-characterized crystalline semiconductor alloy films, however. For example, the crystalline perfection of $Al_xGa_{1-x}As/GaAs$ heterostructures allows one to determine the Al composition, x, to within ± 0.03, averaged over the top $\sim 15\,\text{Å}$ of the film, as a function of time based on (ψ, Δ) measurements versus time at one photon energy (13). This has been achieved through calibrations that have determined $(\varepsilon_1, \varepsilon_2)$ versus x that are valid only for the specific photon energy and growth temperature. This capability in turn allows one to characterize compositional profiles as a function of time (or depth into the film) in cases where the composition is varying intentionally or unintentionally during growth. It should be emphasized that such approaches are well suited to epitaxial growth processes because of their perfection; any deviation from the crystalline perfection in the growth process leads to errors in data interpretation.

Amorphous, microcrystalline, or polycrystalline films inevitably exhibit nucleation and growth phenomena, and roughness on the film surface, typically much greater than the monolayer level. This makes analysis of single-photon energy ellipsometry data more difficult and subject to potential errors and uncertainty. For a semitransparent film that ultimately grows to opacity, one typically compares the trajectory in the ellipsometry angles that is swept out in the (ψ, Δ) plane as a function of time during film growth, to the corresponding trajectory assuming perfect uniform growth. Discrepancies in the data from the model can be considered as being due to deviations from the perfect layer-by-layer process. Then more sophisticated optical models of film growth can be constructed that include, for example, a nucleation process consisting of cluster formation, coalescence, and surface roughness evolution. The resulting simulations can then be compared to the data in order to understand the discrepancies from perfect growth (14–17). Once these discrepancies are understood, the deduced features can be studied as a function of deposition conditions to obtain insights into the film growth process (16).

Because of the complexities associated with the growth process, the ellipsometry data at a single photon energy are often not sufficient to establish unique growth models (18). This is especially true in micro- or polycrystalline materials in which the dielectric function $(\varepsilon_1, \varepsilon_2)$ may be very sensitive to the microstructure through finite size effects. Recent advances in ellipsometric instrumentation have been stimulated in the last few years by the hope of solving these more complex problems, and have culminated in the development of real-time spectroscopic ellipsometry (SE) as a more powerful *in situ* probe of thin-film growth (9, 19, 20).

B. Real-Time Spectroscopic Ellipsometry

Developments in SE as an *ex situ* probe of materials and thin films have roughly paralleled the development of real-time, single-photon energy ellipsometry. The motivating force behind SE was the physical information inherent in the photon energy dependence of $(\varepsilon_1, \varepsilon_2)$ for a thin film or bulk material (21). For example, the plasma-frequency and electron relaxation time can be determined for a bulk free-electron metal, and the interband critical-point energies can be determined for a bulk semiconductor. In fact, through the latter, information on composition, temperature, doping, and defects can be deduced. Furthermore, the added spectroscopic information available allows one to solve thin-film problems more readily. This involves extracting layer thickness information that can lead one to the true dielectric function of the film.

In *ex situ* SE studies, one typically uses a combination broad-band source/spectrometer and photomultiplier tube detector to measure (ψ, Δ) at a number of closely spaced photon energies, typically from the near infrared to the near ultraviolet (21). This generally requires several minutes to scan

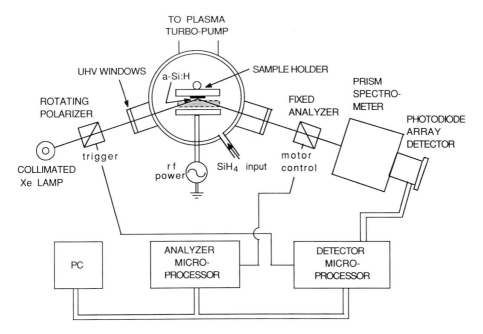

FIG. 1. A schematic of the rotating polarizer multichannel ellipsometer. The deposition system in this case is designed for rf plasma-enhanced chemical vapor deposition in the capacitively coupled, parallel-plate configuration. In this configuration, tetrahedrally bonded thin films such as hydrogenated amorphous silicon (a-Si:H), amorphous silicon–carbon alloys (a-Si$_{1-x}$C$_x$:H), and microcystralline Si (μc-Si:H) are prepared. The windows are oriented for a 70° angle of incidence at the sample surface. After Ref. 22.

the spectrometer over the desired range with a reasonable spectral resolution. If the spectrometer is removed from the source side of the spectroscopic ellipsometer and the photomultiplier is replaced by a spectrograph/photodiode array combination (see Fig. 1), the instrument can then collect spectroscopic data in parallel over a wide photon energy range (20). This is the key development that has led to the capability of collecting spectra from the near infrared to the near ultraviolet in a time as short as 16 ms (22), opening up the possibility of real-time SE.

The purpose of this chapter is to describe the basic operational principle of the real-time spectroscopic ellipsometer (Section II), and then present a number of applications drawn from extensive studies of tetrahedrally bonded thin films (Section III). In many of these applications, real-time SE is providing detailed glimpses into the complex microstructural development of the thin films for the first time. Where relevant, the connections are emphasized between the deposition parameters, the monolayer-scale microstructural evolution, and the ultimate properties of the materials. Section III consists of four subsections describing studies of hydrogenated amorphous silicon (a-Si:H) (23–26), amorphous silicon carbon alloys (a-Si_xC_{1-x}:H) (27), microcrystalline silicon (μc-Si:H) (28, 29), and polycrystalline diamond (30).

In general, the thin-film materials will be presented in this order because the sequence represents increasing complexity of the growth processes, as well as the difficulty of the measurement and interpretation. The studies of a-Si:H and a-$Si_{1-x}C_x$:H are simplified by the fact that the nucleation process occurs on a relatively small scale (first 20 Å of growth), and finite-size effects appear to be negligible. In contrast, μc-Si:H films tend to grow with a low nucleation density, and the dielectric function of the growing crystallites changes continuously as a function of crystallite size. Although such size effects are not significant in diamond, the substrate treatments required to nucleate diamond on non-diamond substrates, and the resulting alterations that the substrate undergoes in this process, push real-time SE to the limits of its capabilities. In this most difficult case, however, it is demonstrated that the real-time SE interpretation is reliable through comparison with scanning electron microscopy.

II. Techniques of Real-Time Spectroscopic Ellipsometry

The polarization states of the incident and reflected beams in an ellipsometry experiment are defined by the ratios E_{pi}/E_{si} and E_{pr}/E_{sr}, respectively. Here E_j denotes the complex electric field including both amplitude and phase, p and s denote the two orthogonal components parallel and perpendicular to the plane of incidence, and i and r denote the incident and reflected beams. In the ellipsometry experiment, one measures the polarization state change that occurs upon reflection from the surface as the ratio between the reflected beam and incident beam polarization states. Thus, the ellipsometry angles are defined in terms of the change in polarization state according to (*10*)

$$\tan\psi \exp(i\Delta) \equiv (E_{\mathrm{pr}}/E_{\mathrm{sr}})/(E_{\mathrm{pi}}/E_{\mathrm{si}}) = (E_{\mathrm{pr}}/E_{\mathrm{pi}})/(E_{\mathrm{sr}}/E_{\mathrm{si}}) = r_{\mathrm{p}}/r_{\mathrm{s}}. \qquad (1)$$

Here $r_{\mathrm{p}} \equiv E_{\mathrm{pr}}/E_{\mathrm{pi}}$ and $r_{\mathrm{s}} \equiv E_{\mathrm{sr}}/E_{\mathrm{si}}$ are the complex amplitude reflection coefficients of the surface for the p and s components. These coefficients depend on the dielectric functions and thicknesses of all the different layers that make up the sample surface, as well as the angle of incidence and the photon energy. Equation (1) defines the real-valued ellipsometry parameters ψ and Δ.

A. Instrumentation

Figure 1 shows the experimental apparatus used to collect ~ 100 (ψ, Δ) values over a photon energy range from 1.5 to 4.5 eV in a time as short as 32 ms during semiconductor film growth by PECVD (*9,20,22*). The instrument consists of a collimated Xe source, a continuously rotating polarizer assembly with a quartz Rochon element, a vacuum preparation and processing system with strain-free windows, a stepping motor-controlled analyzer with a calcite Glan–Taylor element, a prism monochromator, and a 1,024-element photodiode array and controller. Each frequency component v of the white light reaching the sample surface is linearly polarized and continuously rotating at a mechanical frequency of ω, which can be set from 10 to 30 Hz. Upon reflection from the surface, each optical frequency component is converted to elliptically polarized light whose azimuth and ellipticity are modulated at 2ω. The fixed analyzer is used for the polarization state analysis of all optical frequency components simultaneously.

As a result, the time dependence of the spectral irradiance reaching detector element k is given by (*9*)

$$I(hv_k) = I_0(hv_k)[1 + \alpha(hv_k)\cos 2(\omega t - P_{sk}) + \beta(hv_k)\sin 2(\omega t - P_{sk})]. \qquad (2)$$

In most experiments, the detector elements are grouped by eight; thus, k ranges from 1 to 128. In this equation, I_0 is the average irradiance, and α and β are the normalized 2ω Fourier coefficients of the waveform, which all depend on the photon energy, hv_k, accepted by detector element k. Time zero is defined by the phase of the detector readout, and $-P_{sk}$ is the polarizer position at time zero for detector element k, measured with respect to the plane of incidence. Because each detector element is read out sequentially, $P_{sk} = P_0 + (k-1)\delta P$, where δP is the angular rotation of the polarizer during the time required to read one detector element, and P_0 is a constant phase factor. Both δP and P_0 are determined in calibration (31,32).

For each detector element, the readout provides the irradiance striking the detector area, integrated over the time since the previous readout (33). If the array is read out at frequency 8ω, with triggering provided by an encoder mounted on the polarizer motor shaft, then the resulting four spectra S_i ($i = 1, 2, 3, 4$) obtained over an optical cycle can be used to calculate spectra in the three unknowns I_0, α, and β, along with a spectroscopic consistency check. Only the results for α and β are of interest for the purposes of the present chapter (34). The spectra in α and β are obtained from the experimental relations

$$\alpha' = (\pi/2)(S_1 - S_2 - S_3 + S_4)/(S_1 + S_2 + S_3 + S_4), \tag{3a}$$

$$\beta' = (\pi/2)(S_1 + S_2 - S_3 - S_4)/(S_1 + S_2 + S_3 + S_4), \tag{3b}$$

by applying a $2P_{sk}$ rotation transformation:

$$\alpha = \alpha' \cos 2P_{sk} + \beta' \sin 2P_{sk} \tag{4a}$$

$$\beta = -\alpha' \sin 2P_{sk} + \beta' \cos 2P_{sk}. \tag{4b}$$

In Eqs. (3)–(4), the explicit dependences on photon energy have been omitted for simplicity. Next, spectra in α and β provide the spectra in (ψ, Δ) according to

$$\tan \psi = [(1-\alpha)/(1+\alpha)]^{1/2} \tan A, \tag{5}$$

$$\cos \Delta = \beta/(1-\alpha^2)^{1/2}, \tag{6}$$

where A is the fixed analyzer angle measured with respect to the plane of incidence (9).

B. Data Collection and Interpretation

Figure 2 shows a typical example of real-time SE data collected during the preparation of a ~ 110 Å thick a-Si:H film by PECVD onto a native oxide-covered crystalline Si (c-Si) substrate held at 250°C (22). Here, 250 pairs of pseudo-dielectric functions have been plotted versus time in the first 16 s of growth. The pseudo-dielectric function is an alternative way of expressing the ellipsometric data and is deduced directly from

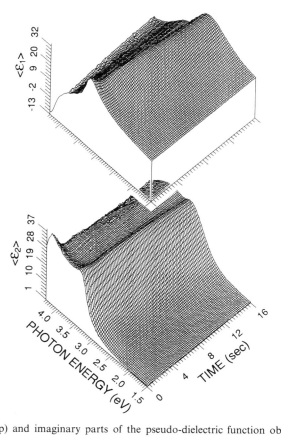

FIG. 2. Real (top) and imaginary parts of the pseudo-dielectric function obtained from real-time measurements during the growth of a-Si:H by PECVD at a high plasma power, yielding a deposition rate of 400 Å/min. The substrate was a native oxide-covered c-Si wafer at 250°C. The acquisition and repetition times for a single pair of spectra were both 64 ms. The plot was constructed from ~250 pairs of spectra, each consisting of ~48 spectral positions between 1.5 and 4.3 eV. After Ref. 22.

$\rho \equiv \tan \psi \exp(i\Delta)$ according to (21)

$$\langle \varepsilon \rangle = \langle \varepsilon_1 \rangle + i \langle \varepsilon_2 \rangle = \varepsilon_a \sin^2 \theta \{ 1 + [(\rho - 1)/(\rho + 1)]^2 \tan^2 \theta \}, \quad (7)$$

where θ is the angle of incidence and ε_a is the dielectric function of the ambient medium. If the solid surface under study presents a single, atomically abrupt interface to the ambient medium, then $\langle \varepsilon \rangle = \varepsilon$, the true dielectric function of the solid.

In collecting the data set of Fig. 2, the rotation frequency of the polarizer was set at 15.6 Hz. In principle, then, a single pair of spectra in (ψ, Δ) can be calculated from four successive detector readouts performed during one-half of the mechanical period, each readout having an exposure time of 8 ms. To improve the signal-to-noise ratio and reduce mechanical alignment errors, however, the raw spectra (S_i, see Eq. (3)) from two successive half-mechanical cycles were accumulated and averaged for a total acquisition time of 64 ms for a single pair of (ψ, Δ) spectra. The repetition time for consecutive (ψ, Δ) spectra was also 64 ms, i.e., no delay was imposed between the collection of one pair of spectra and the next. The a-Si:H film deposition rate in Fig. 2 was ~400 Å/min; thus, 0.4 Å accumulates during the data acquisition time. This also represents the average accumulation between successive measurements of the (ψ, Δ) spectra. As a result, sub-monolayer resolution is achieved in monitoring this growth process. With an acquisition time of 64 ms, monolayer sensitivity is also achieved since the precision in (ψ, Δ) at 2.5 eV is better than 0.03° (22).

At the front of the plot in Fig. 2 at $t = 0$ is the pseudo-dielectric function of the c-Si substrate at 250°C. This differs from the true dielectric function of c-Si owing to the presence of a 23 Å thick native oxide. The two sharp features in $\{\langle\varepsilon_1\rangle, \langle\varepsilon_2\rangle\}$ at 3.3 and 4.2 eV are the E'_0-E_1 and E_2 critical point structures due to transitions near $\Gamma-\Lambda$ and X in the energy band structure. As film growth proceeds, the higher-energy feature dampens quickly with time as the a-Si:H film grows, whereas the low-energy feature remains throughout the growth of the ~110 Å a-Si:H film. These observations indicate that the film is opaque at 4.2 eV, yet still semitransparent at 3.3 eV, when data collection is terminated after 16 s of film growth.

There are two basic approaches for interpretation of $\{\langle\varepsilon_1\rangle, \langle\varepsilon_2\rangle\}$ data sets such as those in Fig. 2: least-squares linear regression analysis (35) and mathematical inversion (36). In the first approach, the true dielectric functions of all the materials in the problem must be known, and a linear regression analysis procedure is used to extract best-fit photon energy–independent free parameters such as film thicknesses and material volume fractions. The number and type of free parameters are established on the basis of a model for the film structure (see next paragraph). In this case, an inspection of the unbiased estimator of the mean square deviation σ, defined by (35)

$$\sigma = (N - p - 1)^{-1/2} \left\{ \sum_{k=1}^{N} (|\tan\psi_e(hv_k) - \tan\psi_c(hv_k)|^2 + |\cos\Delta_e(hv_k) - \cos\Delta_c(hv_k)|^2) \right\}^{1/2}, \quad (8)$$

along with the 90% confidence limits on the deduced free parameter values, allows one to assess the validity of the model for the film structure. In Eq. (8), N is the number of spectral points, p is the number of independent parameters deduced in the analysis, and the subscripts "e" and "c" denote the experimental and best fit calculated ellipsometric spectra. The second approach, mathematical inversion, deals with the opposite situation in which the structural parameters are known (or trial values are chosen for them), and Newton's method is applied to deduce one unknown dielectric function in the model structure from the (ψ, Δ) spectra (36).

The simplest demonstration of both approaches is in the analysis of the c-Si substrate at the front of the plot ($t = 0$) in Fig. 2. In the first step, the (ψ, Δ) spectra are measured on the substrate prior to heating it to the deposition temperature (250°C). Using the known room temperature dielectric functions of the bulk c-Si (37) and vitreous silica (38), the (ψ, Δ) spectra are subjected to linear regression analysis, assuming that the substrate consists of a single homogeneous oxide of unknown thickness on the c-Si. In this analysis, the thickness of the oxide is determined that provides a best-fit simulation of the experimental data (23 Å, in this case). Next, the substrate is heated to the deposition temperature, and (ψ, Δ) are collected again. The thickness and dielectric function of the oxide are assumed to remain unchanged with heating, and the dielectric function of the underlying Si substrate at 250°C is obtained from this second set of (ψ, Δ) spectra by mathematical inversion. This substrate information is now sufficient to undertake optical modeling of the a-Si:H film growth process.

Before proceeding, however, it is helpful to present a brief description of the effective medium theory (EMT) used extensively here in modeling the microstructural evolution of tetrahedrally bonded thin films. A specific EMT developed by Bruggeman (39) has been applied, because it has been demonstrated in earlier *ex situ* studies to provide the best fits, among the simple alternatives, when modeling the microstructure of amorphous (40) and polycrystalline Si (41) films by linear regression analysis. The Bruggeman EMT is expressed by

$$0 = \sum_{i=1}^{N_c} f_i \frac{\varepsilon_i - \varepsilon}{\varepsilon_i + 2\varepsilon} \quad (9)$$

and allows one to calculate the effective complex dielectric function ε of a microscopic composite from the dielectric functions ε_i and the volume fractions f_i of its N_c component materials. The assumptions implicit in Eq. (9) are that the geometry of the aggregate structure is spherical and that only dipole interactions are involved. Furthermore, the application of an EMT in general is based on the assumption that the microstructural

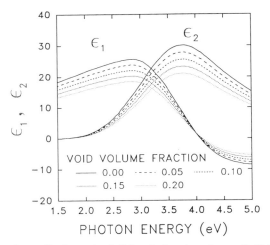

FIG. 3. The experimentally determined dielectric function of an a-Si:H film prepared by PECVD under conditions (250°C substrate temperature and 2 W rf plasma power) that lead to optimum photoelectronic properties and a low volume fraction of voids (solid line). Also shown are the dielectric functions calculated using the Bruggeman effective medium theory for hypothetical materials composed of a mixture of this a-Si:H material and voids, with the volume fraction of the latter component given in the figure. Thus, by its definition, the volume fraction here is measured on a scale relative to that of the optimum material.

inhomogeneity is much smaller than the wavelength of light (42). Usually, the component dielectric functions ε_i are chosen to be the same as those of bulk materials, but such an assumption must be viewed critically, particularly for the crystalline components of the composite.

In modeling the amorphous semiconductor thin films in the nucleation stage of growth, when the films consist of isolated clusters on the substrate that increase in size with time, Eq. (9) is used to calculate the effective dielectric function of the film $\varepsilon = \varepsilon_1 + i\varepsilon_2$. In this calculation, it is assumed that the film consists of a volume fraction f_m of material that simulates the nuclei, with a dielectric function the same as that of bulk material, and a volume fraction $f_v = 1 - f_m$ of free space that simulates the voids between clusters. Figure 3 shows the effective dielectric functions as a function of the void volume fraction, which is controlled by the geometry of the nucleation process. In Fig. 3, the dielectric function for the solid component ($f_v = 0$) is chosen as that of bulk a-Si:H, prepared under conditions leading to a high density, and measured at room temperature in thick-film form. The Bruggeman EMT is also used to model the surface roughness layers on thin films, also as a composite of bulk material and void. In this case the void volume fraction is usually in the range of 0.4 to 0.6. More details concerning the modeling will be presented in the next section.

III. Studies of Tetrahedrally Bonded Thin Films

Tetrahedrally bonded thin films have numerous important technological applications that justify detailed investigation of their processing–property relationships. Hydrogenated amorphous silicon semiconductors have been developed for use in large area display and photovoltaic applications (*43, 44*). The amorphous alloys of Si and C expand the capabilities of amorphous semiconductors and improve the performance of devices made from them (*45*). These wider gap alloys have been employed as intrinsic layers in visible light-emitting diodes (*46*) and as wide gap components of triple $p-i-n$ junction solar cells (*47*). Microcrystalline silicon layers have also found applications as the p and n layers of $p-i-n$ solar cells because they exhibit a higher doping efficiency, electrical conductivity, and transparency in comparison to doped a-Si:H (*48*). Diamond thin films and thick coatings, on the other hand, have a much wider range of applications due the extremes in the properties of diamond (*49, 50*). These applications include high-temperature electronic devices that rely on the high carrier mobility, heat sinks that rely on the high thermal conductivity, and coatings for cutting tools that rely on the hardness of diamond (*51*).

A. Hydrogenated Amorphous Silicon

Because the dielectric function of a-Si:H cannot be expressed in terms of a small number of photon-energy independent parameters (*52*), simple linear regression techniques to analyze real-time SE data are inadequate. Thus, any analysis of data, such as those presented in Fig. 2, must extract both the dielectric function and the microstructure simultaneously. In Part 1 of this subsection, the focus will be on the details of the microstructural evolution of a-Si:H deduced from the real-time SE data. In Part 2, the optical functions will be discussed, particularly since they provide justification for some of the assumptions made in extracting the microstructure. In these two parts, the focus will be on a-Si:H prepared under PECVD conditions that lead to optimum photoelectronic characteristics. The standard deposition configuration for such samples is shown in Fig. 1. The substrates are mounted onto the grounded electrode, which can be heated, and rf power is capacitively coupled to the opposing electrode. The preparation parameter dependence of the microstructural evolution, including the effects of substrate temperature and rf plasma power, will be highlighted in Part 3 of this subsection.

1. Microstructural Analysis from Real-Time Observations

To start the analysis of real time $\{\langle\varepsilon_1\rangle,\langle\varepsilon_2\rangle\}$ data, such as those in Fig. 2, a simple one-layer model for the a-Si:H film is proposed. In this model it is assumed that the film grows as a composite of bulk material and void and is characterized by a single void volume fraction that varies with time. The void volume fraction in the one-layer model represents an average throughout the thickness of the film and simulates the inhomogeneity that develops in the nucleation and coalescence processes. In this model, one needs to determine the dielectric function of the film, as well as its thickness and void volume fraction as a function of time.

First, a single pair of real-time $\{\langle\varepsilon_1\rangle,\langle\varepsilon_2\rangle\}$ spectra is selected, associated with a specific time t_0, later in the growth process. By this time, it is assumed that the void volume fraction in the single layer has stabilized. The stabilized value is defined to be 0 in order to set a relative scale for the void volume fraction. Then a trial thickness value is guessed for the film at $t = t_0$. With this guess, the selected pair of spectra can be numerically inverted to provide a trial dielectric function for the film. The trial dielectric function is correct only if the trial thickness is correct. Next, the trial dielectric function is used in a linear regression analysis of the full real-time data set, in order to extract the thickness and relative void volume fraction as a function of time. At this point, however, the critical parameter is the unbiased estimator of the mean square deviation σ, defined by Eq. (8), which is summed for the individual fits to each pair of experimental (ψ, Δ) spectra versus time. This sum, denoted $\Sigma_t\sigma$, is a measure of the quality of the fit to the two three-dimensional surfaces (see, for example, Fig. 2). This entire process is repeated for different initial guesses of the thickness at $t = t_0$, and the correct guess is the one that minimizes $\Sigma_t\sigma$. This in turn provides the global best fit dielectric function as well as the thickness and relative void fraction versus time.

Figure 4 shows $\sigma(t)$ in the first 40 s of film growth for the one-layer model that provides the global best fit to real-time data collected in the first 250 s of a-Si:H film growth (open points). For this data set, the a-Si:H was deposited onto c-Si at a temperature of 250°C and an rf plasma power of 2 W (52 mW/cm^2 power flux at the grounded electrode). These PECVD conditions lead to optimum electronic quality material at a growth rate of ~ 80 Å/min. In all, 250 pairs of spectra were collected, one every second with

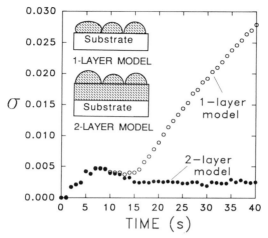

FIG. 4. Unbiased estimator of the mean square deviation plotted as a function of time for linear regression analysis of real-time SE data collected during optimum PECVD a-Si:H deposition onto SiO_2/c-Si at 250°C. Results for one- and two-layer models are included (see inset). Microstructural parameters obtained in the fits using the one-layer model for $t < 12$ s and using the two-layer model for $t \geqslant 12$ s are included in Fig. 6.

an acquisition time of 160 ms, and the pair at $t_0 = 250$ s was selected for the determination of the dielectric function. This pair also set the scale for the relative void volume fraction. It is clear from Fig. 4 that the quality of the fit to the data assuming a one-layer model becomes increasingly poor for deposition times > 15 s.

As a result, a two-layer model for the film was developed in an attempt to improve the agreement between the best fit simulated results and the experimental data. In this case only the top layer of the film is modeled as a composite, consisting of a mixture of voids and the material of the underlying layer. This mixture is an attempt to simulate the nuclei that may form in the early stages of growth, as well as the surface roughness on the film in the later stages of growth. The underlying material simulates a bulk-like layer whose dielectric function is assumed to be thickness-independent. Thus, in this model, three microstructural parameters are assumed to vary as a function of time, the surface layer thickness, d_s, the underlying "bulk" layer thickness, d_b, and the surface layer void volume fraction, f_{vs}. The immediate goal then is to obtain the bulk layer dielectric function (which is also a component of the surface layer) and the three microstructural parameters as a function of time that provide the best fit to the real-time data set. The ultimate goal, however, is to learn about the nucleation and growth processes from the time evolution of the microstructural parameters.

The two-layer modeling approach is simply an extension of the one layer approach (24). Again one pair of experimental spectra at $t = t_0$ is selected from among the full data set, trial guesses are made for the surface and bulk layer thickness values, and f_{sv} is set to 0.50, as might be expected for a surface roughness layer. With these guesses, mathematical inversion of the experimental data is possible to extract the dielectric function of the bulk layer. This dielectric function is used in a linear regression analysis of the real-time data set, and the resulting $\Sigma_t \sigma$ calculated from the individual fits is used as a criterion to establish the correct thickness choices at $t = t_0$. By mapping $\Sigma_t \sigma$ for a two-dimensional grid corresponding to the choices for d_s and d_b, a global minimum in $\Sigma_t \sigma$ is identified. Figure 5 shows results obtained for the same optimum a-Si:H deposition of Fig. 4, using the dielectric function determined at $t_0 = 30\,\text{s}$. The top panel in Fig. 5 shows

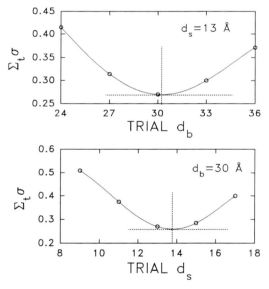

FIG. 5. A plot depicting the two-layer analysis applied to determine the surface roughness and bulk layer thicknesses as well as the bulk layer dielectric function for a very thin film of a-Si:H deposited under optimum PECVD conditions onto a SiO_2/c-Si substrate at 250°C. The ordinate depicts the sum of the unbiased estimators of the mean square deviation obtained in linear regression analysis fits to the spectra collected from 20 s to 100 s. A trial dielectric function in the analysis was determined from the pair of experimental spectra collected at 30 s, using a guess for one thickness (plotted along the abscissa) and a value for the other near the minimum (either $d_b = 30\,\text{Å}$ or $d_s = 13\,\text{Å}$).

$\Sigma_t \sigma$ as a function of the choice of d_b with d_s fixed at 13 Å, and the bottom panel shows $\Sigma_t \sigma$ as a function of the choice of d_s with d_b fixed at 30 Å. On the basis of further iteration, one can conclude that a global best fit occurs for $d_s = 13$ Å and $d_b = 30$ Å.

From these observations, it is concluded that the dielectric function of the bulk layer obtained by inversion of the experimental data at $t_0 = 30$ s with $d_b = 30$ Å and $d_s = 13$ Å is the correct one, at least within the assumptions of the two-layer model. Figure 4 (solid points) shows the quality of the best fit as a function of time in the first 40 s of film growth, for comparison with the results for the one-layer model. It is clear that σ for the two-layer model remains low throughout this time regime; in fact, σ is almost as low as can be expected, given the experimental accuracy and precision of the measurements. In addition, the low σ values and high-quality fits extend to the much later times, as well. It is also interesting to note that the σ values for the one- and two-layer models nearly coincide on the scale of Fig. 4 for times less than 12 s. From this behavior, it is concluded that no advantage is gained by including the second layer of the model in this time regime. Thus, the statistical information suggests a transition from a one- to a two-layer model at this point.

Such a conclusion can also be drawn by inspecting d_s and d_b as a function of time during the a-Si:H film growth, as shown in Fig. 6. Here, it is evident that for times less than 12 s, the bulk layer thickness is less than ~ 2.5 Å, which corresponds to a single monolayer. Thus, it is clear that the one-layer model, with $d_b = 0$, should be used for $t < 12$ s. Figure 6 also includes the thickness d deduced from the one-layer model in the first 11 s. Very good overlap with d_s is obtained, as would be expected since the sub-monolayer values of d_b have a negligible effect on the value of d_s deduced in the two-layer model.

The behavior of d_s and d_b in Fig. 6 is consistent with a nucleation, coalescence, and bulk film growth process. In the early stages of growth ($t < 12$ s), the surface layer is simulating the isolated nuclei that increase in size versus time. When the nuclei reach a critical thickness of 20 Å, they make contact to form the first dense monolayer ($d_b = 2.5$ Å) at $t = 12$–13 s. From this point onward, the incoming flux from the plasma contributes to the bulk layer, and the surface layer now simulates roughness that rides atop the bulk layer during the remainder of the growth process. It is also interesting to note, however, that the surface layer decreases in thickness once the first bulk monolayer is formed, indicating a smoothening effect as the nucleation-related microstructure coalesces into the bulk layer. Once this coalescence is complete and the nucleation-induced surface roughness layer has stabilized, the underlying bulk layer increases linearly with time,

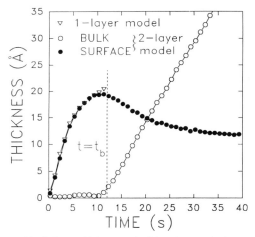

FIG. 6. Surface and bulk layer thicknesses (solid and open circles, respectively) obtained in a linear regression analysis of SE data collected during parallel-plate PECVD of a-Si:H onto SiO_2/c-Si at 250°C (see Fig. 1), assuming a two-layer model for film growth. (See Figs. 4 and 5 for the analysis approach used to establish this model.) The plasma power at the substrate was set at 2 W (52 mW/cm^2 flux), leading to a deposition rate of ~ 80 Å/min. The acquisition and repetition times for the SE spectra were 160 ms and 1 s, respectively. The bulk layer is negligible for $t < t_b$, where $t_b = 12$ s is the time at which the first bulk density monolayer forms ($d_b \sim 2.5$ Å). The layer thickness in a one-layer model is also shown for $t < t_b$ (triangles). The low-density nucleating layer becomes a surface roughness layer when $t \geq t_b$. After Ref. 25.

providing an accurate measure of the deposition rate (~ 80 Å/min for the optimum a-Si:H film of Fig. 6).

Figure 7 shows typical results for f_v plotted versus d for an optimum a-Si:H film, deposited onto native oxide-covered c-Si (open circles). These two parameters are obtained using the one-layer model prior to the divergence of the one- and two-layer $\sigma(t)$ values (see Fig. 4) and provide direct information on the monolayer-level geometry of the nucleation process. In fact, in the first partial monolayer of growth ($d = 2.5$ Å), the film covers 60% of the SiO_2/c-Si surface. As the film thickness increases to 18 Å, prior to the formation of the first bulk-like monolayer, the average void content in the film increases to about 50 vol.%. The overall behavior indicates that flat disks form first on the SiO_2/c-Si substrate, and these evolve into hemispheres with time as d increases. After the formation of the first bulk monolayer, f_{sv} from the two-layer model stabilizes at 0.50 throughout the remainder of the growth process. This value should be viewed with suspicion, since f_{sv} is forced to 0.50 in the initial numerical inversion (e.g., at $t_0 = 30$ s for the sample of Figs. 4–6). However, if other

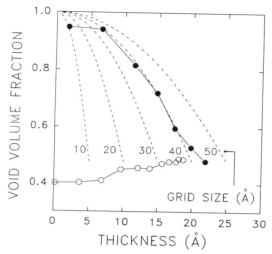

FIG. 7. Void volume fraction versus film thickness deduced in a linear regression analysis of real-time SE data using one-layer models for a-Si:H films prepared by parallel-plate PECVD from pure SiH$_4$ at a plasma power of 2 W and a substrate temperature of 250°C, on native oxide-covered c-Si (open circles) and on Cr (solid circles). These results were obtained in the first ~10 s of film growth before the first bulk monolayer forms. The broken lines are model calculations assuming that hemispherical nuclei form on a square grid, increase in radius, and make contact (at a void volume fraction of 0.48). The values beside the curves indicate the grid size.

values are chosen at this point, the final solution for f_{sv} in the linear regression analysis shows gradual drifts as a function of time in the bulk film growth regime. Such drifts are believed to be artifacts due to an incorrect initial choice for f_{sv}. Any uncertainties associated with the proper stabilized value of f_{sv}, however, has little effect on results such as those in Figs. 6 and 7.

The characteristics of Figs. 6 and 7 provide a sensitive signature of the microstructure throughout the nucleation, coalescence, and growth process. In the nucleation regime, one might expect the observed characteristics to depend on the substrate material. Figure 7 also includes the one-layer model parameters in the initial stage of growth for a-Si:H prepared under the same optimum conditions as for Figs. 4–6, but with a Cr-coated c-Si wafer as a substrate. Although the Cr surface was chemically etched prior to its insertion into the deposition chamber, a thin oxide probably remained on its surface.

The results in Fig. 7 suggest that the a-Si:H exhibits very low coverage of the Cr substrate, <10%, in the first partial monolayer. This behavior is not consistent with the formation of disk-shaped nuclei as in the case of the SiO_2/c-Si substrates, but rather hemispherical clusters. Figure 7 includes a comparison of the measured d and f_v values with those predicted assuming that hemispherical nuclei form on a square grid and increase in size, making contact at $f_v = 0.48$ (broken lines in Fig. 7, with values indicating grid size). The experimental results for a-Si:H growth on the Cr substrate agree with those for a spacing of 40 Å between nucleation sites. This means that the a-Si:H nuclei must make contact upon reaching a thickness of 20 Å, just as in the case of the SiO_2/c-Si substrates. This thickness value is also consistent with the results of the two-layer model for a-Si:H growth on Cr (see next paragraph). It is interesting to note that the film structure from the one-layer model tends toward the same state with $f_v = 0.5$ and $d = 20$ Å, as the initial nuclei make contact, independent of the substrate material. Experimental results similar to those in Fig. 7 for a-Si:H growth on the Cr substrate are also obtained for growth on a Mo-covered c-Si substrate.

The same general characteristics of coalescence and bulk film growth in the two-layer model for optimum a-Si:H on SiO_2/c-Si (see Fig. 6) are also observed when the Cr substrates are used. In fact, even the same quantitative values are obtained—for example, a surface layer thickness of ~20 Å for the formation of the first bulk monolayer and surface smoothing by ~8 Å in the first 50 Å of bulk film growth (see Section III.A.3) (25). It must be noted, however, that to obtain the results described here, the metallic substrate films must be deposited under conditions that lead to ultrasmooth surfaces (e.g., low-pressure ion beam or magnetron sputtering). If these films are prepared by thermal- or electron-beam evaporation, they exhibit rough surfaces that then lead to a rough interface between the metal and the growing a-Si:H film. This interface layer makes microstructural interpretation much more difficult.

The evolution of the volume of film per unit area for a given a-Si:H deposition can be calculated from the evolution of d and f_v in the one-layer regime prior to nuclei contact ($d < 20$ Å in Fig. 7), as well as from the evolution of d_s, d_b, and f_{sv} in the two-layer regime of coalescence and bulk film growth ($d_b > 2.5$ Å in Fig. 6) (26). The film volume per area is also called the mass thickness and is defined by $d_m = (1 - f_v)d$ in the nucleation regime and $d_m = d_b + (1 - f_{sv})d_s$ in the coalescence and bulk growth regime. The rate at which the mass thickness increases describes the rate at which Si atoms are bonded into the film. The results of such calculations for different depositions show variations that are probably due to differences in the details of plasma ignition. This, in turn, leads to differences in the initial flux of reactive radicals, relative to that reached at steady-state in the bulk growth regime.

The plasma ignition problem masks effects due to differences in the sticking probabilities of reactive radicals on the substrate and film surfaces. To avoid this problem, an a-Si:H film prepared at very low rate ($\sim 5\,\text{Å/min}$) onto a c-Si substrate at 250°C with a remote plasma configuration was studied (see Fig. 11 in Section III.A.3). As described in Sec. III.A.3, the signatures of nucleation, coalescence, and bulk film growth for such a film are nearly identical to those in Fig. 6, in spite of the factor of ~ 17 lower growth rate (see Fig. 12 in that section). Figure 8 shows the evolution of the mass thickness for such a film. Although the Si incorporation rate shows complex, nonlinear behavior in the nucleation regime [$d_m = (1 - f_v)d < 10\,\text{Å}$], eventually the incorporation rate becomes linear. Although the first 2.5 Å mass thickness forms very rapidly, the incorporation rate appears to decrease and reach a minimum just before the formation of the first bulk monolayer (at $t = 3.5$ min), as the final uncovered regions of the substrate surface fill in. This latter effect leads to a lower average incorporation rate during nucleation (3.4 Å/min vs. 5 Å/min in the bulk growth regime) and may be due to a barrier to coalescence. One interesting difference exhibited in the corresponding studies of remote PECVD of a-Si:H on Cr substrates is a much slower rate of formation for the first 2.5 Å of film mass.

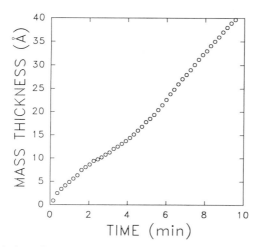

FIG. 8. Evolution of mass thickness with time deduced from the results of one- and two-layer analyses of a-Si:H growth by remote He plasma-enhanced CVD onto a SiO_2/c-Si substrate at 250°C. The mass thickness is defined as the product of the volume fraction of the material and its thickness. In the case of the two-layer model, this product is summed for both layers. The evolution of the structural parameters in the two-layer model for this deposition is provided in Fig. 12. After Ref. 26.

In concluding this part, some comments concerning the physics of the nucleation process with regard to the influence of the substrate material are in order. The origin of the differences between the results in Fig. 7 for SiO_2/c-Si and Cr substrates can be understood from a thermodynamics viewpoint (53). For the case of a-Si:H/SiO_2/c-Si, the sum of the a-Si:H surface and a-Si:H/SiO_2 interface free energies must be less than the native SiO_2 surface free energy. Thus, the total free energy of the system is minimized by a two-dimensional growth process that maximizes interface area. For the case of the a-Si:H/Cr interface, the reverse must be true, and growth is three-dimensional from the start in order to minimize the interface area. This also explains the differences in the rate of formation of the initial partial monolayer on the two different substrates noted at the end of the previous paragraph. There is also a simple explanation for the lack of complete monolayer coverage of SiO_2 by the depositing Si. Coalescence of the monolayer a-Si:H disks may occur at the cost of high surface and interface free energies, which would then favor formation of the second monolayer. The evolution of the mass thickness in Fig. 8 has indicated a barrier to the coalescence of hemispherical structures with $d_{mass} \sim 10$ Å, as well.

2. Optical Functions from Real-Time Observations

The previous part has concentrated solely on the evolution of the microstructure deduced from a least-squares regression analysis of real-time (ψ, Δ) spectra, using one- and two-layer models for the a-Si:H film. The dielectric function of the a-Si:H bulk layer is obtained simultaneously in the two-layer analysis when the bulk and surface roughness layer thickness values are chosen to minimize $\Sigma_t \sigma$ (24). Figure 9 shows the dielectric function of the 30 Å a-Si:H bulk layer (solid line) extracted in the two-layer analysis of optimum PECVD a-Si:H growth associated with Figs. 4–6. Initially, a few points concerning the results in Fig. 9 should be emphasized. First, this is the true dielectric function for the 30 Å a-Si:H layer, having been corrected for the effects of the 13 Å surface roughness extracted in the two-layer model. Second, the results have been obtained during film growth with a 160 ms acquisition time, making this the fastest method available for collecting the data required to extract the optical functions of a thin film. Third, since the dielectric function in Fig. 9 was obtained during film growth under optimum PECVD conditions, the results are representative of a substrate temperature of 250°C.

Before interpreting these optical functions in further detail, the assumptions regarding the a-Si:H dielectric function that were made in the two-layer analysis of Figs. 4–6 will be reexamined. Because of the very low

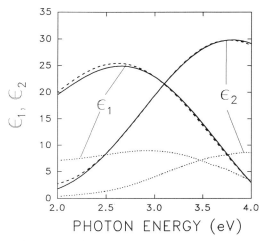

FIG. 9. Dielectric functions obtained from SE data collected with a 160 ms acquisition time during PECVD of optimum a-Si:H onto substrates held at 250°C. Dotted lines: a 21 Å thick cluster film deposited onto a Cr-coated Si wafer and measured just prior to coalescence (see Fig. 7). Solid and broken lines: a bulk film deposited onto SiO_2/c-Si, measured in the early and later stages of growth when the bulk layer thicknesses were 30 Å and 328 Å, respectively. The dielectric function at 30 Å bulk film thickness was obtained in the analysis of Figs. 4–6.

σ values obtained versus time in Fig. 4 for the two-layer model, the assumptions appear to be justified. However, the small deviations from these assumptions provide further information on the growth process and the optical functions.

The first assumption made was that the dielectric function of the bulk layer is independent of thickness throughout the full structural evolution. Again, if this were not true at least to a very good approximation, σ in Fig. 4 would increase over time, yet no significant such effect is observed. As a further test of this assumption, the $\Sigma_t\sigma$ procedure of Fig. 5 was performed, selecting a pair of (ψ,Δ) spectra collected at a later stage in the growth process, namely at 250 s, rather than at 30 s as in Fig. 5. At this time, the bulk and surface roughness layer thickness values that minimized $\Sigma_t\sigma$ were found to be 328 Å and 9 Å, respectively. The resulting bulk layer dielectric function for the film at this later stage is also given in Fig. 9 (dashed lines), for comparison with the result for the 30 Å film. The agreement is quite close over the photon energy range from 2.75 to 4.0 eV, where the sample absorbs most strongly, again providing support for the initial assumption. The discrepancies at lower energies may be caused by small inadequacies in the model for the roughness layer that will be described next.

The second important assumption made in the analysis of Figs. 4–6 was that the roughness layer in the two-layer model and the nucleating layer in the one-layer model can be simulated using the Bruggeman EMT as a mixture of voids with the bulk material, the latter having the dielectric function of Fig. 9. In order to test this, the dielectric function was extracted from (ψ, Δ) spectra collected at 7 s during PECVD of a-Si:H on the Cr-coated Si wafer substrate before the onset of bulk film growth. In this analysis, the film thickness was fixed at 21 Å, the value obtained in the one-layer analysis (see, for example, Fig. 7), and the (ψ, Δ) spectra were numerically inverted to determine $(\varepsilon_1, \varepsilon_2)$. The results will not necessarily be identical to the effective dielectric function of an $\sim 0.5/0.5$ bulk a-Si:H/void mixture because σ does not vanish at this time. (For example, see the results for a-Si:H growth on c-Si at $t = 8$ s in Fig. 4.)

Figure 9 also includes this dielectric function (dotted line) for comparison with the results for the 30 and 328 Å bulk films. First, the very low magnitude of the dielectric function is attributed to the high volume fraction of voids that are present because the initial nuclei have just started to make contact prior to the formation of the first bulk monolayer. In fact, the void volume fraction from the one-layer model is 0.48, consistent with contacting hemispherical structures located on a square grid as noted in Section III.A.1. A more interesting aspect is the peak energy of ε_2, which appears at 4.0 eV for the 21 Å cluster film, ~ 0.2 eV higher than in the corresponding value in the bulk films. An inspection of the calculations of Fig. 3 shows that such an effect cannot be attributed to the presence of the voids.

The two remaining possibilities for the shift in the peak in ε_2 are (i) quantum size effects that may shift the joint density of states for the clusters to higher energies or (ii) the presence of a larger number of Si–H bonds in the clusters than in the bulk material. It is well known that in a-Si:H films, the onset for absorption shifts to higher energies continuously as a function of the bonded H-content (*54*). In more detailed real time SE studies, a-Ge and a-Si cluster films, prepared by sputtering solid targets in pure Ar in the absence of H, show no detectable shifts in the ε_2 peak energies when measured relative to the bulk materials (*55*). Thus, the shift in Fig. 9 for the cluster film relative to the bulk material is attributed to the presence of additional Si–H$_n$ bonds ($1 \leqslant n \leqslant 3$) that presumably cover the surface of the clusters. Some contribution to the higher peak energy may also be due to the near-surface regions of the cluster with a higher concentration of H due to the incomplete cross-linking of the network. The thickness of such a layer is estimated to be 4–8 Å from earlier real-time, single-photon energy ellipsometry (*56*) and mass spectrometry (*57*) studies.

These results indicate that the dielectric function of the 21 Å cluster film, upon removal of the effect of the 0.48 volume fraction voids, would be a better approximation to the dielectric function of the initial growth clusters for use in the one-layer model. In addition, it might be expected that this dielectric function would also provide an improved description of the solid component of the a-Si:H surface roughness layer in the two-layer model. Further analysis of the evolution of the microstructure for a-Si:H growth on c-Si using this refined dielectric function in the one- and two-layer models lead to a small improvement in the final $\Sigma_t\sigma$, but the resulting microstructural parameters differ negligibly from those in Figs. 6 and 7.

In theory, the dielectric functions in Fig. 9 provide information on the concentration of various Si and H bonding units in the film. For example, a tetrahedron model has been developed whereby, given the concentration of the various Si-Si$_{4-n}$H$_n$($0 \leqslant n \leqslant 3$) tetrahedra in the network, the dielectric function can be calculated (58). However, the reverse process of calculating the concentrations of the tetrahedra on the basis of optical measurements such as those in Fig. 9 is not possible because of the number of free parameters involved. Furthermore, it is not clear that the dielectric functions assigned to the three different H-containing tetrahedron ($1 \leqslant n \leqslant 3$), calculated from the experimental results for unhydrogenated a-Si using a theoretical scaling law (59,60), are sufficiently accurate for analysis of this nature. In fact, the standard way of determining the concentration of such bonding configurations is not through optical measurements such as these but through infrared vibrational absorption spectra, calibrated by nuclear resonance reaction techniques (61).

The most important information that can be extracted from the dielectric functions of the a-Si:H films, at least for electronic device applications, is the band gap. The most popular way to determine the band gap is by the Tauc method (62), which involves plotting $(\varepsilon_2 E^2)^{1/2}$ vs. photon energy $E = h\nu$. For parabolic densities of states distributions in both the valence and conduction bands and for an energy-independent momentum matrix element, this plot should be linear, and the energy axis intercept provides the band gap. It has been widely recognized that when high-quality dielectric functions ε_2 are available for a-Si:H samples over a sufficiently wide energy range, the Tauc plot is not linear, suggesting that the initial assumptions are incorrect (63,64). An alternative method involves plotting $\varepsilon_2^{1/2}$ versus E, which should be linear for parabolic densities of states and an energy-independent dipole matrix element (65). When the high-quality experimental data are plotted in this way, much-improved linearity is obtained and a well-defined band gap can be extracted.

Figure 10 shows the results of this latter method for extracting the band gap from the dielectric functions of Fig. 9, obtained in real time. In this case

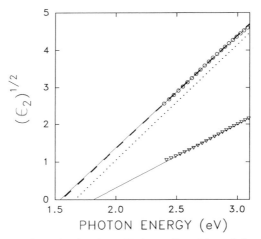

FIG. 10. Band-gap determination from SE data collected in real time for PECVD a-Si:H films whose dielectric functions are given in Fig. 9. Open circles and overlapping solid line: data for 30 Å bulk a-Si:H film and corresponding best fit linear relationship. Broken line: best fit linear relationship that describes the data for the same film at 328 Å. Dotted line: best fit linear relationship for the 328 Å film, measured *in situ* after cooling the sample to room temperature. Triangles and overlapping solid line: data for 21 Å a-Si:H cluster film and best fit linear relationship. The photon energy at which $\varepsilon_2 = 0$ for the best fit linear relationship defines the band gap in each case.

the plot is restricted to the higher energy range in order to avoid the experimental errors that give rise to the deviations in ε_2 between the 30 and 328 Å results in Fig. 9. The open circles show the data for the 30 Å film, with the solid line indicating the extrapolation to the energy axis that provides a band gap of 1.54 eV. The broken line is the extrapolation for the thicker film, leading to a gap value of 1.55 eV, identical within error. Again, it should be noted that these gap values are appropriate for a sample temperature of 250°C.

At the end of the deposition, the a-Si:H sample was cooled to 25°C under high vacuum and remeasured in order to obtain the gap relevant for electronic device operation at room temperature. In an analysis of the resulting (ψ, Δ) data to extract the dielectric function, it was assumed that the bulk and surface roughness thicknesses were the same as those obtained from the two-layer analysis of spectra collected just after the deposition was terminated. Figure 10 also includes the band gap extrapolation relevant for 25°C (dotted line). The room-temperature band gap value of 1.66 eV, obtained *in situ* with a data acquisition time of 160 ms, is in good agreement with the literature result for PECVD a-Si:H (1.64 eV), considering that the latter value was obtained by *ex situ* transmittance and reflectance measurements that may be subject to errors from surface oxides and roughness (65). In addition, the average temperature shift of the band gap, obtained from the real-time and *in situ* results, 4.9×10^{-4} eV/K, is in reasonable agreement with the literature results for the corresponding shift of E_{03}, the energy at which the absorption coefficient is 10^3 cm^{-1} (5.8×10^{-4} eV/K) (66).

Finally, Fig. 10 includes the band gap determination for the 21 Å cluster film of Fig. 9 (open triangles). The band gap is 1.8 eV, 0.25 eV larger than that for the bulk a-Si:H layers measured at the same temperature. This difference is in good agreement with the shift in the peak energy of ε_2 for the cluster film relative to the bulk films (see Fig. 9). This suggests that both differences have the same origin, a higher bonded H concentration associated with the clusters.

Overall, the results of the band gap determinations in Fig. 10 are consistent with expectations and suggest that quite accurate values can be obtained from the high-speed, real-time measurements. These measurements represent the first band gap determinations made from passive, real-time observations during thin film semiconductor growth. Although the band gap of a-Si:H prepared by optimum PECVD is well known, this capability becomes much more important in studies of tetrahedrally bonded amorphous semiconductor alloys such as a-$Si_{1-x}C_x$:H, prepared using mixtures of SiH_4 and CH_4 reactive gases. This subject will be treated in detail in Section III.B.1.

3. Process–Property Relationships in a-SiH

The two-layer model results of Fig. 6 provide a clear signature of the nucleation, coalescence, and bulk growth characteristics of the thin films. The differences in the microstructural evolution of different films can be quantified by two important parameters that provide information on the atomic-scale surface processes (25). First, the surface layer thickness at the time, t_b, when the first bulk monolayer forms, denoted $d_s(t_b)$, corresponds to the critical thickness at which isolated thin-film nuclei make contact. This in turn provides information on the average density of nuclei on the substrate surface. The specific geometry of hemispherical nuclei on a square grid has been found to be consistent with the observed void volume fraction when a-Si:H nuclei make contact (see Fig. 7). For this geometry, the nucleation density N_d can be estimated from $d_s(t_b)$ according to $N_d = [2d_s(t_b)]^{-2}$. Second, the amount of smoothening that the surface layer undergoes after the formation of the first bulk monolayer describes the extent to which the initial growth microstructure generated in the nucleation process coalesces to form a smooth surface.

When comparing the nucleation of thin films under different deposition conditions or for different substrates, a lower value of $d_s(t_b)$ implies a higher nucleation density. When the nucleation process is heterogeneous, this simply implies a higher density of substrate defects that serve as initial nucleation sites. When the nucleation process is homogeneous, however, a higher nucleation density implies a reduced diffusion length for precursors on the substrate surface. To understand this, one may consider a single stable nucleus on the substrate surface. Other nuclei will not form on the substrate surface within the area over which the stable nucleus serves as a sink for diffusing precursors. This area contracts, and hence the nucleation density increases, as the precursor surface diffusion length decreases. Thus, within a homogeneous model, any process that enhances the diffusion length of precursors on the substrate surface is expected to lead to a lower nucleation density.

Similarly, the degree of surface smoothening or roughening after the first bulk monolayer forms can provide information on the diffusion length of precursors on the film surface. In general, thin-film simulations suggest that when the precursor surface diffusion length λ_d is longer than the dominant roughness wavelength λ_r on the surface, then the roughness decreases in amplitude with continued film growth (67). If the reverse is true, then the roughness is enhanced with continued growth. In the nucleation studies, the dominant roughness wavelength is given by the spacing between the nuclei, i.e., approximately $2d_s(t_b)$. Thus, when the nucleation density is low, a relatively large surface diffusion length is needed to smoothen the nucleation-related microstructure. In contrast, when the nucleation density is high, a low diffusion length is often sufficient to smoothen the nucleation-related microstructure.

Figure 6 shows clearly that for optimum electronic quality a-Si:H, the surface roughness layer does not decrease to zero in the coalescence process, but rather stabilizes at $\sim 10\,\text{Å}$. Thus, if the nucleation density is sufficiently low to give a surface roughness layer of $\sim 10\,\text{Å}$ or less, no further smoothening would be expected irrespective of the precursor surface diffusion length. For such films, information on the processes of precursor surface diffusion is inaccessible. In the case of homogeneous nucleation, when a high nucleation density implies a low precursor diffusion length on the substrate, a low diffusion length on the film surface is expected, as well. In the deposition of a-Si:H from the decomposition of SiH_4, this may be true for precursor radicals such as Si, SiH, and SiH_2, which have a wider range of bonding possibilities in comparison to SiH_3 radicals (68–70). Finally, we note that a low diffusion length of the precursors on the film surface may also be detectable through gradual roughening trends over time that are initiated through random fluctuations in the surface profile.

Next, some of these general ideas will be demonstrated through experimental results that show consistent monolayer-layer trends as a function of the conditions or techniques of deposition. First, results for the growth of a-Si:H in a much slower remote PECVD process will be compared with those for growth by conventional parallel-plate PECVD (see Fig. 1). In the remote process, pure SiH_4 gas is injected into the chamber downstream from a He plasma tube as shown in Fig. 11 (71). The excited He atoms from the plasma break up the SiH_4 molecules, and the resulting radicals lead to a-Si:H growth on a substrate, mounted on a holder further downstream.

Figure 12 shows a comparison between the two-layer model results for the remote (R) and optimized parallel-plate (PP) PECVD processes. In both cases the substrate is SiO_2/c-Si, held at 250°C during deposition. The results for conventional PECVD are reproduced from Fig. 6 and are shown as the

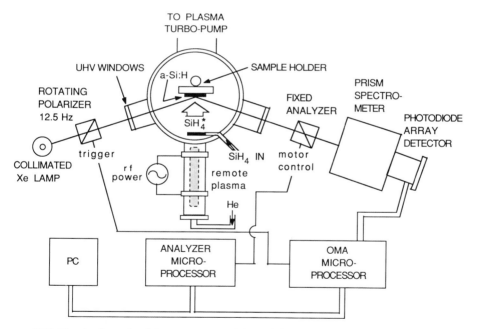

FIG. 11. A schematic of the apparatus used for real-time monitoring of a-Si:H film growth at low rates (~5 Å/min). The deposition system in this case is designed for plasma-enhanced chemical vapor deposition in the remote He rf plasma configuration. In this configuration, the SiH_4 is injected into the chamber downstream from a He plasma tube; the substrate is mounted on a heatable platform located downstream from the SiH_4 injector. The windows on the chamber are oriented for a 70° angle of incidence at the sample surface.

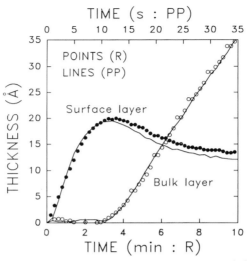

FIG. 12. Surface and bulk layer thicknesses (solid and open circles, respectively) obtained in a linear regression analysis of SE data collected during remote He PECVD of a-Si:H onto SiO_2/c-Si at 250°C (see Fig. 11), assuming a two-layer model for film growth. (See Fig. 8 for the evolution of mass thickness for this model.) The acquisition and repetition times for the SE data were 3.2 s and 14.5 s, respectively. The two-layer model results for remote and direct PECVD are compared by scaling the corresponding data of Fig. 6 for optimum PECVD a-Si:H growth on SiO_2/c-Si by a factor of 17 along the time axis (bold lines, referring to upper scale).

solid lines. Because the growth rate of the bulk layer in the remote process is a factor of ~17 lower than in the conventional process, the time axes differ by this same factor in order to better compare the results. Overall, it is found that with this scaling, the two depositions show nearly identical microstructural evolution, including the same nucleation density and the same coalescence behavior. This result implies that the nucleation and coalescence behavior of a-Si:H prepared by conventional PECVD under optimum conditions is not kinetically limited. It also suggests that the film precursors are the same in optimum parallel-plate PECVD and in the remote PECVD process.

With the high-speed capability of the ellipsometer, we can also compare the behavior shown in Fig. 12 with that obtained at much higher deposition rates (25). Figure 13 shows the results of the two-layer analysis of the real-time $\{\langle\varepsilon_1\rangle,\langle\varepsilon_2\rangle\}$ spectra shown in Fig. 2 for conventional PECVD on a 250°C c-Si substrate. For this deposition, the rf plasma power was a factor of 10 higher than that used for optimum electronic quality a-Si:H (20 W vs. 2 W), but all other conditions were identical. This leads to a bulk layer

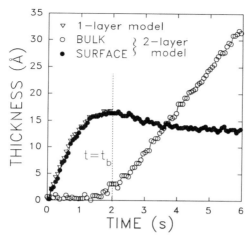

FIG. 13. Surface and bulk layer thicknesses (solid and open circles, respectively) obtained in a linear regression analysis of SE data collected during parallel-plate PECVD of a-Si:H onto SiO_2/c-Si at 250°C (see Fig. 1), assuming a two-layer model for film growth. The plasma power at the substrate was set at 20 W (520 mW/cm^2 flux), leading to a deposition rate of 400 Å/min. The acquisition and repetition times for the SE data were both 64 ms. The bulk layer is negligible for $t < t_b$, where t_b is the time at which the first bulk density monolayer forms ($d_b \sim 2.5$ Å). The layer thickness in a one-layer model is also shown for $t < t_b$ (triangles). After Ref. 25.

growth rate of 400 Å/min, a factor of ~ 5 higher than that for the optimum material, and a factor of ~ 80 higher than that for the remote PECVD material. Two differences are noted between the results of Fig. 13 and Fig. 6 (or Fig. 12, solid lines). First, the thickness at which the nuclei make contact, $d_s(t_b)$, is 3 Å less for the higher-rate film, indicating a slightly higher nucleation density. Second, the amount of surface smoothening that occurs in the first 50 Å of bulk film growth, denoted by Δd_s, is 4 Å for the high-rate deposition compared to 8 Å for the low-rate one. This results in an ultimate roughness layer on the high-rate film that is thicker than that on the optimum film.

These two aspects of film growth at high plasma power, a higher nucleation density and a weaker coalescence effect are what might be expected for kinetically limited film growth. In other words, the precursor diffusion time on either the surface of the substrate or the film is limited by the accumulation time for a single monolayer. However, there are two other possible explanations. First, it is possible that under high plasma power conditions, the distribution of precursors changes and more Si, SiH, and SiH_2 are generated, which exhibit lower diffusion lengths on the film and substrate surfaces (68–70). Alternatively, the higher plasma density for the higher power deposition may result in an enhanced rate of ion impact-induced desorption of H from the growing film surface (72) or a higher density of defects at the substrate surface. These effects would also serve to reduce the diffusion length of precursors on the film and substrate surfaces, respectively.

Next, we consider the effect of substrate temperature, which is another important variable parameter in studying the film growth process (25). Figure 14 shows the results of the two-layer analysis for the growth of a-Si:H by conventional PECVD onto a c-Si substrate held at 120°C. These results are to be compared with those of Fig. 6 or Fig. 12 (solid lines) for deposition at the optimum substrate temperature of 250°C, but under conditions otherwise identical to those in Fig. 14. From Fig. 14, the film thickness is ~ 19 Å when the nuclei first make contact. This value is within 1 Å of that obtained for a-Si:H prepared at the optimum substrate temperature. It is also evident that the surface smoothening effect in the first 50 Å of bulk film growth is weaker for growth at the lower substrate temperature, ~ 3 Å for 120°C as opposed to ~ 8 Å for 250°C. As a result, the ultimate

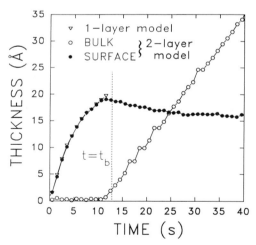

FIG. 14. Surface and bulk layer thicknesses (solid and open circles, respectively) obtained in a linear regression analysis of SE data collected during parallel-plate PECVD of a-Si:H onto SiO_2/c-Si at 120°C (see Fig. 1), assuming a two-layer model for film growth. The plasma power was set at the standard value of 2 W (52 mW/cm^2 flux), leading to a deposition rate of 74 Å/min. The acquisition and repetition times for the SE data were 160 ms and 1 s, respectively. The layer thickness in a one-layer model is also shown for $t < t_b$ (triangles), where t_b is the time at which the first bulk density monolayer forms ($d_b \sim 2.5$ Å). After Ref. 25.

roughness layer on the film after coalescence is thicker at the lower substrate temperature.

The origin of the weaker coalescence behavior at the lower substrate temperature most likely arises from a shorter diffusion length of film precursors on the film surface as the thermally activated diffusion process is frozen out. If this were to occur on the substrate surface, as well, a reduction in the nucleation density and $d_s(t_b)$ would be predicted, in contrast to experimental observations. Thus, the nucleation process may be heterogenous, controlled by the density of defects on the substrate surface, at least under the PECVD conditions of Figs. 6 and 14. This model may also explain the lack of a significant dependence of nucleation density on deposition rate as seen in Figs. 12 and 13.

Figure 15 shows the magnitude of surface layer smoothening in the first 50 Å of bulk layer growth, deduced in analyses of real-time SE data collected during a-Si:H depositions on c-Si (open symbols) and Cr (solid symbols). In most cases, conventional parallel-plate PECVD was used (circles) at

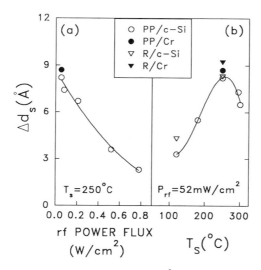

FIG. 15. Surface smoothening in the first 50 Å of bulk film growth for PECVD a-Si:H, plotted as a function of (a) rf plasma power flux at the substrate surface with a fixed substrate temperature of 250°C, and (b) substrate temperature with a fixed rf plasma flux of 52 mW/cm². These results were deduced from the decay of the surface roughness layer in analyses such as those in Figs. 12–14. The key is as follows; PP, parallel-plate PECVD; R, remote He PECVD; Cr, sputtered Cr substrate; c-Si, native oxide-covered c-Si substrate. The solid lines are guides to the eye for the PP/c-Si depositions. After Ref. 25.

different values of (a) rf power flux measured at the substrate surface and (b) substrate temperature. Selected depositions were performed by remote PECVD (triangles) using substrate temperatures of 120°C and 250°C. The first observation to be noted from this plot is the well-defined trends as a function of preparation parameters for a-Si:H growth on c-Si by the conventional PECVD process (lines). These results suggest sub-monolayer-scale sensitivity (± 0.5 Å) to the surface smoothening effect. Because the nucleation density is similar for these films, these trends are directly related to the degree of coalescence of the initial growth microstructure, which also controls the ultimate roughness layer thickness on the films. Although the trends in Fig. 15 have been discussed in terms of changes in the precursor surface diffusion length with plasma power and substrate temperature in conjunction with Figs. 13 and 14, the origin of the weakening of the coalescence effect at the higher substrate temperatures in Fig. 15 is not clear at present. It is possible that this effect arises from an increase in the reaction rate of precursors on the surface, which gives rise to a reduction in the diffusion length of the precursors. It should be noted that the deposition rate increases from 74 Å/min to ~ 100 Å/min as the substrate temperature increases from 120°C to 300°C.

It is interesting to note that the conditions in conventional PECVD widely known to yield optimum electronic quality a-Si;H are those that result in the strongest coalescence behavior in Fig. 15. Apparently, the conditions of lower precursor surface diffusion that lead to incomplete nuclei coalescence on the monolayer scale, a thicker surface roughness layer, and in most cases a higher void volume fraction in the resulting film, also give rise to the bulk defects that limit the electronic device quality of the material. Thus, through real-time SE measurements, processing–property correlations have been established on a monolayer scale, and these correlations provide insights into the origins of poor electronic performance.

To expand this picture, Fig. 16 shows the void volume fraction plotted versus deposition temperature. The void fraction for each film was obtained on a relative scale by fitting the dielectric function from the two-layer model using linear regression analysis and the Bruggeman EMT. A reference dielectric function for the film deposited at 300°C with the lowest void fraction was used in Eq. (9), along with that of voids, and the volume fractions of the components were determined in the analysis. The results at lower temperature ($T \leqslant 250$°C) show that the void volume fraction is reduced when the smoothening effect of coalescence is enhanced. Such an effect is predicted on the basis of continuum models of film growth (67). In fact, the presence of voids may account for the poor electronic performance

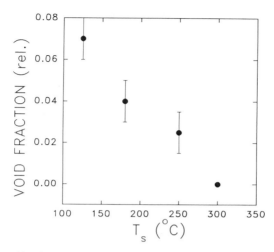

FIG. 16. Relative void volume fraction in the bulk layer for the PECVD a-Si:H films of Fig. 15 plotted versus the substrate temperature.

of the a-Si:H prepared at low substrate temperature. A higher volume fraction of voids may result in a reduced microscopic mobility for excited carriers. The voids may also lead to more extensive exponential tails in the density of states that extend from the valence and conduction band edges into the gap, resulting in a lower effective mobility (73). At the highest substrate temperature, the void volume fraction continues to decrease even though the coalescence effect in Fig. 15 is weakened. This suggests that the weakening in the coalescence does not result in extended void structures, and that the poor electronic performance of such films is associated with point defects such as threefold coordinated Si atoms.

To complete this subsection, in Fig. 17, we show the results of a two-layer analysis for pure a-Si prepared by magnetron sputtering a Si target in Ar, for comparison with those in Figs. 6 and 12–14 (23). In this process, the precursors are Si atoms, and there is no H in the process to passivate the surface during growth. Thus, any adatom surface diffusion arises from impact and thermal energies. For the deposition of Fig. 17, the sputtering conditions were set to obtain the highest density for the bulk layer, as determined from linear regression analysis with the EMT. In fact, for the lowest possible Ar sputtering pressure and a relatively high substrate temperature of 300°C, the void volume fraction of the resulting a-Si film comparable to that of the highest density PECVD a-Si:H in Fig. 16. The

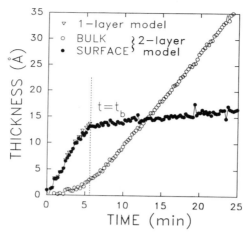

FIG. 17. Surface and bulk layer thicknesses (solid and open circles, respectively) obtained in a linear regression analysis of SE data collected during low-pressure magnetron sputtering of pure a-Si onto SiO_2/c-Si at 300°C, assuming a two-layer model for film growth. The acquisition and repetition times for the SE data were 3.2 s and 14.5 s, respectively. The layer thickness in a one-layer model is also shown for $t < t_b$ (triangles), where t_b is the time at which the first bulk density monolayer forms ($d_b \sim 2.5$ Å). After Ref. 23.

low pressure is expected to enhance adatom surface diffusion, since the impinging Si atoms are not thermalized in transit from the target to the substrate.

In spite of this choice deposition parameters, the difference between the physical vapor deposition (PVD) process and the PECVD process is significant, especially in view of the differences among the PECVD films of Fig. 15. First, the thickness at which initial nuclei make contact is 13 Å, compared to 20 Å for optimum PECVD. This difference is well outside possible experimental error and leads to a sizable difference in the nucleation density, 1.5×10^{13} cm^{-2} for PVD versus 6×10^{12} cm^{-2} for optimum PECVD. Second, once the nuclei make contact and the first bulk monolayer forms, the residual nucleation-related microstructure for the PVD a-Si in Fig. 17 tends to be enhanced with further film growth. Although the higher nucleation density on the substrate could be related to the generation of additional nucleation sites by the energetic impact of Si, the observations overall are attributed to a much lower diffusion length of Si adatoms on both substrate and film surfaces in comparison to any of the PECVD films. This behavior is expected considering the important role that H plays in the PECVD of a-Si:H.

B. Hydrogenated Amorphous Silicon–Carbon Alloys

The powerful capabilities described in subsection A for a-Si:H become even more important in studies of complex materials such as the amorphous tetrahedrally bonded alloys. First, because two reactive gases are required to form the alloys, one has the capability of varying the band gap of the alloy continuously over a wide range by varying the flow ratio of the two gases. This aspect of alloy preparation will be discussed in the first part of this subsection. Second, experience has shown that a wider range of preparation variables needs to be considered for alloy optimization, since the precursors associated with the two different elements may undergo different surface processes and incorporate into the film differently. This aspect will be discussed in the second part of this subsection.

As an example of this second aspect, consider the case of hydrogenated silicon–carbon alloys, a-Si$_{1-x}$C$_x$:H, with low x (45). In the growth of this material, a silyl (SiH$_3$) radical precursor that forms a single Si–Si bond with the surface at a temperature of 250°C can readily cross-link with neighboring SiH$_n$ units, leading to the release of H$_2$. In contrast, a methyl (CH$_3$) radical precursor bonded to Si at the surface is much more stable due to the strong C–H bonds; thus, cross-linking to the neighboring SiH$_n$ units will not occur. This can lead to a methylated form of amorphous Si in which methyl groups rather than H atoms serve to relax the strain in the tetrahedral network (74). In order to avoid this undesirable structure, it is necessary to enhance the exchange of H atoms between the surface (and in particular the C–H bonds) and the gas phase, which may then promote the cross-linking of C-based precursors. The exchange of H can be enhanced through ionic bombardment and through additional atomic H in the growth environment. Enhanced bombardment can be achieved by reducing the gas pressure and increasing the plasma power, and H can be added to the gas phase by diluting the reactive feed gases with molecular hydrogen in the growth process. In the second part of this subsection, we will focus on the role of H$_2$-dilution in the monolayer-scale surface processes during alloy growth.

1. Optical Gaps from Real-Time Observations

The real-time data collection and analysis techniques described in the previous subsection have also been applied to a-Si$_{1-x}$C$_x$:H. In preparing such films, the initial step is to establish the connection between CH$_4$:SiH$_4$ gas flow ratio and the band gap. Figure 18 shows the band gap determination from real-time SE data, according to the methods of Section III.A.2 (see

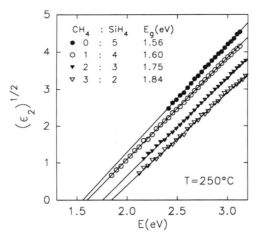

FIG. 18. Analysis used to determine the band gap deduced from real-time observations (160 ms acquisition time) for ~150 Å thick a-Si$_{1-x}$C$_x$:H films prepared onto c-Si substrates with a fixed H$_2$:(SiH$_4$ + CH$_4$) ratio of 4:1 and a variable CH$_4$:SiH$_4$ flow ratio. The band gap is obtained by an extrapolation of the linear behavior in $\varepsilon_2^{1/2}$ to zero ordinate. These results are appropriate for a substrate temperature of 250°C.

Fig. 10). For this set of samples, the substrate temperature was 250°C, the rf power flux was 2.5 times higher than that of optimum PECVD a-Si:H (5 W vs. 2 W), and H$_2$ dilution was used with a H$_2$:(CH$_4$ + SiH$_4$) flow ratio of 4:1. The rationale behind the latter two parameter choices is an attempt to promote C-atom cross-linking in the network. The CH$_4$:SiH$_4$ flow ratio for the four samples was varied from 0:5 to 3:2, maintaining the total gas flow constant.

As noted earlier, the band gap determination method of Fig. 18 has a number of important advantages over the usual methods of *ex situ* transmission and reflection spectroscopy. First, because the former measurements are performed *in situ*, the effect of native oxide overlayers on the determination of the optical functions need not be considered. Second, ε_2 is obtained in the $\Sigma_i\sigma$ minimization procedure of Section III.A.1, assuming a two-layer model; thus, the effect of surface roughness layers is included in the analysis. Third, because the film thickness at the time of data collection in Fig. 18 was ~150 Å, this approach to determining the energy gap can be applied even for ultrathin doped amorphous films that are often employed in electronic devices. Fourth, the (ψ, Δ) spectra that were analyzed to obtain the results in Fig. 18 were acquired in 160 ms during deposition. This high speed is an advantage if the ultimate goal is to extract the band gap itself, rather than just the raw data, through real-time analysis. Finally, because the band gap is obtained by a reflection technique, it is possible to measure the band gaps of a series of films, sequentially-deposited to opacity on the same substrate. Such an approach allows one to scan preparation parameter space expeditiously and find conditions that give the desired band gap.

The extrapolations of the linear behavior in $\varepsilon_2^{1/2}$ versus energy to zero ordinate provide the band gaps, which are tabulated in Fig. 18 and plotted as a function of the $CH_4:SiH_4$ flow ratio in Fig. 19. Also shown in Fig. 19 are the gaps obtained from measurements performed *in situ* after depositing the film and cooling it to room temperature. These values are of greater interest for device operation and differ from those obtained from real-time measurements in accordance with an average temperature shift of the gap that ranges from 4.5×10^{-4} eV/K for a-Si:H to $\sim 7 \times 10^{-4}$ eV/K for the higher C-content alloys. Applying correlations established elsewhere, this range in the room-temperature band gap corresponds to a range in x in a-$Si_{1-x}C_x$:H from 0 to ~ 0.2 (*45*). Based on the results in Fig. 19, if one

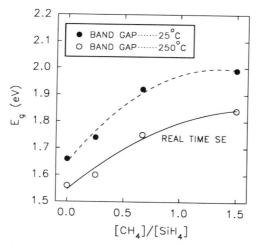

FIG. 19. Optical gaps for the a-$Si_{1-x}C_x$:H films of Fig. 18, prepared with a fixed H_2-dilution ratio of 4:1, plotted as a function of the $CH_4:SiH_4$ flow ratio. The plot includes gaps deduced from real-time measurements (see Fig. 18), which are characteristic of a sample temperature of 250°C (open points). The gaps deduced from measurements performed after cooling the sample to 25°C (solid points) are included as well.

requires a material with a room-temperature band gap of 1.9 eV, for example, then the $CH_4:SiH_4$ flow ratio should be chosen as 2:3. In the next part, we will focus on the ability of H_2 dilution to improve the properties of alloys with a gap of 1.9 eV.

2. Role of H_2-Dilution of Reactive Gases

Figure 20 shows the results of a two-layer analysis of real-time SE data for two PECVD a-$Si_{1-x}C_x$:H films prepared with $H_2:(CH_4 + SiH_4)$ gas flow ratios, R, of 4 and 20. The substrate temperature and plasma power were set at the values used in the depositions of Fig. 19, and the $CH_4:SiH_4$ ratio was fixed at 2:3 to obtain room-temperature band gaps of ~ 1.9 eV. The data presentation is extended over a longer time scale than similar plots in Figs. 12–14 in order to better compare the longer-term surface microstructural evolution in the two cases.

An inspection of the results in Fig. 20 reveal a number of differences between the two depositions. First, the higher H_2 dilution level leads to a lower deposition rate, as indicated by the slope of the bulk layer thickness

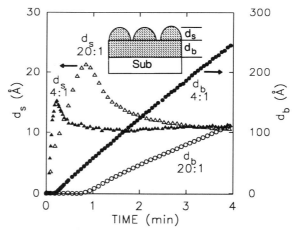

FIG. 20. Microstructural analysis deduced from real-time SE data for two a-$Si_{1-x}C_x$:H films prepared by parallel-plate PECVD onto SiO_2/c-Si substrates at 250°C, one with a 4:1 H_2-dilution ratio (solid symbols) as in Fig. 18 and the other with a 20:1 ratio (open symbols). The triangles and circles denote the thicknesses of the low-density nucleating layer, d_s, and the bulk layer, d_b, respectively (see inset). The latter is negligible for $t < t_b$, where t_b is the time at which the first bulk density monolayer forms ($d_b \sim 2.5$ Å). The low-density nucleating layer becomes a surface roughness layer when $t \geqslant t_b$.

versus time (33 Å/min for $R = 20$ compared with 65 Å/min for $R = 4$). Second, the thickness at which the nuclei make contact is significantly larger at the high dilution level (21 Å for $R = 20$ compared with 15 Å for $R = 4$). Although both films exhibit smoothening behavior after the first bulk monolayer forms, for the film with $R = 20$ the smoothening trend continues throughout the deposition, whereas for $R = 4$ surface roughening is observed for bulk layer thicknesses >100 Å.

Figure 21 presents quantitative microstructural characteristics for a number of a-Si$_{1-x}$C$_x$:H depositions plotted as a function of the H$_2$ dilution ratio. These parameters include the thickness at which nuclei make contact

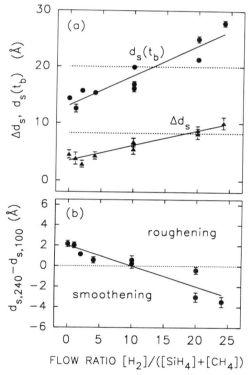

FIG. 21. Monolayer-scale characteristics of the growth of a-Si$_{1-x}$C$_x$:H films extracted from data such as those of Fig. 20 and plotted versus H$_2$-dilution ratio. (a) Thickness of the surface layer at time t_b when the first bulk monolayer forms [$d_s(t_b)$] and the smoothening in the first 50 Å of bulk film growth [Δd_s]. (b) Longer-term trends in the evolution of the surface morphology, given by the difference between the roughness layer thicknesses measured at bulk layer thicknesses of 240 Å and 100 Å. Positive and negative values indicate surface roughening and smoothening, respectively.

(top panel), the amount of surface smoothening in the first 50 Å of bulk film growth (center, as in Fig. 15 for a-Si:H), and the difference between roughness thicknesses at bulk layer thicknesses of 240 and 100 Å (bottom panel). These three characteristics exhibit clear trends for $0 \leqslant R \leqslant 24$, indicating that the differences in Fig. 20 are representative of the general features of H_2-dilution. The behavior in Fig. 21 can be considered in two separate regimes: a low H_2 dilution regime with $0 < R < 5$ and a high dilution regime with $R \geqslant 20$.

In the low dilution regime, the thickness of the nuclei upon contact, $d_s(t_b)$, is < 16 Å. This is smaller than the value for optimum a-Si:H from pure SiH_4 (20 Å; see horizontal broken line in Fig. 21), indicating a higher nucleation density for these alloys ($N_d > 1.0 \times 10^{13}$ cm^{-2} vs. 6×10^{12} cm^{-2}). In fact, the nucleation density for a-Si$_{1-x}$C$_x$:H with $R = 0$ is nearly the same as that observed for sputtered a-Si without H ($N_d = 1.5 \times 10^{13}$ cm^{-2}). This behavior suggests that some C-bearing precursor in the plasma exhibits a smaller surface diffusion length on the substrate, bonding more strongly to it in comparison to the components of a pure SiH_4 plasma. In addition, for $R < 5$ in Fig. 21 (center), there is a weak smoothening effect by 3–4 Å in the coalescence process. Although the magnitude of this effect appears to be limited by the minimum roughness layer thickness of 10 Å, the smoothening trend reverses and the roughness thickness increases with time over the 100–200 Å bulk layer thickness scale.

In the high H_2 dilution regime, corresponding to the samples with $R = 20$ and 24, the nucleation density becomes lower than that for optimum a-Si:H prepared from pure SiH_4, whereas the smoothening effect becomes comparable to that for a-Si:H ($\Delta d_s \sim 8$ Å; see horizontal line in Fig. 21 at center). In addition, the longer-term trend in the evolution of the surface morphology is toward continued smoothening. The reduction in nucleation density means that the H in the plasma is very effective at either passivating sites for nucleation that exist on the substrate surface or scouring the substrate of the low diffusion-length, C-bearing precursors that otherwise bond readily to the substrate. It is also possible that H_2-dilution changes the distribution of species in the plasma, eliminating those precursors that exhibit a short diffusion length on the substrate surface. The large smoothening effect at the higher dilution can be attributed to the corresponding conditioning of the film surface or plasma by the excess H, leading to the accumulation of material through the incorporation of precursors with large diffusion lengths. This is also consistent with the transition from a roughening to smoothening trend with increased H_2-dilution at the bottom of Fig. 21.

For a dilution ratio of $R = 30$, some of the trends in the microstructural quantification of Fig. 21 reverse—in particular, the longer-term surface smoothening trend. For $R = 30$, the growth conditions are approaching those under which μc-Si:H is prepared (see Section III.C). In this regime, the granular structure of the film appears to be enhanced at the near-surface, leading to greater roughening, because the bonds connecting neighboring microstructural units may not be stable in the extremely high H_2-dilution

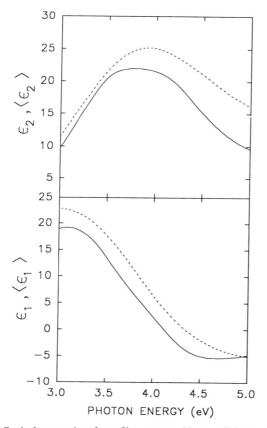

FIG. 22. Optical properties of two films prepared by parallel-plate PECVD onto SiO_2/c-Si substrates at 250°C using a CH_4:SiH_4 ratio of 2:3. The solid lines correspond to the pseudo-dielectric function for a 1,215 Å thick, two-phase film prepared with a 80:1 H_2-dilution ratio. This film consists of a mixture of Si crystallites in a matrix of a-$Si_{1-x}C_x$:H. The pseudo-dielectric function is plotted because the spectra were not corrected for surface roughness. The broken lines correspond to the true dielectric function ε for a 7,240 Å thick, single-phase film of a- $Si_{1-x}C_x$:H, prepared with a 20:1 H_2-dilution ratio. In both cases the films were opaque over the plotted photon energy range, and the measurements were performed *ex situ* with a serially scanning ellipsometer.

environment. For $R = 40$, there is evidence of microcrystallites in thick films, and for $R = 80$, microcrystallinity can be detected early in the growth process. As an example, Figure 22 shows the optical properties of alloy films prepared with $R = 20$ and $R = 80$, but with all other conditions identical. The $R = 80$ film exhibits a double-peaked structure in $\langle \varepsilon_2 \rangle$ that is characteristic of the $E_0'-E_1$ and E_2 transitions in Si microcrystallites. In fact, a more detailed analysis suggests that this material is composed of a physical mixture of crystallites along with a second phase of a-$Si_{1-x}C_x$:H. Because the energy positions of the $E_0'-E_1$ and E_2 transitions in the crystallites are not at the same position as those for bulk Si, further investigation into the structure of the crystallites is required. A model study, better suited to providing insights into the nature of pure microcrystalline Si, will be presented in Section III.C.

The relative void volume fractions for the films of Fig. 21 have been determined from analyses of the dielectric functions, which in turn were obtained from real-time SE data through the $\Sigma_i \sigma$ minimization procedure (see Section II.A.1). In the void fraction analyses, the dielectric functions were modeled as physical mixtures of void and a thicker a-$Si_{1-x}C_x$:H material prepared with $R = 20$, that served as a high-density reference. The final results of the analysis as a function of the H_2-dilution ratio appear in Fig. 23. From these results, one can conclude that the favorable surface processes induced by excess atomic H in the gas phase, characterized by a reduction in the nucleation density and an enhancement in the coalescence

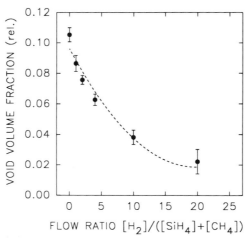

FIG. 23. Relative void volume fraction in the bulk layer for the final PECVD a-$Si_{1-x}C_x$:H films of Fig. 21 plotted versus the H_2-dilution ratio used in PECVD.

and long-term microstructural stability, also result in films with an improved bulk microstructure. The electronic properties of such films have also been measured and reveal that, as a function of H_2-dilution from $R = 0$ to 20, the density of mid-gap defect states decreases, the inverse slope of the Urbach edge decreases, and the electron photoresponse (mobility-lifetime product) improves (75).

Thus, as in the case of unalloyed a-Si:H (see Fig. 15), there appears to be a correlation between the observation of a favorable nucleation–coalescence sequence and the ultimate quality of the material for devices. In the case of a-Si:H, it was proposed that precursor surface diffusion promotes the coalescence of the film and prevents the development of extended microstructural defects, and it is these defects, or localized defects associated with them, that limit photoexcited carrier transport in most cases. Such relationships between monolayer-scale processes and the ultimate photoelectronic properties probably account for the observations in the a-$Si_{1-x}C_x$:H films as well (76). However, one must also consider the possible role of gas-phase atomic H, in eliminating point defects on the surface directly, without the need for surface diffusion. Such defects may include Si–CH_3 bonding structures that may become trapped in the film at low dilution levels, as discussed earlier (74), and also sp^2 bonded C that is converted to sp^3 in the presence of atomic H. Such defects have been detected in a-$Si_{1-x}C_x$:H prepared in the absence of dilution by infrared transmittance and nuclear magnetic resonance measurements, respectively (77, 78).

C. Microcrystalline Silicon

The term microcrystalline Si (denoted μc-Si:H) has been applied to films that can be prepared from Si sources in the presence of a high gas phase concentration of atomic H, usually from a plasma (79, 80). Such films generally can be deposited on a wide variety of substrates without special seeding procedures (see Section III.D for diamond) and consist of crystallites ranging from tens of angstroms to tenths of microns in size. Depending on the deposition conditions, some films consist of a two-phase structure of crystalline grains within an amorphous matrix. For the deposition system of Fig. 1, μc-Si:H can be prepared at 250°C by heavy dilution of SiH_4 in H_2, operating at a relatively high power to enhance the growth rate (81).

In general, one can understand the growth of the crystalline phase in terms of a chemical equilibrium between the gas phase and the solid surface (79). Because the Si-bearing precursors are continually being deposited onto and removed from the substrate surface, only solid Si in the crystalline structure is sufficiently stable to form nuclei in this environment. However, an amorphous phase will often form in the regions between crystalline grains in the coalescence and bulk growth regime. Real-time spectroscopic ellipsometry is an ideal probe of the nucleation and growth of μc-Si:H. The presence of microcrystals is revealed by the E'_0-E_1 and E_2 transitions in the near ultraviolet near 3.3 and 4.2 eV (82), and the amorphous component is revealed by a broad absorption band extending to photon energies below the lowest direct gap in c-Si. In this subsection, we focus on the evolution of the microstructure, optical, and electronic characteristics of μc-Si:H films consisting of isolated crystalline nuclei in the absence of the amorphous component (28,29). Although the initial goal was to characterize the microstructure during nucleation, interesting information on the effects of crystallite size on the electronic transitions also can be extracted from the real-time SE results.

1. *Optical and Microstructural Analysis from Real-Time Observations*

In Section III.A.1, it was found that for an a-Si:H film consisting of isolated nuclei, a one-layer model is a suitable approximation to the structure, at least for the purposes of interpreting the optical properties. After the nuclei make contact, however, two layers are required in the optical model. One of the two layers represents the underlying bulk film, and the other represents the roughness layer originating from the nucleation process. Thus, in studies of μc-Si:H in the nucleation stage, a one-layer model is also used.

In Section III.A.2, the optical functions of a 21 Å thick layer of a-Si:H nuclei could be interpreted in terms of a physical mixture of bulk material and voids with a relatively small modification due to excess H at the nuclei surfaces. The lack of size-dependent effects can be understood because the mean free path of electrons in the amorphous solid is very short and the optical properties are determined by the local bonding. In contrast, the optical properties of Si microcrystallites are influenced by the long-range periodicity and are expected to depend more sensitively on the size of the crystallites. As a result, in an analysis of μc-Si:H film growth, the thickness and dielectric function of the nucleating film are extracted independently at each time. To do this, the correct physical thickness, d, is chosen by trial and error as the one that provides a dielectric function obtained by inversion that is free of artifacts (83). Such artifacts arise from features in the dielectric function of the substrate, Cr at 2.7 eV in this case, and are introduced only when the thickness of the μc-Si:H film is chosen incorrectly.

Figures 24 (top; solid circles) and 25 provide the thickness as a function

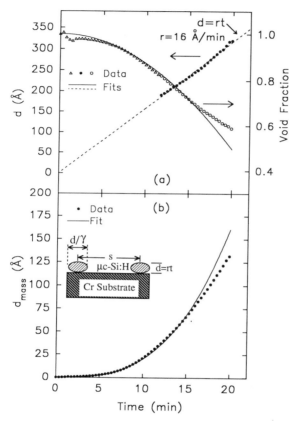

FIG. 24. Time evolution of the microstructure of a discontinuous film of Si microcrystallites, deduced from an analysis of (ψ, Δ) spectra collected in real time during growth. The film was prepared by parallel-plate PECVD from a 1:80 $SiH_4:H_2$ flow ratio onto a Cr substrate at 250°C. The physical thickness d (solid points) and the void volume fraction f_v (open points) are presented in (a). The broken line is the best fit to d assuming a constant deposition rate r. The values of f_v denoted by open triangles were derived using an EMT and the single-crystal Si optical functions at 250°C; for the open circles an analytical formula for the c-Si optical functions was employed (see fits in Fig. 25). The mass thickness, or Si volume per unit area $d(1 - f_v)$, is presented in (b). The solid line in (b) represents the best fit to the data assuming three-dimensional, isolated particle growth, as shown schematically in the inset. The solid line in (a) is derived from the fit in (b) in accordance with $f_v = 1 - (d_{mass}/rt)$, where t is the elapsed time. After Ref. 29.

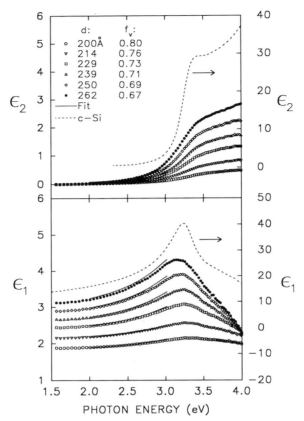

FIG. 25. Effective dielectric functions extracted from (ψ, Δ) spectra collected during the growth of a discontinuous film of Si microcrystallites at selected times identified by the physical thicknesses d, listed at upper left. The substrate is an opaque layer of chromium on glass, held at 250°C. A one-layer model was used for the film in order to extract d. The solid lines are three parameter fits to the effective dielectric function for $2 \leqslant hv \leqslant 3$ eV using a two-component version of the Bruggeman effective medium theory. One component is c-Si, whose dielectric function is calculated from a simple analytical formula, and the other is void, whose volume fraction f_v is the free parameter of greatest interest here (listed alongside the corresponding thickness). The dielectric function of single-crystal Si at 250°C is also presented for comparison (broken lines). After Ref. 29.

of time and dielectric functions at selected times that result from this analysis procedure. The μc-Si:H film in this case was prepared onto Cr-coated glass with a substrate temperature of 250°C, a $SiH_4:H_2$ gas flow ratio of 1:80 and an rf power level 10 times higher than that used in optimum a-Si:H deposition (20 W versus 2 W). The analysis could only be performed for times $\geqslant 12$ min, corresponding to physical thicknesses > 190 Å. Because of the low volume of material in the film at shorter times, the dielectric function of the μc-Si:H is too weak for successful application of the artifact minimization procedure. From Fig. 24 the physical thickness increases linearly with time at a rate of 16 Å/min, and when extrapolated backward in time intercepts the origin, indicating that there is no induction time for nuclei formation.

There are a number of points that can be made with respect to the dielectric functions in Fig. 25, which are appropriate for the substrate temperature of 250°C. First, the imaginary parts of the dielectric functions at each thickness approach zero at the lowest energies. This behavior is expected below the lowest direct gap of c-Si and provides further support for the validity of the artifact minimization procedure used to determine d and $(\varepsilon_1, \varepsilon_2)$. Second, the very low amplitudes of the dielectric function of the film, in comparison with the dielectric function of bulk single-crystal Si (see broken lines in Fig. 25), reflect well-spaced nuclei with a very high intervening void volume fraction. The increasing amplitude with thickness results from the reduction in void volume fraction as the particles increase in size and approach one another. Third, the presence of the c-Si E_1 critical point structure near 3.3 eV and the lack of significant absorption strength from 2.0 to 2.5 eV, relative to that at 4 eV, show that the film is composed of Si microcrystallites without any significant amorphous phase. Buried beneath the dominant features in Fig. 25 are subtle effects due to crystallite size that will be discussed in the next part.

In order to better understand the microstructural development, the void volume fractions can be extracted from the effective dielectric functions shown in Fig. 25. Thus, the effective dielectric functions have been fitted assuming a simplified structural modeling consisting of a mixture of void and particles. In this linear regression analysis fitting procedure, which employs the Bruggeman EMT, the dielectric functions of the particles themselves, $(\varepsilon_{1p}, \varepsilon_{2p})$, are assumed to obey a simple analytical expression for $2 \leqslant h\nu \leqslant 3$ eV with a minimum number of free parameters. The expression is based on the assumption that the particle dielectric functions are dominated by the onset of a two-dimensional M_0-type critical point below the E_1 feature, i.e.,

$$\varepsilon_p(h\nu) = -A[\chi(h\nu)]^{-2} \ln\{1 - [\chi(h\nu)]^2\}, \qquad (10)$$

where $\chi(h\nu) = (h\nu + i\Gamma)/E_1$ (84). In this expression A, E_1, and Γ are the amplitude, transition energy, and broadening parameter of the critical point. By fixing A at the bulk c-Si value, the dielectric function spectra can be fit using only three free parameters, E_1, Γ, and f_v, the void volume fraction.

The results of this fitting procedure are shown in Fig. 25 as the solid lines, and Fig. 24 includes the results for the void fraction (top panel, open circles). If the effects of particle size on $(\varepsilon_{1p}, \varepsilon_{2p})$ are neglected, and the experimental dielectric functions in Fig. 25 are fitted assuming a mixture of bulk c-Si and void, then the void volume fractions that result are the same as those obtained from the analytical formula for $(\varepsilon_{1p}, \varepsilon_{2p})$ within ~ 0.01. This implies that the effects of particle size on the dielectric functions are small enough that they do not invalidate the void fraction determination. The void fraction determination can be extended to shorter times in Fig. 24 by assuming thicknesses given by the extrapolated linear rate and determining the void volume fraction by this latter technique. The results are given by the triangles in the top panel of Fig. 24. Thus, the near-continuity of f_v versus time in the region near 12 min, where the transition between the two modeling approaches is made, supports the validity of both.

In the lower panel of Fig. 24 the mass thickness, defined by $d(1 - f_v)$, is plotted using the results from the top panel of the figure. The overall behavior in Fig. 24 can be understood on the basis of simple three-dimensional particle growth (see inset, Fig. 24). If the physical thickness is a linear function of time t and the microcrystallites are assumed to nucleate on a square grid with spacing s, then the mass thickness is given by $d_{\text{mass}} = (cr^3/s^2)t^3$. Here r is the physical thickness rate (16 Å/min), and c is a unitless constant that depends on the shape of the crystallites, assumed to be fixed over time. The expression $c = \pi h/6\gamma^2$ covers the simplest cases of spheroidal ($h = 1$) and hemispheroidal ($h = 4$) crystallites. In this expression $\gamma = d/b$ is the axial ratio of the spheroid, where b is the length of the major axis assumed to be parallel to the substrate plane (see inset in lower panel of Fig. 24).

A model of this sort is in fact reasonable because of the linearity of the physical thickness over time. Such behavior would be expected if the dominant pathway for incorporation of Si in the discontinuous film is through attachment to the nuclei that exist from $t = 0$, rather through the formation of new nuclei with a nonzero induction time. The solid line in the lower panel of Fig. 24 is a fit to this simple three-dimensional nuclei growth model, using the results for $d < 265$ Å, to prevent potential biasing by the contact of neighboring microcrystallites. The void fraction corresponding to this fit is given as the solid line in the top panel of Fig. 24. Overall, the fits are reasonable, considering the simplicity of the model. If the crystallites are assumed to be hemispherical in shape the spacing would be 650 Å. This corresponds to a nucleation density of $2.4 \times 10^{10}\,\text{cm}^{-2}$, which is a factor of 250 lower than that for optimum a-Si:H on similarly prepared Cr substrates.

The deviations in d_{mass} from the cubic time dependence above 265 Å may result from the onset of partial aggregation of neighboring crystallites. It may also be due to thin-film self-shadowing effects in which neighboring crystallites block the flux of film precursors from reaching the near-substrate regions of the crystallites. This would result in a change in the particle shape and in a tendency toward columnar crystallite formation in the pre-coalescence stage. When the crystallites make contact, the mass thickness should increase linearly with time, corresponding to two-dimensional growth.

2. Size Effects in Microcrystallites

In order to study the effects of size on the dielectric function of the microcrystallites, $(\varepsilon_{1p}, \varepsilon_{2p})$, one might consider the energy and broadening parameters deduced in the fits of the dielectric functions of Fig. 25 to Eq. (10). Although the simple analytical approximation to the dielectric function is appropriate for extracting the void volume fraction, based on the reduction in the amplitudes of the dielectric functions of Fig. 25 relative to that of bulk c-Si, the best fit values of E_1 and Γ are not sufficiently accurate. The problem arises because other known contributions to the analytical expression for the c-Si dielectric function are neglected over the $2 \leqslant hv \leqslant 3$ eV photon energy range.

In order to determine accurate parameter values that characterize the electronic structure of the crystallites, standard critical-point line-shape analysis is performed on $(\varepsilon_{1p}, \varepsilon_{2p})$ (85). To do this, the known void volume fractions of Fig. 24 are extracted from the effective dielectric functions of Fig. 25 by inversion of the Bruggeman EMT equation (Eq. (9)), in order to deduce the evolution of the dielectric functions $(\varepsilon_{1p}, \varepsilon_{2p})$ of the crystallites. The resulting spectra are similar in shape to those of Fig. 25, but with significantly increased amplitudes. To be consistent with earlier studies of bulk c-Si, a purely excitonic line shape is assumed for ε_p in the neighborhood of the E_1 critical point, i.e.,

$$d^2\varepsilon_p/d(hv)^2 = 2A[\exp(i\phi)](hv - E_1 + i\Gamma)^{-3}, \qquad (11)$$

where ϕ is the phase associated with the E_1 optical transitions (85). Figure 26 shows typical second derivative spectra for μc-Si:H at thicknesses of 191 Å and 250 Å, derived from the real-time observations (lower pair of broken lines, in top and bottom panels). To obtain these results, standard procedures for smoothing and differentiation are employed.

Figure 26 also shows spectra for static bulk c-Si (upper broken lines, in the top and bottom panels), measured with the multichannel ellipsometer

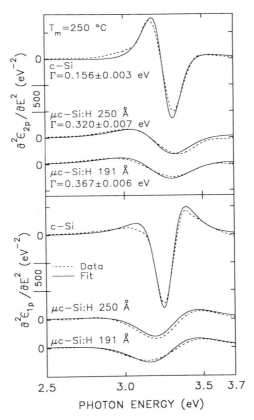

FIG. 26. Critical-point analysis applied to the E_1 transitions associated with the dielectric functions of Si microcrystallites at physical thicknesses of 250 Å and 191 Å (lower curves). The critical-point amplitude, energy, broadening, and phase are determined from the fit (solid lines) to the second derivative spectra (broken lines). Also shown are the corresponding results for bulk single-crystal Si (upper curves). All second derivative spectra are offset for clarity. In all cases the measurement temperature is 250°C, and an excitonic line shape (see Eq. (11)) is assumed. After Ref. 29.

under identical conditions, namely, with an acquisition time and sample temperature of 3.8 s and 250°C, respectively. The parameters obtained by fitting the results to Eq. (11) $[(E_1, \Gamma, \phi) = (3.268 \text{ eV}, 0.16 \text{ eV}, -65°)]$ are in good agreement with the corresponding values $[(E_1, \Gamma, \phi) = (3.273 \text{ eV}, 0.17 \text{ eV}, -40°)]$ obtained by a higher-precision, serially scanning spectroscopic ellipsometer (85) that typically requires many minutes for data collection. This agreement supports the validity of the line-shape analysis of real-time SE data, in spite of the lower precision and energy resolution of the multichannel ellipsometer as compared to a serially scanning instrument. An obvious conclusion from Fig. 26 is the significant broadening of the critical point feature for the microcrystallites in comparison to the bulk c-Si counterpart.

Figure 27 shows the best fit energy, broadening, and phase parameters for the crystallites as a function of the physical thickness from 190 Å to 250 Å, obtained in least-squares fits to Eq. (11) (see solid lines in Fig. 26). The error bars denote 90% confidence limits obtained in the fits. The real-time SE data are analyzed only in the three-dimensional growth regime ($d \leq 250$ Å) in order to be able to associate crystallite size with physical thickness. As noted earlier, however, there is insufficient material (i.e., $d_{\text{mass}} < 35$ Å) for

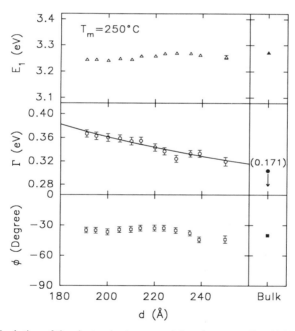

FIG. 27. Evolution of the electronic structure of the microcrystalline Si film of Figs. 24–26 as a function of physical thickness at a measurement temperature of 250°C. These results were obtained from fits such as those in Fig. 26. From top to bottom, the critical point energy, broadening parameter, and phase are plotted, and values for single-crystal Si at 250°C are plotted at the far right. The broadening parameter for single-crystal Si is 0.171 eV (off-scale, at right-center). The error bars are the 90% confidence limits in the critical point analysis. After Ref. 29.

physical thicknesses less than 190 Å to enable one to extract ε_p spectra of sufficient quality for line-shape analysis. From Fig. 27, it can be seen that the energy and phase parameters are constant, within consideration of potential experimental errors. Furthermore, the average values of the two parameters $(E_{1,a}, \phi_a) = (3.26\,\text{eV}, -36°)$ are very close to the results obtained on bulk c-Si with the higher-precision instrument.

The broadening parameter in Fig. 27, on the other hand, decreases from 0.37 eV to 0.32 eV with increasing thickness over the accessible range and is a factor of two greater than that obtained for bulk c-Si (0.16–0.17 eV). This behavior is consistent with a size dependence in the dielectric function of the microcrystallites controlled by scattering of electrons at microcrystallite surfaces which limits the excited state lifetime. However, the crystallites are large enough so that the physical origin and optical transition energy associated with the E_1 critical point are not altered. The size effect can be understood from the following expression relating the broadening parameter to the lifetime, based on the uncertainty principle (86):

$$\Gamma(d) = h/\tau(d) = \Gamma_{\text{bulk}} + h[v/\lambda(d)]. \tag{12}$$

Here, τ is the excited state lifetime, h is Planck's constant Γ_{bulk} is the broadening parameter of bulk single c-Si, and v and λ are the speed and mean free path of electrons in the crystallite that participate in the E_1 transitions, i.e., those with wave vector near the Λ point in the band structure. Equation (12) is based on the assumption that the Si bulk and crystallite scattering mechanisms are independent. In addition, in Eq. (12), an explicit film thickness (d) or crystallite size dependence in λ has been included, recognizing that electron scattering at crystallite surfaces may be the dominant mechanism that increases Γ for the crystallite from Γ_{bulk} for the single crystal. This physical mechanism can be included by substituting $\lambda = d/2$, an expression that has been used extensively for scattering of free electrons in spherical metal particles (87).

To assess the validity of Eq. (12) with this substitution, Γ from Fig. 27 is replotted versus d^{-1} in Fig. 28. Approximate linear behavior is apparent, which is not unexpected given the relatively narrow range in d^{-1}. However, the best fit linear relation extrapolates to 0.165 eV, which is the expected value of Γ_{bulk} from the line-shape analysis of c-Si in Fig. 26. Thus, consistency with Eq. (12) is obtained. As another check of the model, the electron speed can be calculated from Fig. 28 as 4.7×10^7 cm/s. This is reasonable considering that the speed of electrons travelling in the $\langle 111 \rangle$ direction with k at the Λ point is $\sim 2.6 \times 10^7$ cm/s, as calculated from the Si band structure (88). The origin of the discrepancy is unclear; it may be due to the participation of other electronic transitions or to errors in the

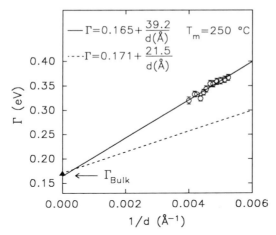

FIG. 28. Broadening parameter for the E_1 optical transitions of microcrystalline Si at 250°C, replotted from Fig. 27 as a function of inverse physical thickness (open points). The solid line is a linear fit to these results, suggesting agreement with a finite-size effect theory, in accordance with Eq. (12). The triangular point is the broadening parameter for single-crystal Si at 250°C (not included in the fit). The broken line is calculated from Eq. (12) assuming that the mean free path is $d/2$, the electron speed is 2.6×10^7 cm/s (as estimated from the Si band structure), and the bulk broadening parameter is 0.171 eV. After Ref. 29.

oversimplified mean free path relationship, $\lambda = d/2$, that arise from a nonspherical particle shape.

Based on this study, it can be concluded that the film analyzed in Figs. 24–28 consists of high-quality crystallites. Otherwise, defects internal to the crystallites rather than the particle surfaces would limit the mean free path of electrons. In fact, it has been found that the results for μc-Si:H films prepared by rf PECVD at higher plasma power do not obey the simple relationship of Eq. (12). This suggests that for such films electron scattering is limited by internal defects. Thus, the approach developed here can be used to characterize the crystalline quality of isolated microcrystallites, providing their density on the surface is sufficiently high for accurate line-shape analysis.

This work also provides insights into the optical functions of continuous micro- and polycrystalline Si films. A number of studies have shown that the near-ultraviolet optical functions of these materials can be simulated accurately with the Bruggeman EMT as three-component microscopic composites consisting of a-Si, c-Si, and void (*17, 18, 89*). In such studies, the SE data collected *ex situ* on the films are subjected to linear regression analysis in order to extract the volume fractions of the independent components as well as the thickness and composition of roughness and/or oxide overlayers in a two-layer model for the films.

In spite of these successful fits, attempts to correlate the resulting bulk film structural information with that deduced from Raman spectroscopy have revealed an ambiguity in the interpretation (90). Some μc-Si:H films that exhibit a considerable a-Si component from SE have shown no such detectable phase from Raman spectroscopy. This result suggests that the a-Si component in the SE analysis may serve the additional role of simulating the broadening of the optical transitions that arises from a reduction in the excited-state lifetime of electrons due to scattering from grain boundaries or defects within the grains. These effects are not included in the standard effective medium theory approach, which employs the assumption that the optical functions of the c-Si component in micro- and polycrystalline Si are identical to those of single crystal Si. Thus, an analysis of the dielectric functions for materials that appear to be two-phase, such as the a-SiC:H/μc-Si composite films discussed in Section III.B.2, must be viewed critically in light of the potential artifacts that lead to an overestimation of the amorphous component. These issues will be discussed further in Section III.D as they relate to a study of the substrates used in diamond film growth.

D. Diamond

Diamond thin films can also be prepared by CVD methods, and the most common approach is similar to that used to obtain μc-Si:H, namely decomposition of CH_4 in the presence of large quantities of atomic H (91). In addition to promoting chemical equilibrium between the gas phase and solid surface, as in the case of μc-Si:H, the excess H ensures that solid carbon forms in the tetrahedral sp^3 bonding configuration. Because the C–C and C–H bonds are stronger in comparison to Si–Si and Si–H bonds, a much higher substrate temperature (700–1,000°C) is needed to achieve equilibrium and prevent the development of an amorphous phase. Generally, the conventional rf plasma generated in the parallel-plate configuration (see Fig. 1), using CH_4 highly diluted in H_2, does not provide sufficient atomic H for high-rate diamond deposition. Alternative enhanced CVD methods, using heated-filament (92, 93) or microwave plasma excitation (94) are more effective for H_2 dissociation, and are the most common methods for preparing diamond. In this subsection, a real-time SE study of the nucleation of diamond using the heated filament excitation method will be discussed (30).

Real-time SE does not appear to be as powerful for probing diamond growth as it is for probing Si growth. One reason for this is that, depending on the substrate seeding and gas excitation method, the diamond phase may develop with nucleation density low enough so that the nuclei spacing becomes comparable to the wavelengths of the probe beam. In this case, the film loses its specularity and scatters the incident light, leading to systematic errors in the SE data that are often difficult to detect. Also, in contrast to Si, the onset of direct transitions in diamond occurs above the high-energy limit of the real-time spectroscopic ellipsometer. Thus, over the accessible energy range, $\varepsilon_2 = 0$ for pure diamond films.

Thus, there exist no optical features due to interband transitions that allow one to distinguish pure diamond from tetrahedrally or sp^3-bonded amorphous C, as are available to distinguish c-Si from a-Si. Tetrahedrally-bonded a-C films also contain an admixture of trigonally or sp^2-bonded C, generating a weak, nonzero ε_2 spectrum in the visible that can be detected by SE (95). However, fine-grained nano- or microcrystalline diamond also exhibits a nonzero ε_2, very similar in form to that observed in the a-C films (96), arising from graphitic defects at grain boundaries (97). Thus, even the presence of weak optical absorption does not allow one to distinguish the amorphous from crystalline sp^3-bonded films solely on the basis of SE. However, for films that are known to be crystalline—typically through scanning electron microscropy, which can reveal faceting, or Raman spectroscopy, which can reveal the sharp $1,332 \, cm^{-1}$ mode of diamond (49, 50)—SE can be applied to quantify the volume fractions of diamond, sp^2-bonded C, and void (98). In this subsection, the initial nucleation regime of diamond will be studied in which the film consists of isolated diamond crystallites.

1. Substrate Treatment and Annealing

In the diamond growth process, one can envision how the diamond network can propagate in the excited $CH_4:H_2$ gas mixture. The atomic H in the gas ensures that any C=C bonds that may form on the surface will be readily saturated with the insertion of H atoms. In addition, any C atoms that bond into the network at non-crystalline sites are less stable and either relax to the crystalline site assisted by H, diffuse to a new site, or leave the surface entirely. It is much less clear how the diamond phase can form in the first place on a non-diamond substrate. Most often it is found that substrates must be seeded, for example, by polishing with diamond powder, so that diamond nucleation will occur immediately in the growth environment. Otherwise, depending on the energetics of the excited gas, the diamond phase may require an extended period for nucleation to occur, or the diamond phase may not form at all. Apparently, nanometer-scale particles of diamond are embedded into the surface in the various seeding treatments, and these serve as nucleation sites (99).

In this first part, the effects of the seeding procedure on the structure of the Si substrate used for diamond deposition will be described. Figure 29 (top) shows room-temperature optical spectra of a Si wafer surface, mechanically polished using <1 μm diamond powder. After treatment, the excess

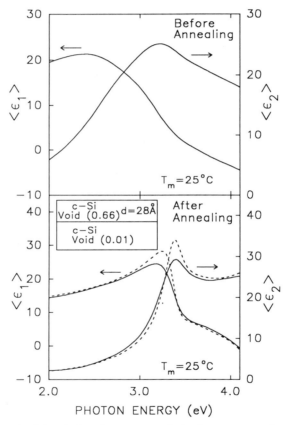

FIG. 29. Pseudo-dielectric function spectra obtained *in situ* on diamond-polished Si substrates at 25°C before (top) and after (bottom) annealing at 520°C (solid lines). The broken lines in the lower panel of the figure are simulated spectra using the best fit structural model shown in the inset. The values in parentheses in the inset are the void volume fractions for the two layers used in the simulation. The agreement between the data and simulation is very good except in the vicinity of the E'_0–E_1 critical point feature. After Ref. 100.

diamond powder was removed from the surface with a methanol rinse. The spectra in Fig. 29 were collected using the multichannel ellipsometer with a 3.2 s acquisition time and are expressed as the pseudo-dielectric function. Comparing these data with the dielectric function of bulk single-crystal Si (see Fig. 25), one can observe that polishing nearly obliterates the E'_0-E_1 critical point feature of the c-Si optical functions near 3.3 eV, indicating a highly defective structure.

In heating the seeded substrate to the deposition temperature for diamond, much of the damage anneals out at $\sim 400-500°C$. Figure 29 (lower panel) includes the corresponding pseudo-dielectric function spectra (solid lines) for the seeded sample measured at room temperature after annealing under vacuum at 520°C. Real-time SE was used during annealing to ensure a stabilized structure prior to cooling the sample. These results are also characteristic of the substrate surface at a diamond deposition temperature of 700°C (see next part), since no further structural changes occur with additional annealing between 520°C and 700°C. A linear regression analysis fit to the (ψ, Δ) spectra for the annealed substrate in Fig. 29 (broken line), using the bulk Si dielectric function in the Bruggeman EMT, suggests the presence of a 28 Å microscopic surface roughness layer on the substrate (see inset in Fig. 29). However, no amorphous phase or residual diamond seeds are detectable; the deviations in the subsurface structure of the annealed sample from that of undamaged Si can be accounted for with 0.01 volume fraction of voids. The fit to these data using the structural model depicted in the inset is quite good except in the neighborhood of the E'_0-E_1 transitions.

To obtain more information on the reason for the poor fit here, the 28 Å surface roughness layer is removed analytically by mathematical inversion, and a line-shape analysis is performed on the critical point feature of the resulting bulk material dielectric function, as described in Section III.C.2. From this sample, lineshape parameters $(E_1, \Gamma, \phi) = (3.373\text{ eV}, 0.180\text{ eV}, -36°)$ are obtained, as compared to $(E_1, \Gamma, \phi) = (3.377\text{ eV}, 0.120\text{ eV}, -37°)$ for bulk c-Si at 25°C. Agreement in E and ϕ indicates that the nature of the electronic transitions is the same for the annealed and single-crystal samples; however, the larger Γ for the annealed sample indicates a reduced electron lifetime. If one proposes that this effect is due to grain boundary scattering, then a grain size of ~ 650 Å can be estimated, based on the correlation in Fig. 28, extrapolated to 25°C (85).

It is of interest at this point to consider the potential of SE to detect diamond seeds embedded in the Si. If one considers the high-energy range of the spectra in Fig. 29 (bottom panel), where the penetration depth of the light is ~ 100 Å, then one can make the crude assumption that the volume of a single detectable seed within this depth corresponds to that of a 100 Å diameter sphere. With this geometry, then the upper limit volume fraction of 0.005 for the diamond component in the material, deduced from the microstructural analysis of the lower spectra in Fig. 29, leads to a lower limit for the seed spacing of $\sim 1,000$ Å. If all these seeds lead to diamond nuclei in the growth process, then an upper limit on the thin-film nucleation density of 10^{10} cm^{-2} would be observed. This result is consistent with the value of $1-2 \times 10^9$ cm^{-2} measured both by SEM and by real-time SE as described in Part 2 of this subsection. In addition, one would expect the growth of single-crystal Si grains during annealing to be limited by the presence of the diamond seeds, and therefore the Si grain size would be smaller than the spacing between seeds. This is also consistent with the experimental Si grain size deduced in the line-shape analysis of Fig. 29. In summary, all experimental observations suggest that the density of diamond seeds embedded into the Si wafer as a result of the treatment process is too low to be detected in these optical experiments.

2. Microstructural Analysis from Real-Time Observations

Next, a real-time SE study of diamond growth on the treated c-Si substrate of Fig. 29 will be discussed in detail (30). Figure 30 shows the experimental configuration used for diamond growth and real-time monitoring by SE. The ellipsometer configuration is the same as that in Figs. 1 and 11; however, the gas excitation method is different. As shown in Fig. 30, during the growth process a $CH_4:H_2$ gas mixture with a flow ratio of 1:80 and total flow of 81 standard cm^3/min was directed through a heated tungsten filament. In this case the filament was a coil of 0.5 mm wire, positioned 8 mm from the substrate, and heated to 1,950°C by applying 175 W power. The treated substrate was mounted on a graphite heater to intercept the excited gas downstream from the filament. The substrate was maintained at 700°C during the growth process, and the total gas pressure was 9 Torr.

For the problem of diamond deposition, real-time SE has a number of uses other than characterizing nucleation and growth (100). First, it was found that a filament pre-carburization procedure was necessary to avoid metallic W contamination of the substrate/diamond film interface in the early stages of film growth. Pre-carburization involves heating the filament for about 1 h in a 1:80 flow ratio of $CH_4:H_2$, at the same power used for diamond deposition and with the substrate out of position. If this procedure is not followed, and an untreated substrate is used to avoid the complications of concurrent diamond growth, a ~ 35 Å discontinuous metallic layer of W or WC_x stabilizes on the heated substrate (700°C) in the first 15

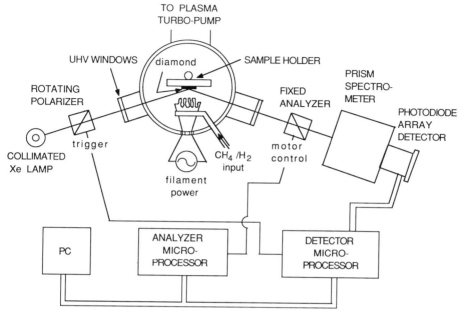

FIG. 30. A schematic of the apparatus used to monitor diamond film growth in real time. The deposition system in this case is designed for filament-enhanced chemical vapor deposition. In this technique the gas mixture, consisting of CH_4 highly diluted in H_2, is excited by passing it through a tungsten filament resistively heated to 1,950°C. The substrate is mounted 8 mm from the filament on a graphite plate that is heatable to 800°C. The windows on the chamber are oriented for a 70° angle of incidence at the sample surface.

minutes after filament ignition. Second, real-time SE can be used to calibrate the true temperature of the substrate surface. An untreated substrate is used for the calibration, and the E_1 transition energy from a line-shape analysis of the second derivative of the dielectric function provides the temperature of the top 250 Å of the surface, through the relationship $T[K] = \{3.486 - E_1[eV]\}/4.07 \times 10^{-4}$ (85). Near the true diamond growth temperature of 700°C, the true temperature of the surface is typically ~50°C lower than that given by a thermocouple embedded in the heater block. This error results from radiation losses and is reduced to ~20°C when the filament is ignited.

Figure 31 shows ~350 pseudo-dielectric function spectra collected in real time during diamond growth using a pre-carburized filament. Each pair of spectra was acquired in 3.2 s, and the time between successive spectra was 17 s. This was sufficient for monolayer resolution at the diamond growth

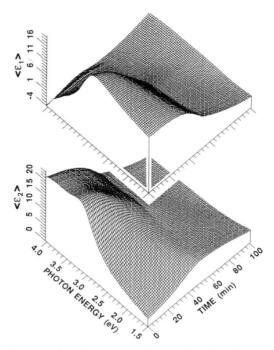

FIG. 31. Real (top) and imaginary parts of the pseudo-dielectric function obtained at 17 s intervals during filament-enhanced CVD of diamond onto a seeded c-Si substrate at a temperature of 700°C. The acquisition time for a single pair of spectra is 3.2 s. The plot was constructed from ∼350 pairs of spectra, each consisting of ∼80 spectral positions between 1.5 and 4.0 eV. At $t = 15$ min, the $CH_4:H_2$ gas flow ratio was decreased from its initial value of 1:40, which stimulated nucleation, to 1:80, which led to a higher quality for the resulting diamond film. After Ref. 30.

rate obtained in this study. The filament was ignited in pure H_2, and upon temperature stabilization at a true surface value of 700°C, real-time SE data acquisition was initiated just before the introduction of CH_4. Initially, the $CH_4:H_2$ flow ratio was set at 1:40 in order to stimulate nucleation. The real-time SE output allowed adjustment of the flow ratio down to 1:80 after several minutes when diamond growth was detected by a uniform decrease in the Δ spectrum. The 1:80 gas flow ratio provides higher-quality diamond with a low sp^2C volume fraction, but leads to a nucleation delay of an hour or more.

Linear regression analysis methods employing the Bruggeman EMT have been used to interpret the spectra of Fig. 31. In an attempt to minimize errors due to light scattering, which increases with increasing photon energy, the analysis was restricted to the energy range from 1.5 to 3.0 eV. In the analysis, the bulk optical functions of the material components of the film structure, including c-Si, diamond, and sp^2C, should be known for the temperature of 700°C. The true dielectric function of the substrate at 700°C is extracted by mathematical inversion from the initial $\langle \varepsilon \rangle$ spectra of Fig. 31, using the microstructure deduced in an analysis of the substrate at room temperature after first an anneal and then an excited H_2 exposure. These two steps simulate the diamond growth sequence prior to the introduction of CH_4. For the diamond component of the film, which is non-absorbing ($\varepsilon_2 = 0$) over the spectral range of interest, the effect of temperature on ε_1 can normally be neglected in thin films consisting of isolated diamond nuclei. This simplification leads to a maximum void volume fraction error of 0.01 in the structural analysis of the film studied here. Such an error is at the level of the 90% confidence limits obtained in the analysis (see Fig. 32). Finally, for sp^2C, the featureless optical functions of glassy carbon are used. These data were also collected at 25°C, and the corresponding error does not appear to be important in this particular study.

Figure 32 shows the two-layer structural model required to fit the time evolution of the raw spectra in Fig. 31, as well as *ex situ* SE data measured after cooling the final sample. The top layer is a physical mixture of diamond and void in order to simulate the isolated particles, observed directly in this case by *ex situ* SEM and verified to be diamond crystallites by noticeable faceting and by Raman spectroscopy. The underlying layer is a three-component mixture of diamond, c-Si, and sp^2C that simulates an absorbing roughness and carbidic surface material. This model indicates

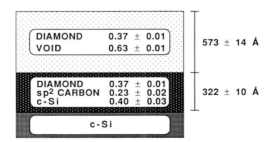

FIG. 32. Two-layer structural model and parameters deduced in a linear regression analysis of the final spectra for the diamond deposition of Fig. 31. The 90% confidence limits on the free parameters are included. The top layer is a diamond/void mixture that simulates isolated diamond crystallites that nucleate on the substrate surface. The lower layer is a three-component mixture that simulates an absorbing roughness and carbidic surface material. After Ref. 30.

that diamond nucleates over a small fraction of the surface, possibly where embedded diamond seeds act as nucleation sites, whereas a complex reaction between the substrate and the gas-phase species occurs over the remainder of the surface. The parameters shown in Fig. 32 were obtained from an analysis of the final spectra of Fig. 31. In an analysis of the full real-time data set, the volume fractions of diamond in the two layers are equated to reduce correlations that occur for the thinnest films.

In the structural model of Fig. 32, correlations also exist between the thickness and the diamond volume fraction in the top layer for the earliest stages of growth. This arises from an inability to extract the thickness and dielectric function ε independently for thin transparent films (<100 Å) when ε is close to that of the ambient. In an analysis of the thin surface layer, we found that over a reasonable range of fixed values for the diamond volume fraction in both layers, the deduced total diamond mass thickness was nearly constant. Thus, the mass thickness of diamond is well defined by the data and can be extracted for values as low as ~ 5 Å. Figure 33 shows the final result for the evolution of the total mass thickness of diamond, including the component in the top layer as well as the interface layer. The mass thickness evolution for the latter contribution alone is also given in the

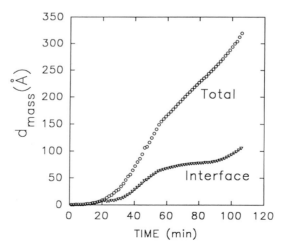

FIG. 33. Diamond mass thickness, i.e., the product of the physical thickness and the diamond volume fraction, plotted as a function of time during diamond deposition. These results were deduced from an analysis of the data of Fig. 31, using the structural model of Fig. 32. The circles and triangles represent the total mass thickness and the contribution from the substrate interface layer, respectively. After Ref. 30.

figure. The results of Fig. 33 suggest three-dimensional growth of diamond crystallites, apparently followed by two-dimensional growth after about an hour. As noted in Section III.C.1, the latter transition may occur even prior to nuclei contact and coalescence, as the near-substrate regions of the crystallites are shadowed by the neighboring crystallites. The stabilization of the diamond mass thickness in the interface layer after about 60 min is also consistent with such an effect.

In the case of diamond growth, the nucleation density is low enough so that the results of the real-time SE analysis at the end of deposition (see Fig. 32) can be compared with direct measurements by SEM, performed *ex situ* after removing the film from the chamber. Figure 34 shows the size distribution of crystallites deduced from plan-view micrographs. For crystallite diameters $\geqslant 1{,}000$ Å, the results can be fitted to a log-normal distribution (*101*), and the best fit statistical median occurs at 1,560 Å. Crystallites with diameters < 500 Å are not included in Fig. 34, as they contribute negligibly to the parameter of interest, the mass thickness. Edge-on micrographs revealed aspect ratios of crystallite diameter (in Fig. 34) to height of 1.4 to 1.7, and cross-sectional shapes between the extremes of rectangle and ellipse. From this information, and from the distribution in Fig. 34, the mass thickness can be estimated from SEM as 335 ± 70 Å, a value in agreement with real-time ellipsometry at the end of deposition, 330 ± 10 Å (see Fig. 33). This agreement tends to suggest that the assumptions made in the interpretation of the real-time SE data do not distort the

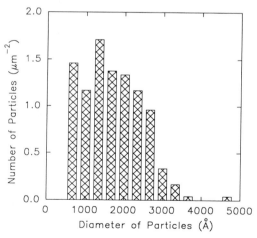

FIG. 34. Diamond particle size distribution measured in the plane of the substrate for the final diamond film of Figs. 31–33, obtained by scanning electron microscopy. After Ref. 30.

results. For example, it appears that diffuse scattering can be neglected for the purposes of diamond mass thickness determination.

If the median particle diameter in the film plane obtained by SEM is scaled by the aspect ratio, then a median thickness for diamond in the range of $1,015 \pm 100$ Å is estimated. This value at least overlaps the 895 ± 25 Å value obtained from real-time SE analysis, including both layers in Fig. 32. Thus, for a film consisting of isolated particles with a broad distribution of particle sizes, the thickness deduced by real-time SE appears to approximate the median particle height. If the quantification by SEM is ignored for the moment, and we were to apply a nucleation model consisting of hemispherical particles on a square grid (as might be assumed if SEM was

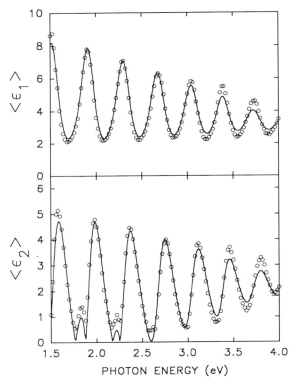

FIG. 35. Experimental pseudo-dielectric function spectra (points) obtained from *ex situ* measurements of a thick diamond film prepared by heated filament-enhanced CVD. These data were collected with a serially scanning spectroscopic ellipsometer. The solid line is the best-fit simulation resulting from a linear regression analysis of the experimental data using the three-layer structure and free parameters provided in Fig. 36. After Ref. 102.

unavailable), then a nucleation density of $(2.2 \pm 0.2) \times 10^9 \text{ cm}^{-2}$ would be obtained. This value is to be compared with the SEM result of $(1.2 \pm 0.5) \times 10^9 \text{ cm}^{-2}$. The agreement with the hemispherical assumption is reasonably good, since this shape closely approximates that of the crystallites. In any event, the small discrepancy arises from the assumption of the incorrect crystallite shape and disappears when the correct shape is used. This agreement promotes greater confidence in optically deduced nucleation information, such as that in Sections III.A.1, III.B.2, and III.C.1.

Finally, in Figure 35 experimental SE data, expressed as the pseudo-dielectric function (points), are presented for a much thicker diamond film prepared by the filament excitation process (102). These data were obtained *ex situ* with a serially scanning instrument after cooling the film and removing it from the deposition chamber. Linear regression analysis has been applied to obtain information on the sample microstructure; the best fit to the data and the resulting parameters are shown in Figs. 35 (solid line) and 36, respectively. A satisfactory fit has been obtained with a three-layer structure for the film, including a ~200 Å thick surface roughness layer, a ~6,700 Å bulk layer, and a ~1,250 Å low-density interface layer.

This three-layer model provides a highly simplified view of a very complex film microstructure. The near-substrate diamond layer is expected to exhibit a void density that decreases with increasing distance from the substrate–film interface. Thus, the interface between the near-substrate layer and the denser, bulk film is not expected to be sharp as in Fig. 36, but smeared out

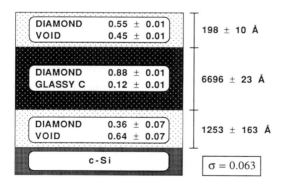

FIG. 36. Schematic of the three-layer film microstructure, the best fit parameters, and 90% confidence limits obtained in a linear regression analysis of the experimental data of Fig. 35. The layers simulate a complex film microstructure, including a dense bulk film sandwiched between a near-substrate interface of incompletely coalesced diamond crystallites and a roughness layer of diamond crystallites protruding from the surface. After Ref. 102.

over a depth of hundreds of angstroms, as columns in the shape of inverted cones gradually make contact. It is interesting that in the region where contact has been made and bulk-like film growth occurs, a significant volume fraction of sp^2C defects must be included in the model for a good fit. These defects appear to occur in the grain boundary regions between interlocking diamond crystallites, whose size is estimated to be ~ 700 Å from x-ray diffraction measurements. The top surface roughness layer arises from the individual crystallites protruding above the film surface.

Apparently, the voids between nuclei, observed in the early stages of growth in the real-time SE study (see Fig. 32), are not forced out of the network by uniformly enlarging nuclei, but are trapped at the substrate–film interface and remain within the final film. This observation from Fig. 36 is consistent with the transition to linearity in the mass thickness increase in Fig. 33, signifying that the nuclei begin to grow two-dimensionally (i.e., preferentially normal to the plane of the film) even before coalescence. This behavior is distinctly different from the growth pattern of the amorphous films of Sections III.A and III.B, but has been observed in thicker μc-Si:H films. The high volume fraction of void trapped in the near-interface layer appears to be a consequence of a very low nucleation density. Recent real-time, single-photon energy ellipsometry studies of diamond growth by microwave PECVD suggest that a much higher nucleation density can be achieved with this technique, provided specific substrate seeding recipes are followed (103). Such films exhibit a much lower near-substrate interface void fraction and a more uniform void density distribution, when deposited as thick layers, than those prepared by the heated filament process onto similarly seeded substrates (98). Future investigations will concentrate on the effect of the preparation process and parameters, as well as the substrate treatment procedure, on the nucleation density and ultimate uniformity of diamond films.

IV. Summary

Over the years, ellipsometry has been developed as a powerful probe of the preparation and properties of thin films. In ellipsometry, one measures the magnitude and phase, $\tan \psi$ and Δ, of the complex amplitude reflection ratio (p-to-s). Together these parameters characterize the change in polarization state incurred when an incident polarized light beam is reflected from the surface of a specular solid. Since the development of automatic instrumentation for ellipsometry, the technique has generally progressed along two different lines. Spectroscopic ellipsometry involves measuring ψ and Δ continuously versus photon energy from the near infrared to the near ultraviolet. Real-time ellipsometry involves measuring ψ and Δ continuously versus time at one photon energy during film growth with a resolution in thickness at the monolayer level.

The first application, spectroscopic ellipsometry (SE), has been advanced to measure static surfaces and has been employed to determine the optical properties of bulk materials and thin films, as well as the structure of complex, multilayered samples. Because ellipsometry involves polarization state detection and provides two parameters (ψ, Δ), it maintains an important advantage over conventional reflectance measurements owing to its ability to determine both real and imaginary parts of the dielectric function of bulk materials, even from measurements at one photon energy or over a restricted energy range. From the dielectric function measured versus photon energy for a bulk crystalline solid, the electronic transitions can be tabulated, including the characteristic energies and line-shapes, which provide information on the nature of the transitions and the electronic band structure of the solid. From these results, very useful additional information on alloy composition, dopant and defect density, and surface temperature can be extracted, providing an appropriate data base exists.

SE also has important advantages over alternative optical techniques for the characterization of thin films. In such cases, three unknowns are required to fully solve the optical problem, namely the real and imaginary parts of the dielectric function of the film as well as its thickness. Thus, a solution is not possible, in general, from an ellipsometry measurement at a single photon energy. However, the continuous ellipsometric spectra are often sufficient to identify the correct thickness by eliminating unphysical spectral artifacts that appear in trial dielectric functions when an incorrect thickness is chosen. (The trial dielectric function is deduced by mathematical inversion of (ψ, Δ) spectra once a trial thickness value is chosen.) In another application of SE, one can characterize multilayered structures in which the individual layers may exhibit complex microstructures, as long as the dielectric functions of all the components of the structure are known. This method applies an effective medium theory and least-squares regression analysis, using the thicknesses and component volume fractions as free parameters.

The second application made possible by the development of automatic instrumentation, real-time ellipsometry, has been advanced to monitor thin-film deposition. Until recently, for thin-film deposition rates on the order of a monolayer per second, (ψ, Δ) measurements were only possible at one photon energy owing to the time required to scan the spectrometer. For thin-film systems exhibiting ideal growth behavior, such as epitaxial, ternary-alloy semiconductors, (ψ, Δ) data collected at a single photon energy in real time are sufficient to extract the depth profile of the alloy composition that characterizes the final sample. For amorphous, microcrystalline, or polycrystalline thin films, however, growth is often far from ideal. In addition, the dielectric function at a single photon energy is not a definitive characteristic of the material composition, in contrast to the situation for single crystals. For example, for binary-alloy amorphous semiconductors, both the alloy composition and the void volume fraction can vary, making it difficult to characterize the material from real-time measurements at a single photon energy. In situations in which only the film microstructure (i.e., the void volume fraction) evolves with thickness, real-time ellipsometry can provide information on the growth process, as long as the dielectric function can be linked to the microstructure through an effective medium theory. This approach tends to work reasonably well for simple amorphous semiconductors such as hydrogenated amorphous silicon. In the case of micro- and polycrystalline films, however, the dielectric function may depend on time during growth, not only through changes in the film composition, but also through grain-size effects.

More recently, the new technique of real-time spectroscopic ellipsometry has been developed, making effective use of the advances over the years in both spectroscopic and real-time ellipsometry. By combining the important attributes of the two approaches, more difficult thin-film growth problems can be solved. The key to the development of real-time SE is the replacement of the spectrometer and photomultiplier tube detector in the conventional instruments with a spectrograph and rapid-scanning photodiode array detector. Because the array is an integrating detector, raw data at the different photon energies are collected in parallel, in contrast to the earlier serially scanning instruments. With parallel detection one can collect a 100-point pair of (ψ, Δ) spectra from 1.5 eV to 4.3 eV with a minimum acquisition time of 32 ms. This is only about a factor of three slower than the minimum acquisition time for (ψ, Δ) at a single photon energy, using a photomultiplier tube detector in an otherwise similarly designed rotating polarizer-type instrument. The novel real-time spectroscopic ellipsometer has been applied most widely in studying the growth of various technologically important, tetrahedrally bonded films, including hydrogenated amorphous silicon (a-Si:H) and silicon–carbon alloys (a-Si$_{1-x}$C$_x$:H), microcrystalline silicon (μc-Si:H), and diamond, all prepared by enhanced chemical vapor deposition (CVD) methods.

In the case of amorphous semiconductor growth by plasma-enhanced CVD on substrates having atomic-scale smoothness, information on both microstructural evolution and the optical properties can be obtained from real-time SE. The statistical information that results from a least-squares regression analysis of the microstructural evolution provides strong evidence for a transition during film growth from a one-layer optical model for the film to a two-layer model. The one-layer model simulates isolated thin film nuclei that form in the initial stages of growth; the two-layer model simulates a bulk-like layer with a residual surface roughness layer on top. The latter two layers result when nuclei coalesce. The transition between the two models occurs when the nuclei make contact and the first bulk-like monolayer forms at the substrate interface.

From these models, a wealth of microstructural information can be extracted, including (i) the void volume fraction and thickness versus time in the nucleation regime, providing information on the nuclei geometry; (ii) the nuclei thickness when the first bulk-like monolayer forms, providing information on nucleation density; (iii) the evolution of the bulk layer thickness with time in the later stages of growth, providing the deposition rate; (iv) the evolution of the surface roughness layer thickness after nuclei make contact, providing information on the degree of coalescence and precursor surface diffusion; (v) the mass thickness versus time, providing information on the rate of incorporation of precursors in the different regimes of nucleation, coalescence, and bulk film growth; and (vi) the relative void volume fraction in the final bulk film, providing information on the ultimate material microstructural quality.

The effect of substrate material has been investigated, simply to determine which microstructural features are intrinsic to the growth process and which are extrinsic, depending on the choice of substrate. In such studies, the substrate must be atomically smooth to avoid the complexities associated with the conversion of a substrate surface roughness layer to a interface roughness layer upon film growth. Although the geometry of the nuclei is sensitive to the substrate, the microstructural evolution after nuclei make contact is found to be independent of the substrate, at least for the materials investigated to this point. The effect of deposition conditions on the microstructural characteristics have also been investigated and have provided insights into the monolayer-scale surface processes that control the ultimate properties of the material.

For a-Si:H, it has been found that the conditions of substrate temperature and plasma power for which the maximum surface smoothening is observed upon nuclei contact also correspond to those for which the optimum photoresponse is obtained for the final films. It has been proposed that the larger surface smoothening effect is characteristic of an enhanced diffusion length for precursors on the film surface. This presumably leads to reduced disorder in the resulting a-Si:H network, with a minimum of point defects or extended microstructural defects that limit the photocarrier lifetime and mobility in electronic devices. For a-Si$_{1-x}$C$_x$:H, the use of hydrogen dilution of the reactive gases in the growth process has been found to lead to a lower nucleation density, greater smoothening during coalescence, and a reduced bulk film void volume fraction. These trends also appear to be related to enhanced precursor diffusion on substrate and film surfaces, as well as enhanced hydrogen exchange between the film surface and the gas phase. These desirable features again lead to films with improved electronic performance.

The ability of real-time SE to determine the optical properties of amorphous semiconductor thin films is equally important, since the design and ultimate performance of devices such as solar cells are sensitively dependent on the optical absorption coefficient of the films. The dielectric function of the a-Si:H bulk layer in the two-layer model and the nucleating a-Si:H in the one-layer model can be determined from real-time observations of layers as thin as 20 Å. This in turn provides the band gaps of the layers. Although the dielectric function and band gap of the bulk a-Si:H layer is independent of thickness, the nucleating clusters themselves exhibit a band gap that is ~ 0.25 eV wider than that of the bulk film. The difference is attributed to the additional hydrogen concentration bonded at the cluster surfaces. Although the band gap of the bulk a-Si:H layer is influenced by its H concentration as well, better control of the gap over a wide range is obtained with a-Si$_{1-x}$C$_x$:H, by varying the CH$_4$:SiH$_4$ reactive gas ratio. For such alloys, real-time SE is especially important given the high sensitivity of the optical properties of the film to the preparation conditions.

For enhanced CVD μc-Si:H and diamond films, the evolution of the microstructure and optical properties has been investigated during the nucleation process when the film consists of isolated individual crystallites. For μc-Si:H, an observed cubic time dependence of the mass thickness in the early stages of film growth is consistent with a three-dimensional growth process, i.e., isolated crystallites that increase in size linearly with time. Because the spectral range of the ellipsometer overlaps the E_1 optical transitions in the Si microcrystallites, information on the electronic characteristics can be extracted as a continuous function of the crystallite size. This capability is limited by a low nucleation density that leads to a very low volume of material in the nucleating film (~ 30 Å mass thickness at a physical thickness of 200 Å). Over the accessible range of film thickness (200–250 Å) in these studies, the energy and phase associated with the E_1 transitions are independent of crystallite size. However, the broadening parameter decreases with increasing thickness in accordance with the scattering of electrons at crystallite surfaces. For diamond films, a similar analysis is not possible because the electronic transitions are above the high-energy limit of the ellipsometer.

For the extremely complex diamond growth process by heated filament-assisted CVD, however, real-time SE has had a number of other important applications. For example, using real-time SE (i) information has been obtained on the annealing that diamond-seeded substrates undergo when they are heated to the temperature required for diamond growth; (ii) techniques have been developed that minimize tungsten contamination from the filament at the diamond/substrate interface; (iii) calibrations have been performed that determine the temperature of the top 250 Å of the substrate under diamond growth conditions; and (iv) alterations in gas flow conditions have been implemented in response to diamond growth for a reduction in the nuclei induction time. With such procedures in place, real-time SE monitoring in the early growth stages provides the nucleation density and evolution of the mass thickness. The values obtained for these parameters at the very end of the growth process are found to be in excellent agreement with *ex situ* scanning electron microscopy. Such agreement supports the validity of the real-time SE techniques of thin film microstructural analysis and lends confidence in the results of other studies such as a-Si:H nucleation, in which direct microstructural characterization is extremely difficult.

In conclusion, it is emphasized that although many very effective surface spectroscopies exist, real-time spectroscopic ellipsometry is fundamentally different not only because of its speed, but also because (i) it requires no *in vacuo* instrumentation; (ii) it operates passively during deposition; (iii) it can probe surfaces even in the most adverse environments such as chemical vapor deposition; and (iv) it can also be used to extract rudimentary information such as thickness and film composition, as well as to obtain a basic understanding of material growth processes and optical functions. It is expected to play an important future role in solving a wide variety of problems in materials science and optical physics.

Acknowledgments

The authors gratefully acknowledge research support from the National Science Foundation under Grant Nos. DMR-8957159, DMR-8901031, and DMR-9217169; the National Renewable Energy Laboratory under Subcontract No. XG-1-10063-10; the Electric Power Research Institute; the State of

Pennsylvania under the Ben Franklin Centers of Excellence Program; the AT&T Foundation; and the Office of Naval Research (with funding from the Strategic Defense Initiative's Office of Innovative Science and Technology) under contract No. N00014-86-K-0443. The authors also appreciate the encouragement of their collaborators in this research, Profs. R. Messier, K. Vedam, and C. R. Wronski.

References

1. N. M. Bashara, A. B. Buckman, and A. C. Hall, ed., "Proceedings of the Symposium on Recent Developments in Ellipsometry," North-Holland, Amsterdam, 1969; also published as *Surf. Sci.* **16** (1969).
2. N. M. Bashara and R. M. A. Azzam, ed., "Ellipsometry: Proceedings of the Third International Conference on Ellipsometry," North-Holland, Amsterdam, 1976; also published as *Surf. Sci.* **56** (1976).
3. R. H. Muller, R. M. A. Azzam, and D. E. Aspnes, ed., "Ellipsometry: Proceedings of the Fourth International Conference on Ellipsometry," North-Holland, Amsterdam, 1980; also published as *Surf. Sci.* **96** (1980).
4. F. Abeles, ed., "Ellipsometry and other Optical Methods for Surface and Thin Film Analysis," Les Editions de Physique, Les Ulis, France, 1983; also published as *J. Phys. (Paris)* **44**, C-10 (1983).
5. "The Proceedings of the First International Conference on Spectroscopic Ellipsometry," Paris, France, January 1993; also published as *Thin Solid Films* **233** and **234** (1993).
6. J. B. Theeten, *Surf. Sci.* **96**, 275 (1980).
7. R. H. Muller, *in* "Techniques for Characterization of Electrodes and Electrochemical Processes" (R. Varma and J. R. Selman, ed.), John Wiley & Sons, New York, 1991, p. 31.
8. S. Gottesfeld, *in* "Electroanalytical Chemistry: A Series of Advances, Volume 15" (A. J. Bard, ed.), Marcel Dekker, New York, 1989, p. 143.
9. R. W. Collins, *Rev. Sci. Instrum.* **61**, 2029 (1990).
10. R. M. A. Azzam and N. M. Bashara, "Ellipsometry and Polarized Light," North-Holland, Amsterdam, 1977.
11. A. Moritani, Y. Okuda, H. Kubo, and J. Nakai, *Appl. Opt.* **22**, 2429 (1983).
12. J. B. Theeten and D. E. Aspnes, *Ann. Rev. Mater. Sci.* **11**, 97 (1981).
13. D. E. Aspnes, W. E. Quinn, and S. Gregory, *Appl. Phys. Lett.* **57**, 2707 (1990).
14. F. Hottier and R. Cadoret, *J. Cryst. Growth* **56**, 304 (1982).
15. B. Drevillon, *J. Non-cryst. Solids* **114**, 139 (1989).
16. R. W. Collins, in "Amorphous Silicon and Related Materials," Vol. B (H. Fritzsche, ed.), World Scientific, Singapore, 1989, p. 1003.
17. G. Feng, N. Maley, and J. R. Abelson, *Appl. Phys. Lett.* **59**, 330 (1991).
18. R. W. Collins and B. Y. Yang, *J. Vac. Sci. Technol.* **B7**, 1155 (1989).
19. R. H. Muller and J. C. Farmer, *Rev. Sci. Instrum.* **55**, 371 (1984).
20. Y.-T. Kim, R. W. Collins, and K. Vedam, *Surf. Sci.* **223**, 341 (1990).
21. D. E. Aspnes, *in* "Optical Properties of Solids: New Developments" (B. O. Seraphin, ed.), North-Holland, Amsterdam, 1976, p. 799.

22. I. An, Y. M. Li, H. V. Nguyen, and R. W. Collins, *Rev. Sci. Instrum.* **63**, 3842 (1992).
23. I. An, H. V. Nguyen, N. V. Nguyen, and R. W. Collins, *Phys. Rev. Lett.* **65**, 2274 (1990).
24. I. An, Y. M. Li, C. R. Wronski, H. V. Nguyen, and R. W. Collins, *Appl. Phys. Lett.* **59**, 2543 (1991).
25. Y. M. Li, I. An, N. V. Nguyen, C. R. Wronski, and R. W. Collins, *Phys. Rev. Lett.* **68**, 2814 (1992).
26. Y. M. Li, I. An, H. V. Nguyen, C. R. Wronski, and R. W. Collins, *J. Non-cryst. Solids* **137&138**, 787 (1991).
27. Y. Lu, I. An, M. Gunes, M. Wakagi, C. R. Wronski, and R. W. Collins, *Appl. Phys. Lett.* **63**, 2228 (1993).
28. H. V. Nguyen, I. An, Y. M. Li, C. R. Wronski, and R. W. Collins, *Mater. Res. Soc. Symp. Proc.* **258**, 235 (1992).
29. H. V. Nguyen and R. W. Collins, *Phys. Rev. B* **47**, 1911 (1993).
30. R. W. Collins, Y. Cong, H. V. Nguyen, I. An, T. Badzian, R. Messier, and K. Vedam, *J. Appl. Phys.* **71**, 5287 (1992).
31. N. V. Nguyen, B. S. Pudliner, I. An, and R. W. Collins, *J. Opt. Soc. Am.* **A8**, 919 (1991).
32. I. An, Y. Cong, N. V. Nguyen, B. S. Pudliner, and R. W. Collins, *Thin Solid Films* **206**, 300 (1991).
33. I. An and R. W. Collins, *Rev. Sci. Instrum.* **62**, 1904 (1991).
34. G. Laurence, F. Hottier, and J. Hallais, *Rev. Phys. Appl. (Paris)* **16**, 579 (1981).
35. D. E. Aspnes, *Proc. Soc. Photo. Instrum. Eng.* **276**, 188 (1981).
36. W. G. Oldham, *Surf. Sci.* **16**, 97 (1969).
37. D. E. Aspnes and A. A. Studna, *Phys. Rev. B* **27**, 985 (1983).
38. I. H. Malitson, *J. Opt. Soc. Am.* **55**, 1205 (1965).
39. D. A. G. Bruggeman, *Ann. Phys. (Liepzig)* **24**, 636 (1935).
40. D. E. Aspnes, J. B. Theeten, and F. Hottier, *Phys. Rev. B* **20**, 3292 (1979).
41. D. E. Aspnes, *Thin Solid Films* **89**, 249 (1982).
42. G. A. Niklasson, C. G. Granqvist, and O. Hunderi, *Appl. Opt.* **20**, 26 (1981).
43. J. Kanicki, ed., "Amorphous and Microcrystalline Semiconductor Devices: Optoelectronic Devices," Artech House, Norwood, Massachusetts, 1991.
44. M. J. Thompson, Y. Hamakawa, P. G. LeComber, A. Madan, and E. Schiff, "Amorphous Silicon Technology—1992," Materials Research Society, Pittsburgh, 1992.
45. J. Bullot and M. P. Schmidt, *Phys. Status Solidi B* **143**, 345 (1987).
46. D. Kruangam, T. Endo, W. G. Pu, H. Okamoto, and Y. Hamakawa, *Jpn. J. Appl. Phys.* **24**, L806 (1985).
47. Y. Kuwano, H. Nishiwaki, S. Tsuda, T. Fukatsu, K. Enomoto, Y. Nakashima, and H. Tarui, in "The Proceedings of the 16th IEEE Photovoltaic Specialists Conference," IEEE, New York, 1982, p. 1338.
48. S. Guha, J. Yang, P. Nath, and M. Hack, *Appl. Phys. Lett.* **49**, 218 (1986).
49. J. C. Angus and C. C. Hayman, *Science* **241**, 913 (1988).
50. W. A. Yarbrough and R. Messier, *Science* **247**, 605 (1990).
51. S. Kurita, in "New Diamond, 1990," Japan New Diamond Forum, Tokyo, Japan, 1990, p. 113.
52. A. R. Forouhi and I. Bloomer, *Phys. Rev. B* **34**, 7018 (1986).
53. J. A. Venables, G. D. T. Spiller, and M. Hanbucken, *Rep. Prog. Phys.* **47**, 399 (1984).
54. W. Paul and D. A. Anderson, *Solar Energy Mater.* **5**, 229 (1980).
55. R. W. Collins, I. An, H. V. Nguyen, and T. Gu, *Thin Solid Films* **206**, 374 (1991).
56. R. W. Collins and J. M. Cavese, *J. Non-cryst. Solids* **97/98**, 1439 (1987).
57. A. Gallagher and J. Scott, *Solar Cells* **21**, 147 (1987).

58. K. Mui and F. W. Smith, *Phys. Rev. B* **38**, 10623 (1988).
59. J. C. Phillips, *Phys. Rev. Lett.* **20**, 550 (1968).
60. J. A. Van Vechten, *Phys. Rev.* **182**, 891 (1969).
61. M. Cardona, *Phys. Status Solidi B* **118**, 463 (1983).
62. J. Tauc, R. Grigorovici, and A.Vancu, *Phys. Status Solidi* **15**, 627 (1966).
63. V. Vorlicek, M. Zavetova, S. K. Pavlov, and L. Pajasova, *J. Non-cryst. Solids* **45**, 289 (1981).
64. R. H. Klazes, M. H. L. M. van den Broek, J. Bezemer, and S. Radelaar, *Philos. Mag. B* **45**, 377 (1982).
65. G. D. Cody, in "Semiconductors and Semimetals," Vol. 21 B (J. I. Pankove, ed.), Academic Press, New York, 1984, p. 11.
66. G. Weiser and H. Mell, *J. Non-cryst. Solids* **114**, 298 (1989).
67. A. Mazor, D. J. Srolovitz, P. S. Hagan, and B. G. Bukiet, *Phys. Rev. Lett.* **60**, 424 (1988).
68. R. Robertson and A. Gallagher, *J. Appl. Phys.* **59**, 3402 (1986).
69. A. Gallagher, *Mater. Res. Soc. Symp. Proc.* **70**, 3 (1986).
70. K. Tanaka and A. Matsuda, *Mater. Sci. Rep.* **2**, 139 (1987).
71. G. Lucovsky, D. V. Tsu, R. A. Rudder, and R. J. Markunas, in "Thin Film Processes II" (J. L. Vossen and W. Kern, eds.), Academic Press, New York, 1991, p. 565.
72. S. Veprek, F.-A. Sarrott, S. Rambert, and E. Taglauer, *J. Vac. Sci. Technol.* **A7**, 2614 (1989).
73. R. M. A. Dawson, C. M. Fortmann, M. Gunes, Y. M. Li, S. S. Nag, R. W. Collins, and C. R. Wronski, *Appl. Phys. Lett.* **63**, 955 (1993).
74. I. Solomon, M. P. Schmidt, and H. Tran-Quoc, *Phys. Rev. B* **38**, 9895 (1988).
75. Y. Lu, I. An, M. Gunes, M. Wakagi, C. R. Wronski, and R. W. Collins, *Mater. Res. Soc. Symp. Proc.* (in press, 1994).
76. A. Matsuda and K. Tanaka, *J. Non-cryst. Solids* **97&98**, 1367 (1987).
77. A. H. Mahan, B. von Roedern, D. L. Williamson, and A. Madan, *J. Appl. Phys.* **57**, 2717 (1985).
78. M. A. Petrich, K. K. Gleason, and J. A. Reimer, *Phys. Rev. B* **36**, 9722 (1987).
79. S. Veprek, *Mater. Res. Soc. Symp. Proc.* **164**, 39 (1990).
80. C. C. Tsai, in "Amorphous Silicon and Related Materials," Vol. A (H. Fritzsche, ed.), World Scientific, Singapore, 1989, p. 123.
81. S. Usui and M. Kikuchi, *J. Non-cryst. Solids* **34**, 1 (1979).
82. R. W. Collins, W. J. Biter, A. H. Clark, and H. Windischmann, *Thin Solid Films* **129**, 127 (1985).
83. H. Arwin and D. E. Aspnes, *Thin Solid Films* **113**, 101 (1984).
84. T. Aoki and S. Adachi, *J. Appl. Phys.* **69**, 1574 (1991).
85. P. Lautenschlager, M. Garriga, L. Viña, and M. Cardona, *Phys. Rev. B* **36**, 4821 (1987).
86. G. F. Feng and R. Zallen, *Phys. Rev. B* **40**, 1064 (1989).
87. U. Kreibig, *J. Phys. F* **4**, 999 (1974).
88. J. R. Chelikowsky and M. L. Cohen, *Phys. Rev. B* **14**, 556 (1976).
89. B. G. Bagley, D. E. Aspnes, A. C. Adams, and C. J. Mogab, *Appl. Phys. Lett.* **38**, 56 (1981).
90. R. W. Collins, H. Windischmann, J. M. Cavese, and J. Gonzalez-Hernandez, *J. Appl. Phys.* **58**, 954 (1985).
91. B. V. Spitsyn, L. L. Bouilov, and B. V. Derjaguin, *J. Cryst. Growth* **52**, 219 (1981).
92. S. Matsumoto, Y. Sato, M. Kamo, and N. Setaka, *Jpn. J. Appl. Phys.* **21**, L183 (1982).
93. S. Matsumoto, Y. Sato, M. Tsutsumi, and N. Setaka, *J. Mater. Sci.* **17**, 3106 (1982).
94. M. Kamo, Y. Sato, S. Matsumoto, and N. Setaka, *J. Cryst. Growth* **62**, 642 (1983).
95. D. L. Pappas, K. L. Saenger, J. Bruley, W. Krakow, T. Gu, R. W. Collins, and J. J. Cuomo, *J. Appl. Phys.* **71**, 5675 (1992).
96. Y. Cong, R. W. Collins, R. Messier, K. Vedam, G. F. Epps, and H. Windischmann, *in*

"New Diamond Science and Technology" (R. Messier, J. Glass, J. Butler, and R. Roy, eds.), Materials Research Society, Pittsburgh, 1991, p. 735.
97. A. R. Badzian, T. Badzian, and D. Pickrell, *Proc. Soc. Photo-opt. Instrum. Eng.* **969**, 14 (1988).
98. Y. Cong, R. W. Collins, G. F. Epps, and H. Windischmann, *Appl. Phys. Lett.* **58**, 819 (1991).
99. W. Yarbrough, *J. Am. Ceram. Soc.* **75**, 3179 (1992).
100. Y. Cong, Ilsin An, H. V. Nguyen, K. Vedam, R. Messier, and R. W. Collins, *Surface and Coatings Technology* **49**, 381 (1991).
101. C. G. Granqvist and R. A. Buhrman, *Appl. Phys. Lett.* **27**, 693 (1975).
102. R. W. Collins, Y. Cong, Y.-T. Kim, K. Vedam, Y. Liou, A. Inspektor, and R. Messier, *Thin Solid Films* **181**, 565 (1989).
103. Y. Hayashi, W. Drawl, R. W. Collins, and R. Messier, *Appl. Phys. Lett.* **60**, 2868 (1992).

Real-Time Spectroscopic Ellipsometry Studies of the Nucleation, Growth, and Optical Functions of Thin Films, Part II: Aluminum

HIEN V. NGUYEN, ILSIN AN, AND ROBERT W. COLLINS

Department of Physics and Materials Research Laboratory,
The Pennsylvania State University, University Park, Pennsylvania

I. Introduction 127
II. Theoretical Background 130
    A. The Optical Functions of Bulk Aluminum 131
    B. Size Effects on the Optical Functions 133
    C. The Generalized Effective Medium Theory 139
III. Experimental Apparatus for Real-Time Monitoring 147
IV. Results and Discussion 149
    A. Overview of the Optical Functions of Aluminum
       Thin Films in the Nucleation and Growth Stages 150
    B. Analysis of the Optical Functions of Continuous Films 162
    C. Analysis of the Optical Functions of Particle Films 168
V. Summary 183
    Acknowledgments 187
    References 187

I. Introduction

For more than a century, the optical properties of ultrathin metal films, thick metal films, and bulk metals have been the subject of extensive investigation. In particular, an important goal has been to explain the significant differences in optical properties among these different forms. The problem was first encountered by Faraday, who reported brilliant colors associated wih ultrathin metal films when illuminating them by white light (*1*). He ascribed the effect to a discontinuous morphology of the films. At the beginning of the 20th century, Maxwell-Garnett proposed a theoretical explanation for these colors in terms of optical absorption bands that are not present in the thick films or bulk materials (*2*). This explanation involved a simple morphological model for the film, in which it is treated as a slab of composite material consisting of spherical metal particles embedded in the ambient medium. The particles were further assumed to exhibit the optical properties of bulk material and to interact in accordance with the Lorenz local-field prescription. The Maxwell-Garnett theory in its original form was able to explain the general features of the optical properties of ultrathin metal films, and later it was verified by electron microscopy that such films were in fact discontinuous, consisting of substrate-supported, isolated particles (3).

Since the development of the simple Maxwell-Garnett effective medium theory (EMT), several refinements for calculating the effective optical functions of discontinuous, or particle, films have been proposed. Attempts to justify such refinements often involved comparing the measured optical properties, such as the transmittance spectrum, with those predicted from a morphological model based on direct observations by electron microscopy. The most common generalized Maxwell-Garnett-type EMT is based on a model for the film as a two-dimensional array of dipoles, representing substrate-supported, spheroidal particles having symmetry axes normal to the substrate surface (4–7). In one refinement, the fields due to the images of the dipoles in the substrate surface were also included in the calculation (8).

In one of the most detailed studies, electron microscopy was used to obtain the morphological parameters of gold particle films, which were represented as a two-dimensional array of spheroids with symmetry axes parallel to the substrate surface (9). These parameters included the mean eccentricity of the particles, the distribution of their semiaxes, and the average interparticle spacing. This information was used as input to a generalized Maxwell-Garnett EMT similar to that of Ref. 7 in order to calculate the effective optical functions of the particle films. In this case, the optical functions used for the solid component in the EMT were not those of bulk gold, but rather, were modified by assuming that the free-electron relaxation time was limited by scattering at particle surfaces. Reasonable agreement was obtained between the normal incidence transmittance spectrum obtained experimentally and that predicted from the calculated effective optical functions, using a single variable parameter in the EMT that describes the distribution of particle separations (7). From this agreement, the authors concluded that the morphology of metal particle films completely governs their optical functions. This conclusion was at odds with earlier careful studies finding that the optical functions of silver particle films could be influenced by defects internal to the particles, and the presence of these defects is very sensitive to the preparation technique (10).

More recent studies of the optical properties of metal films over the last decade have exploited developments in spectroscopic ellipsometry (SE) in order to better characterize and understand the relationship between the preparation process and the surface and subsurface ("bulk") morphology of thick metal films (11). For films that are opaque over the near-infrared to near-ultraviolet photon energy range, SE provides an important advantage over reflectance measurements in its ability to extract both real and imaginary parts of the optical functions directly, as long as the metal surface is free of overlayers (12). With the recent development of the rapid-scanning multichannel ellipsometer, the optical functions of thin films can now be determined from high-speed measurements performed *in situ* and in real time during the growth process (13,14). As described in the present chapter, this capability has been exploited in a study of the evolution of the optical functions of thin aluminum films, not just in the bulk film growth stage, but also throughout the stages of nucleation and coalescence (15–17). From the evolution of the optical functions, insights into the evolution of the morphological and electronic characteristics of the film can be extracted. This is an important breakthrough because the optical approach, in contrast to direct probes of morphology, is non-destructive, non-invasive, and requires no specialized equipment internal to the deposition system.

This investigation relies heavily on the foundation developed by previous workers with respect to (i) the optical functions of bulk aluminum (18), (ii) size effects on the optical functions of metal particles in ultrathin films (19) or metal grains in thicker, bulk-like films (20), and (iii) effective medium theories that permit one to calculate the effective optical functions of a film that is density-deficient because of the nucleation process in ultrathin films or the grain structure in thicker films (7, 8, 21, 22). Section II of this chapter provides an overview of each of these three topics. In Section III, the experimental apparatus used for the real-time measurements will be described. For more detailed information on the multichannel ellipsometer, the reader is referred to the previous chapter and to the references (13,14). In the first part of Section IV, a preliminary linear regression analysis interpretation (23) has been described. In this interpretation, the bulk dielectric function of aluminum is employed as the solid component in a very simple effective medium theory in an attempt to explain the evolution of the real-time SE data collected during aluminum preparation throughout the nucleation, coalescence, and bulk film growth regimes. The inability to obtain close fits to the experimental data, particularly in the nucleation and coalescence regimes, points out the need for a more sophisticated approach employing size-dependent optical functions, and/or a more complex effective medium theory.

Thus, the dielectric functions are extracted directly from the data in order to interpret correctly the physical processes that affect the observed optical properties. The interpretation is described in detail in Sections IV.B and IV.C. Subsection B focuses on the results obtained after the percolation threshold when the film is continuous. Subsection C focuses on the effective dielectric functions obtained before the coalescence process when the film consists of isolated particles and exhibits a well-defined absorption band, as first observed by Faraday (*1*). Because a morphological model is used in the data analysis that neglects the statistical distribution in the size and shape of the aluminum grains and particles, less emphasis is placed on the interpretation of the deduced morphological information. However, it must be demonstrated that the simplified morphological model does not compromise the validity of the interesting information extracted on the electronic behavior.

II. Theoretical Background

Analysis of the evolution of the optical functions of aluminum from nucleation and coalescence to bulk film growth requires three components. First, a model must be developed that expresses the optical functions of bulk aluminum in terms of photon energy-independent parameters related to the energy band structure, the occupancy of electronic states, the accessible electronic transitions, and the relaxation time for optically excited electrons. Second, the model should also be able to describe the effect on the optical functions as the size of the solid decreases to the nanometer scale. In this way one can model the intrinsic optical functions of small particles in the nucleation regime or crystalline grains that form after coalescence. Third, a model for the morphology of the film must be developed that identifies the critical parameters that influence the optical properties. As part of this third component, an appropriate effective medium theory must be utilized that allows one to calculate the effective optical functions of metal films from the models for the morphology and for the optical functions of particles and grains. Thus, there are two sets of optical functions of interest in the nucleation regime when the film consists of isolated particles. The first set consists of the intrinsic optical functions of the particles themselves, and the second set consists of the effective optical functions of the particle film that includes the particles and intervening void space. Each of the three modeling components will be described in turn in Subsections A, B, and C.

A. The Optical Functions of Bulk Aluminum

Although aluminum is a technologically important material, used widely in microelectronics and optics, the choice of aluminum for this case study has been motivated by two other important reasons, as well. First, it is a simple metal (without *d*-bands), so that it has been possible to develop a theoretical expression for its dielectric function in terms of photon energy-independent parameters (*18*). Second, a pronounced absorption feature near 1.5 eV and its high-energy tail span the accessible spectral range of the ellipsometer (*24*). Thus, information on the electronic characteristics of aluminum can be obtained from both intraband (free-electron) and interband contributions to the dielectric function.

The optical properties of thick ("bulk") aluminum films have been widely studied because of the information that they provide on the energy band structure of simple polyvalent metals (*24*). Calculations of the band structure of Al indicate that the energy bands are nearly parallel when mapped in the high-symmetry families of planes parallel to the hexagonal (111) and square (200) Brillouin zone faces. These parallel bands are separated by $2|U_{111}| = 0.49$ eV and $2|U_{200}| = 1.53$ eV, energy gap values estimated from an analysis of de Haas–van Alphen measurements, and give rise to two interband features in the dielectric function of Al (*25*). In these expressions, $U_\mathbf{K}$ is the Fourier coefficient of the crystal potential for the reciprocal lattice vector $\mathbf{K}$, corresponding to the appropriate zone face. Reported dielectric functions for Al from the near- to mid-infrared region show a strong asymmetrical $2|U_{200}|$ peak and a barely visible $2|U_{111}|$ feature in the imaginary part. The latter is less visible because of the stronger background from the intraband component that increases with decreasing photon energy, and is observed more clearly in measurements at low temperature (*26*).

Mathematically, the dielectric function of aluminum can be expressed as a sum of the intraband (D: Drude) and interband (PB: parallel-band) components:

$$\varepsilon(\omega) = \varepsilon_\mathrm{D}(\omega) + \varepsilon_\mathrm{PB}(\omega), \tag{1}$$

where ω is the optical angular frequency. Contributions to the dielectric function due to transitions associated with other regions of **k**-space, i.e., non-parallel-band or "normal-band" transitions, are thought to be small in comparison (*18*).

The intraband contribution is described by the Drude model for a free-electron gas:

$$\varepsilon_D(\omega) = \varepsilon_0 + \{\Omega_p/[\omega^2 + (i\omega/\tau_D)]\}, \tag{2}$$

where ε_0 is associated with the core interband transitions at high frequency, Ω_p is the (phenomenological) plasma frequency for the intraband transitions, and τ_D is the intraband relaxation time (27). For thick ("bulk") film aluminum at room temperature, the following values describe

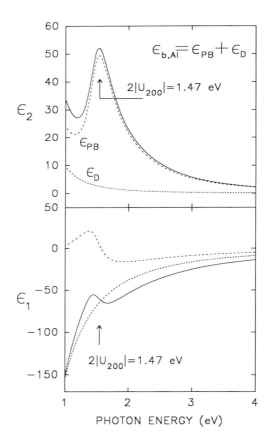

FIG. 1. Dielectric function for bulk aluminum (solid lines) consisting of a sum of the intraband (short broken lines) and interband (long broken lines) contributions. These results were calculated from the Drude free-electron and Ashcroft–Sturm formulas (Eqs. (2)–(3)) using parameters determined by fitting experimental results. The dominant interband contribution in ε_2 arises from the (200) parallel-band transition located at 1.47 eV.

the dielectric function obtained exprimentally: $\varepsilon_0 = 1.03$, $\hbar\Omega_p = 12.5\,\text{eV}$, and $\tau_D = 10.6 \times 10^{-15}\,\text{s}$ (24,28). The intraband contribution from 1 to 4 eV, calculated using these parameters, is given as the short-dashed lines in Fig. 1.

An expression for the interband contribution has been formulated by Ashcroft and Sturm, but is too lengthy to be reproduced here. Formally, the expression can be written as

$$\varepsilon_{PB}(\omega) = (4\pi i \sigma_a/\omega) \sum_{\mathbf{K}} (a_0 K) J_{\mathbf{K}}(\omega, |U_{\mathbf{K}}|, \tau_{\mathbf{K}}), \qquad (3)$$

where a_0 is the Bohr radius, $\sigma_a = (e^2/12a_0 h)$, and $U_{\mathbf{K}}$ and $\tau_{\mathbf{K}}$ are the Fourier coefficient of the crystal potential and the relaxation time for the parallel-band transitions associated with the Brillouin zone face of reciprocal lattice vector $\mathbf{K}$ (18). The sum in Eq. (3) is performed over the two parallel-band transitions (200) and (111), and $J_{\mathbf{K}}$ are two complex-valued functions that appear in the original publication. The final interband contribution, calculated with the bulk values of $2|U_{111}| = 0.49\,\text{eV}$ (25), $2|U_{200}| = 1.47\,\text{eV}$, and $\tau_{111} = \tau_{200} = 4.1 \times 10^{-15}\,\text{s}$ (see Section IV.A), is given as the long-dashed lines in Fig. 1.

The sum of the two contributions is also given in Fig. 1 as the solid line in accordance with Eq. (1). For the most part, the parameters used to generate $\varepsilon(\omega)$ in Fig. 1 have been obtained by fitting optical measurements on thick, opaque films. Such samples exhibit higher-quality surfaces than true bulk solids prepared by polishing procedures (24). As a result, the term "bulk" has been used here to refer to the thick-film optical properties.

B. Size Effects on the Optical Functions

In a bulk crystalline solid, scattering processes that involve phonons or defects and impurities control the electron relaxation time and mean free path. For small particles obtained in the early stages of film growth, however, the bulk electron mean free path may be comparable to or even larger than the particle dimension. Thus, for particles that retain the crystalline quality of the bulk material, the mean free path will be reduced by the scattering of electrons at the surfaces of the particles. The resulting reduction in electron relaxation time leads to a change in the optical properties of the material as described by Eqs. (1–3). To quantify such an effect, the following expression for the relaxation time corresponding to the

jth set of electronic transitions ($j = D$, 111, or 200) has been developed:

$$\tau_j^{-1} = \tau_{j,b}^{-1} + (v_j/\lambda_j). \tag{4}$$

where $\tau_{j,b}$ is the relaxation time associated with the bulk material, v_j is the velocity of the excited electron, and λ_j is the mean free path (*19,29*).

For intraband electronic transitions within an isolated spherical particle of radius R, Eq. (4) is applied with $j = D$, $v_D = v_F$, the Fermi velocity (2.02 × 10^8 cm/s for aluminum), $\tau_{D,b} = 1.06 \times 10^{-14}$ s, and $\lambda_D = R$, as long as scattering from the surface is isotropic and all scattering mechanisms are independent. Even if such an expression is found to be valid for intraband transitions, it is not necessarily correct for interband transitions, and its validity must be tested. Because of the curvature of the upper band, $v_{200} = \hbar^{-1}|\nabla_k E_k|$ is not constant over the full region of **k**-space covering the interband electronic transitions that end at the unoccupied states just above the Fermi level, as well as those that start from the occupied states just below the Fermi level. Figure 2 demonstrates this situation, providing a schematic of the band structure and possible interband transitions for wave vectors within a plane parallel to the (200) Brillouin zone face.

For intraband transitions within the grains of a polycrystalline matrix, the following related expression has also been used:

$$\tau_D^{-1} = \tau_{D,b}^{-1} + \{3\Re v_F/2(1-\Re)\lambda_{D,g}\}, \tag{5}$$

where $\Re \ll 1$ is the reflection coefficient for free electrons at the grain

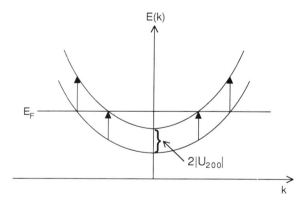

FIG. 2. Schematic of allowed electronic states and Fermi level position for aluminum, plotted in a plane parallel to the (200) Brillouin zone face. The vertical arrows denote possible parallel-band transitions. U_{200} is the Fourier coefficient of the crystal potential for the reciprocal lattice vector perpendicular to the plane.

boundaries (20). The subscript on λ_D is used to distinguish this mean free path for the grains ("g") from the corresponding one for particles in Eq. (4). Again, for spherical grains of radius R_g and isotropic boundary scattering $\lambda_{D,g} = R_g$. Resistivity measurements have suggested that, for aluminum, the choice $\Re = 0.15$ should be made in Eq. (5) (30). Although Eq. (5) has also been appropriated to characterize the relaxation time associated with interband transitions (20), again it is not clear in this case that the same values of electron velocity, mean free path, and reflection coefficient are applicable.

The simple approach, whereby the intrinsic dielectric function of aluminum particles (or grains) is obtained by adjusting the relaxation times in the dielectric function of the bulk material, is a classical one. A quantum-mechanical analogue, for the interband component of the dielectric function, is based on the uncertainty principle. It leads to a similar result, as can be seen by multiplying Eq. (4) by h and replacing h/τ_j by Γ_j, the width of the jth interband feature (31). The resulting behavior has been demonstrated for the direct transitions in semiconductor particles as a function of their size (32). However, when the particles become small enough (while retaining the single-crystalline structure) that the surface scattering rate far exceeds the scattering rate internal to the particle, i.e., $(\pi v_j/R) \gg \tau_{j,b}^{-1}$, then it has been suggested that the simple approach fails (33). Then a complete reexamination of the particle dielectric function in terms of quantum size effects is required (34). This has been performed for free electrons in aluminum using a particle-in-a-cube model, and the strongest discrete transition is predicted at an energy given by $E_Q = [(2m_c + 1)/m_c^2]E_F$, observable when m_c is an integer given by $k_F a_c/\pi$ (33). In these expressions, E_F and k_F are the Fermi energy and wave vector (11.63 eV and 1.73 Å^{-1} for aluminum), and a_c is the cube size. Thus, such transitions should be observable as they shift to lower energies with increasing particle size if $(\pi v_j/R) \gg \tau_{j,b}^{-1}$ and if the distribution in particle size is sufficiently narrow. Further comment on such effects will appear later in Section IV.C.

For the purposes of the present background discussion, the quantum size effects will be neglected, and dielectric functions of aluminum, as modified by the classical description (Eq. (4)), will be presented. This approach does appear to explain the experimental results obtained in this study, as will be shown in Sections IV.B and C. Figures 3 and 4 reveal the effect of the mean free path on the intraband and interband contributions, respectively, calculated according to Eq. (4) with $v_j = v_F$ and the bulk values of τ_j provided earlier. In this case, the mean free path may be considered an effective one to cover the possibilities of surface, boundary, or defect scattering. For example, for free electron scattering at grain boundaries in a continuous

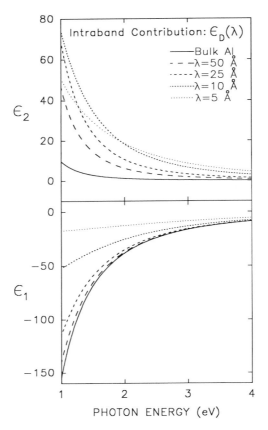

FIG. 3. The intraband contribution (broken lines) to the dielectric function of aluminum assuming various values of the electron mean free path λ, which reduces the free-electron relaxation time from the bulk value in accordance with Eq. (4). The intraband contribution to the bulk dielectric function (Eq. (2)) is also shown (solid lines).

film, the appropriate mean free path is given by Eq. (5) as $\lambda_{D,g} = 3\Re\lambda_D/2(1-\Re)$, where λ_D is the effective mean free path based on Eq. (4). The bulk contributions ($\lambda_j \to \infty$) from Fig. 1 are also included in Figs. 3 and 4 for comparison.

In the intraband contribution in Fig. 3, calculated by combining Eqs. (2) and (4), a reduction in electron mean free path results in an increase in ε_2 for photon energies from the near infrared to the near ultraviolet. In contrast, the magnitude of ε_1 decreases, becoming less dependent on

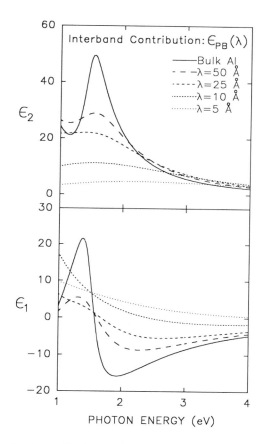

FIG. 4. The interband contribution to the dielectric function of aluminum including the (200) and (111) sets of parallel-band transitions. For the broken lines, various values of the electron mean free path λ have been assumed, which reduces the (111) and (200) relaxation times from the bulk values in accordance with Eq. (4). The interband contribution to the bulk dielectric function (Eq. (3)) is also included (solid lines).

photon energy, with the strongest effect occurring at very short mean free paths. The combined effects are expected on the basis of Eq. (2) as $\omega\tau_D$ decreases to the order of unity. From the interband contribution in Fig. 4, calculated by combining Eqs. (3) and (4) (and assuming $\lambda_{PB} \equiv \lambda_{111} = \lambda_{200}$), one can see that a reduction in mean free path broadens the (200) parallel-band feature near 1.5 eV, which becomes unresolvable as λ_{PB} decreases below 10 Å.

The detectable aluminum particles, prepared by evaporation or sputtering and studied prior to coalescence, exhibit radii typically from 5 to 30 Å. For a common value of the mean free path on this scale, the net dielectric functions, calculated from Eqs. (1–4) with $\lambda_D = \lambda_{PB}$, are nearly featureless, as shown in Fig. 5. In spite of this, ε_2 exhibits roughly equal intraband and interband contributions near the center of the spectral range used in multichannel ellipsometry (2.5 eV). Thus, correct descriptions of both con-

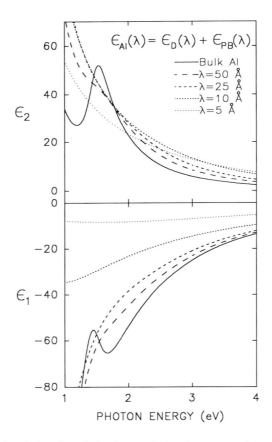

FIG. 5. Dielectric function of aluminum calculated as a sum of intraband and interband contributons (Eq. (1)–(3)), assuming various values of a common mean free path λ that reduces the relaxation times associated with the free-electron, and (111) and (200) parallel-band electron transitions in accordance with Eq. (4). For comparison, the results from Fig. 1 for the dielectric function of bulk aluminum are included (solid lines).

tributions are required in order to understand the optical properties of the particles. It should be noted that the interband contribution is dominated by the (200) transitions, and that of the (111) transitions is much weaker over the energy range accessible to multichannel ellipsometry. As a result, the free parameters for the (111) transitions in Eq. (4) can be equated to those used for the (200) transition with little consequence.

C. The Generalized Effective Medium Theory

Thin films prepared from the vapor phase evolve through different stages of microstructure (*35*). In the initial nucleation stage, isolated particles form and increase in size as a function of time. The particles eventually make contact, and the film enters a coalescence stage during which the void spaces between contacting particles begin to fill with depositing material. If the deposition conditions are suitable, a continuous film with a density close to that of the bulk material develops and increases in thickness. Usually a surface roughness layer rides on the top of the continuous film (*36*), which in some cases is a remnant of the clustering in the nucleation process. The effective dielectric functions of the film in these regimes are distinct, both because of the possible particle and grain size effects as described in Section II.B, but also because of the differences in the morphological characteristics. In order to relate the effective dielectric functions and the morphology, different effective medium theories (EMTs) must be applied, depending on the solid material and film geometry. These theories allow one to calculate the effective dielectric function of the film from the dielectric function of the solid material and the volume fraction of void space that is present in the film (*4–8,21,22*).

A significant body of previous experience exists in this field; however, a review is beyond the scope of this chapter. Generally, it has been found that a theory based on the Maxwell-Garnett type is suitable for metal particle films, as it predicts the observed resonance absorption behavior (*37*). The dielectric functions of continuous metal films with a small void volume fraction relative to the particle films, as well as those of thin surface roughness layers, have been modeled successfully with the Bruggeman EMT (*11*). The EMT of Sen et al. (*38*) has been proposed for modeling the dielectric functions of continuous metal films with a cermet-type grain structure (*39,40*). In this section, the focus will be on a description of the generalized Maxwell-Garnett EMT applicable to metal particle films; the Bruggeman theory has been treated in the previous chapter.

As noted in Section I, the original simple EMT developed by Maxwell-Garnett treated the film as a three-dimensional cubic array of spherical metal particles of intrinsic dielectric function ε_p, embedded within a host of dielectric function ε_h, and the Lorentz local-field correction was employed (2). This led to an isotropic effective dielectric function for the film. Yamaguchi and co-workers used a more realistic geometry for the film, a two-dimensional square array of identical, spheroidal particles having their axes of symmetry perpendicular to the substrate surface, as shown in Fig. 6 (6, 8, 41). The local fields were calculated exactly for this geometry, and the fields due to the mirror images of the particles in the substrate surface were also included. The resulting directionally dependent depolarization fields in this case lead to an anisotropic effective dielectric function for the film of the form

$$\varepsilon_{\parallel} = \varepsilon_h + \{Q(\varepsilon_p - \varepsilon_h)/[\varepsilon_h + F_{\parallel}(\varepsilon_p - \varepsilon_h)]\} \tag{6}$$

$$\varepsilon_{\perp} = \varepsilon_h + \{Q(\varepsilon_p - \varepsilon_h)/[\varepsilon_h + (F_{\perp} - Q)(\varepsilon_p - \varepsilon_h)]\}, \tag{7}$$

characteristic of a uniaxial film with the optic axis normal to the plane of the substrate. In these equations, Q is the volume fraction of the film occupied by metal particles of dielectric function ε_p, and ε_h is the dielectric function of the intervening host material. $F_{\parallel}$ and $F_{\perp}$ are apparent particle depolarization factors given by

$$F_{\parallel} = f_{\parallel} - \{0.716Qd/\xi(\varepsilon_s + 1)a\} - \{\gamma^2(\varepsilon_s - 1)/24(\varepsilon_s + 1)\}, \tag{8}$$

$$F_{\perp} = f_{\perp} + \{1.432\varepsilon_s Qd/\xi(\varepsilon_s + 1)a\} - \{\gamma^2(\varepsilon_s - 1)/12(\varepsilon_s + 1)\}. \tag{9}$$

Here $f_{\parallel}$ and $f_{\perp}$ are the true depolarization shape-factors of the spheroidal particle that depend only on the spheroid axial ratio $\gamma \equiv b/c$ and obey the rule $2f_{\parallel} + f_{\perp} = 1$ (42, 43). In the axial ratio, b and c are the semimajor and semiminor axes parallel and perpendicular to the substrate surface, respectively. The second and third terms on the right in Eqs. (8) and (9) represent static dipole interactions among particles and their images in the substrate surface, respectively (8). In these terms, a is the particle spacing, $d = 2c$ is the film thickness, and ξ is a parameter that allows one to generalize the equations to geometrical arrangements of particles other than the square array of Fig. 6 (7). For the square array, $\xi = 1$, and the thickness-to-spacing ratio d/a can be eliminated from the equations using $d/a = (6Q/\pi\gamma^2)^{1/2}$. As a result, within the geometry of Fig. 6, $F_{\parallel}$ and $F_{\perp}$ are functions only of Q and γ.

Equations (6) and (7) have been described within the framework of the

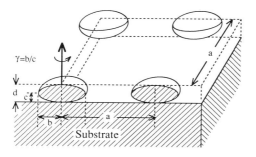

FIG. 6. Morphological model for a particle film, consisting of identical, substrate-supported spheroids arranged on a square grid of spacing a. The axial ratio of the spheroidal particle is $\gamma \equiv b/c$, where b and c are the semimajor and semiminor axes, parallel and perpendicular to the substrate plane, respectively. In this geometry, the film thickness is $d = 2c$. The generalized Maxwell-Garnett theory of Eqs. (6)–(9) employs this geometry when $\xi = 1$ in Eqs. (8) and (9).

geometrical situation depicted in Fig. 6. However, they can also be viewed as more general equations in which the apparent depolarization factors $F_\parallel$ and $F_\perp$ are chosen based on the assumed geometry and interparticle interaction (as long as the assumed geometry leads to a film that is either isotropic or uniaxial with its optical axis normal to the film plane) (44). Thus, they can be labeled as "interaction parameters." For example, $F_\parallel = (1 - Q)/3$ and $F_\perp = (1 + 2Q)/3$ for the simple Maxwell-Garnett theory with the Lorentz local-field interaction. With this substitution in Eqs. (6) and (7), isotropy is regained. It should be noted that in general if $\delta_A \equiv |F_\parallel - (F_\perp - Q)| = 0$, then the effective dielectric function of Eqs. (6) and (7) becomes isotropic.

Figure 7 shows the real (ε_1) and imaginary (ε_2) parts of the principal components of the effective dielectric functions calculated from Eqs. (1)–(3) and (6)–(9) for five hypothetical 50 Å thick aluminum particle films. For the five films, the same particle axial ratio of $\gamma = 2.5$ was assumed, while the spacing between the spheroidal particles located on a square grid was varied from $a = 130$ Å to $a = 150$ Å. In the calculations, $\varepsilon_h = 1$ for ambient vacuum, $\varepsilon_s = 2.17$ for an SiO_2 substrate, and ε_p was set equal to the bulk aluminum dielectric function of Eqs. (1)–(3) and Fig. 1. The bulk dielectric function is used only for illustrative purposes here. An analysis of experimental data presented in Section IV shows that it is necessary to use a modified dielectric function for the particles.

A number of points can be made upon inspection of Fig. 7. First,

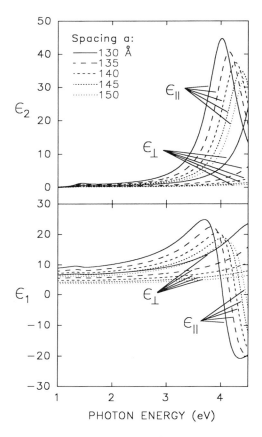

FIG. 7. Effective dielectric functions of 50 Å thick aluminum particle films, simulated according to the morphological model of Fig. 6 and the generalized Maxwell-Garnett effective medium theory of Eqs. (6)–(9). Here $\varepsilon_\parallel$ and $\varepsilon_\perp$ are the principal components of the effective dielectric function parallel and perpendicular to the substrate plane. The morphological model used here consists of identical aluminum spheroids arranged on a square grid. For the different calculations the grid spacing, a, ranges from 130 Å to 150 Å. The axial ratio of the spheroids is fixed at 2.5, and the bulk dielectric function of aluminum is assumed for the particles.

resonance absorption features appear in both $\varepsilon_{2\parallel}$ and $\varepsilon_{2\perp}$ that are not present in ε_{2p}, the dielectric function of the particles. This behavior can be considered to result from the excitation of dipolar plasmon–polaritons (45). Second, the resonance energy in $\varepsilon_{2\parallel}$ is significantly lower than that in $\varepsilon_{2\perp}$

for all films, but the anisotropy weakens for the films of increasing particle density (and aluminum volume fraction). The former behavior is inherent in the form of Eqs. (6)–(9) when $(F_\perp - Q) > F_\parallel$. Third, both resonances in $\varepsilon_{2\parallel}$ and $\varepsilon_{2\perp}$ increase in strength as they shift to lower energies. This overall behavior results from an increase in the dipole interaction between particles as their density increases, characterized by a decrease in $F_\parallel$ and an

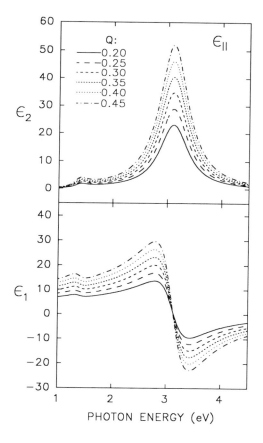

FIG. 8. Effective dielectric functions for aluminum particle films calculated with different values of the metal volume fraction Q according to Eq. (6). Here $\varepsilon_\parallel$ is the principal component of the effective dielectric function parallel to the substrate plane. In the calculation, the bulk dielectric function for the aluminum particles is assumed, and the interaction parameter is fixed at $F_\parallel = 0.04$.

increase in $F_\perp$. Finally, the (200) parallel-band feature is significantly damped in the effective dielectric functions since oscillator strength has been transferred from the intraband and interband transitions to the plasmon–polarization transitions.

Next, the morphological and electronic features that control the reso-

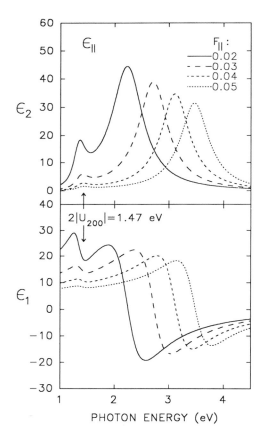

FIG. 9. Effective dielectric functions for aluminum particle films calculated with different values of the interaction parameter $F_\parallel$ according to Eq. (6). Here $\varepsilon_\parallel$ is the principal component of the effective dielectric function parallel to the substrate plane. In the calculation, the bulk dielectric function for the aluminum particles is assumed, and the aluminum volume fraction is fixed at $Q = 0.3$. Note that oscillator strength is coupled into (200) parallel-band transitions only when the contribution to $\varepsilon_{2\parallel}$ from the resonance band is significant near 1.5 eV.

nance behavior observed in Fig. 7 will be explored in further detail. In this case, $\varepsilon_\parallel$ will be the focus of attention. To perform such a study, Eqs. (1)–(4) and (6) have been combined in order to express $\varepsilon_\parallel$ in terms of the

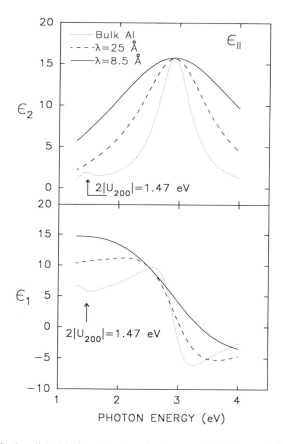

FIG. 10. Effective dielectric functions for aluminum particle films calculated according to Eq. (6) with different values of the common mean free path defined by Eq. (4). The mean free path serves to reduce the relaxation times associated with the free-electron and (111) and (200) parallel-band electron transitions. Here $\varepsilon_\parallel$ is the principal component of the effective dielectric function parallel to the substrate plane. Values of the aluminum volume fraction and the interaction parameter $(Q, F_\parallel)$ are adjusted to yield the same peak magnitude and energy position in $\varepsilon_{2\parallel}$. For $\lambda = 8.5\,\text{Å}$ (solid line), $\lambda = 25\,\text{Å}$ (dashed line), and for $\lambda \to \infty$ (corresponding to bulk aluminum; dotted line), $(Q, F_\parallel)$ are chosen to be $(0.58, 0.044)$, $(0.28, 0.037)$, and $(0.13, 0.035)$, respectively.

three variable parameters, the aluminum volume fraction Q, the interaction parameter $F_\parallel$ (depending on Q and γ), and a single value of the mean free path $\lambda = \lambda_D = \lambda_{PB}$ that is common to both the intraband and interband transitions. This latter value allows the individual relaxation times to be calculated through Eq. (4).

Figure 8 shows the effect on $\varepsilon_\parallel$ of varying the aluminum volume fraction, with $F_\parallel$ fixed at 0.04, and in the limit $\lambda \to \infty$ (meaning that bulk values of τ_j are used in Eqs. (2) and (3)). It is evident that an increase in Q increases the amplitude of the resonance feature without changing its energy position or width. Figure 9 shows the effect of varying $F_\parallel$, with Q fixed at 0.30, and in the limit $\lambda \to \infty$. In this case, the resonance feature shifts to increasing energy and decreases in amplitude as $F_\parallel$ increases, as is also observed in Fig. 7. However, very little change in the width of the resonance feature is observed among the spectra in Fig. 9. Finally, Fig. 10 shows the effect of varying the common mean free path λ, while adjusting $F_\parallel$ and Q independently in order to match the peak position and amplitude of the resonance. The purpose of this approach, in view of the results of Figs. 8 and 9, is to demonstrate that only changes in λ lead to changes in the width of the resonance feature. Conversely, one can conclude that if the width of the resonance feature remains unchanged, the mean free path, related to the electronic properties, also remains unchanged, at least within the simplified morphological model of Fig. 6.

The direct connection between the electronic properties and the width of the resonance band has been noted earlier, from an inspection of the expression that results when Eq. (2) for the intraband component is substituted directly into the simple Maxwell-Garnett EMT (5). In this case, the width of the resonance feature is found to be proportional to ε_{2p}, evaluated at the resonant energy. The origin of the effect in Fig. 10 is similar. As λ increases from 8.5 Å to large values, then ε_{2p} decreases for photon energies between 2 and 4 eV (see Fig. 5), leading to a narrowing of the resonance band.

The results presented in Figs. 8–10 suggest that a linear regression analysis fit of either of the two principal components of the effective dielectric function of a particle film, using a theoretical expression that combines Eqs. (1)–(4) and Eq. (6) or (7), will provide information on both the microstructural properties of the film (Q, $F_\parallel$, $F_\perp$) and the electronic properties of the particles (λ). If the primary interest is the electronic properties, then such an approach is likely to succeed due to the singular effect of λ on the width of the resonance feature. In contrast, the morphological parameters tend to exhibit greater correlation and therefore may be less certain. It should also be noted that if any erroneous assumptions are made regarding the optical response of the particles, it may not be possible to find the correct morphological parameters and obtain a good fit to the effective dielectric function.

III. Experimental Apparatus for Real-Time Monitoring

Figure 11 shows a schematic of the multichannel spectroscopic ellipsometer used for real-time monitoring of aluminum films throughout the nucleation, coalescence, and bulk growth stages. The instrument is the same as that described in the previous chapter, consisting of a Xe arc source,

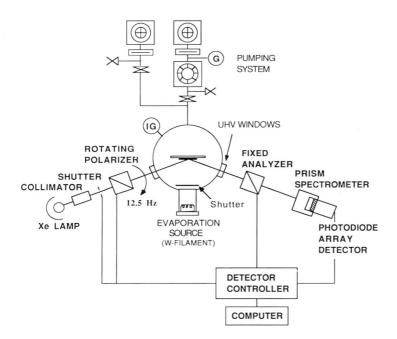

FIG. 11. Schematic diagram of the rotating-polarizer ellipsometer equipped with a multichannel detection system. With this instrument, full spectra in the ellipsometry parameters (ψ, Δ) from 1.3 to 4.0 eV have been collected in real time during the deposition of aluminum thin films by sputtering and evaporation. In these studies, the minum acquisition time for the spectra is 0.32 s. The system configuration for evaporation is shown here.

collimating optics, rotating polarizer (rotation frequency: $\omega_p/2\pi = 12.5\,\text{Hz}$), sample in vacuum chamber, fixed analyzer, spectrograph, 1,024-element Si photodiode array detector, and detector control system (13). The detector elements are grouped by eight to obtain 128 spectral positions over the array surface, leading to 83 points over the useful spectral range from 1.3 to 4.0 eV. The angle of incidence and the fixed analyzer angle were set at 70° and 30°, respectively.

The goal of the experiment is to determine spectra in the ellipsometry angles (ψ, Δ), defined by $\tan\psi \exp(i\Delta) \equiv r_p/r_s$, where r_p and r_s are the complex amplitude reflection coefficients of the surface for p and s linear polarization states (46). Here, p and s denote the directions of electric field vibration parallel and perpendicular to the plane of incidence. The photodiode array is read out four times per optical cycle of the rotating polarizer (one-half mechanical cycle), and the resulting four spectra provide the normalized $2\omega_p$ Fourier coefficients from which spectra in (ψ, Δ) can be calculated, after calibration of the instrument (47). Thus, at the polarizer frequency of 12.5 Hz, the minimum time for a full spectroscopic measurement of (ψ, Δ) is 40 ms (14). In the studies of aluminum, (ψ, Δ) spectra have been obtained as an average of 8 to 20 consecutive optical cycles. This leads to data acquisition and repetition times ranging from 0.32 to 0.8 s and from 1.25 to 5.5 s, respectively, depending on the deposition rate.

The aluminum films studied by real-time SE were deposited by evaporation from a tungsten wire, as shown in Fig. 11. By adjusting the power applied to the filament, deposition rates from 40 Å/min to 280 Å/min were obtained, as determined from analysis of the real-time SE results. For this range of deposition rates, the background pressure-to-rate ratio was maintained at $\sim 10^{-9}$ Torr min/Å, which appears to result in negligible contamination, at least for the purposes of optical property measurements (48). Selected depositions were also performed by dc magnetron sputtering an aluminum target in pure Ar gas; however, these depositions were studied in the particle regime only. The Ar pressure during sputtering was 5×10^{-3} Torr, and the dc power was set at 50 W to obtain a low deposition rate of 16 Å/min and at 100 W to obtain a higher rate of 36 Å/min. The substrates were unheated Si wafers, covered with a 50–150 Å thermally grown oxide layer. The oxide thickness was determined from an analysis of the full substrate spectra prior to aluminum growth. In this procedure, the known dielectric functions of crystalline silicon (49) and vitreous SiO_2 (50) were employed in a linear regression analysis fitting routine in order to determine the oxide thickness that provides a minimum in the unbiased estimator of the mean square deviation σ between experimental and calculated $(\tan\psi, \cos\Delta)$ spectra (23).

The issue of a possible chemical reaction between the SiO_2/Si substrate and the aluminum film is an important one. If such a reaction were to occur, analysis of the real-time SE results would be extremely complicated. At substrate temperatures > 200°C, a chemical reaction between SiO_2 substrates and aluminum films is known to lead to Al_2O_3 and an underlying layer of Si (51–53). The possibility of such a reaction needs to be assessed for the room-temperature depositions discussed in Section IV, given the potential sub-monolayer sensitivity of SE. To accomplish this assessment, representative evaporated and sputtered films were removed from the deposition system and dissolved from their substrates with HCl diluted in H_2O. The exposed substrate surfaces were studied by SE. An analysis of the results indicated that the SiO_2 thicknesses remained unchanged within ± 1 Å from those measured prior to film growth, and no Si layer was detected on the surface. This suggests that it is not necessary to include reacted interface layers in the modeling of the real-time SE data. This conclusion is also consistent with the results of grazing internal reflection spectroscopy and current–voltage measurements on $Al/SiO_2/Si$ tunnel diodes (54). These latter studies showed that the Al/SiO_x interface of a metal–insulator–semiconductor (MIS) diode formed at room temperature exhibited the equivalent of less than 0.6 monolayers of Al–O bonds. Oxidation of the aluminum at the interface was observed when the MIS diode was annealed above 200°C.

IV. Results and Discussion

Representative real-time spectroscopic ellipsometry results obtained for aluminum evaporations performed at low (43 Å/min) and high (280 Å/min) rates were analyzed in detail and will be discussed extensively in this section. The lower rate leads to a lower nucleation density, as would be expected on the basis of homogeneous nucleation, whereby the capture radius for each nucleus expands as the time available for surface diffusion increases (35). The low nucleation density ensures a wider range of thickness during which the growth of isolated particles can be studied. Because of the high nucleation density for the higher deposition-rate film, the coalescence process occurs more abruptly since there appears to be a narrower distribution of interparticle spacings under these conditions. As a result, optimum morphological uniformity and a low void fraction in the final film are obtained. Based on these observations, it was found to be more convenient to characterize the high-rate film in detail in the coalescence and bulk film growth stages (Subsection B), and the low rate film in the particle stage (Subsection C).

A. Overview of the Optical Functions of Aluminum Thin Films in the Nucleation and Growth Stages

Figures 12 and 13 show typical experimental and simulated spectra for the real $\langle \varepsilon_1 \rangle$ and imaginary $\langle \varepsilon_2 \rangle$ parts of the pseudo-dielectric function for the aluminum film evaporated at high rate (280 Å/min). The pseudo-dielectric function is calculated directly from (ψ, Δ) using the Fresnel equations for a single interface between the ambient and an opaque solid (*12, 46*). This expression of the data is used because, late in the deposition, when the sample is opaque and closely approximates a single interface, the pseudo-dielectric function closely approximates the true dielectric function of the aluminum. Differences between the two exist only because of the presence of a microscopic roughness layer on the surface of the film (*36*), as well as possible undetected heterogeneities through the film thickness. At earlier times in the deposition when the aluminum is semitransparent, the pseudo-dielectric function is a weighted average of the dielectric functions of the Al, SiO_2, and Si, with weighting that depends on the layer thicknesses, angle of incidence, and photon energy.

The experimental three-dimensional surfaces in the upper parts of Figs. 12 and 13 were constructed from 121 spectra, each having 83 spectral positions. Thus, a total of 20K data points was collected during the 150 s deposition. The orientations of the surfaces in $\langle \varepsilon_1 \rangle$ and $\langle \varepsilon_2 \rangle$ are chosen differently, in order to provide the best visibility for the initial spectra. The substrate $(\langle \varepsilon_1 \rangle, \langle \varepsilon_2 \rangle)$ spectra, characteristic of 71 Å of SiO_2 on silicon, are visible at time zero prior to any aluminum deposition. The sharp features at 3.35 eV are associated with the E'_0–E_1 transitions in the underlying silicon substrate (*49*). As the deposition proceeds, the accumulation of aluminum rapidly dampens out the silicon features, and the (200) parallel-band features build up and dominate the spectra after 1 min. At the end of the deposition period shown, the aluminum film is virtually opaque over the full spectral range. The thickness of the aluminum at this point is ~700 Å, as estimated from the average deposition rate obtained by SE analysis in the regime of semitransparency.

The simulated three-dimensional surfaces in the lower parts of Figs. 12 and 13 were calculated according to an optical model characterized by the uniform, layer-by-layer growth of aluminum onto the same substrate structure, 71 Å SiO_2 on Si. In this calculation, a thickness-independent aluminum dielectric function has been employed. In order to extract the appropriate dielectric function for the simulations, the effects of an assumed 10 Å microscopic roughness layer have been analytically removed (by mathematical inversion) from the final $(\langle \varepsilon_1 \rangle, \langle \varepsilon_2 \rangle)$ spectra for the 700 Å aluminum film in Figs. 12 and 13 (see later discussion in this section). The deposition rate in the simulation is assumed to be constant, using the same average value as that obtained in the analysis of the experimental data.

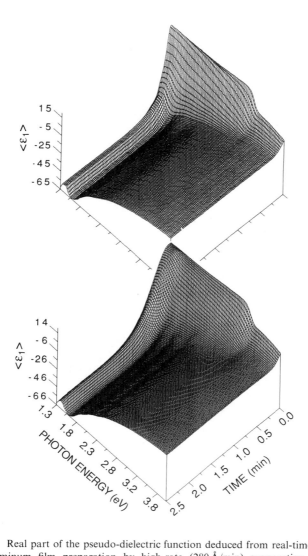

FIG. 12. Real part of the pseudo-dielectric function deduced from real-time measurements during aluminum film preparation by high-rate (280 Å/min) evaporation onto a room-temperature silicon wafer covered with a 71 Å thermally grown oxide layer (top). A total of 121 spectra, each consisting of 83 data points, were collected over a deposition time of 2.5 minutes. Simulated spectra appear at the bottom for the hypothetical optical model in which an aluminum film grows perfectly uniformly on a substrate of the same structure, assuming a thickness-independent aluminum dielectric function characteristic of bulk material.

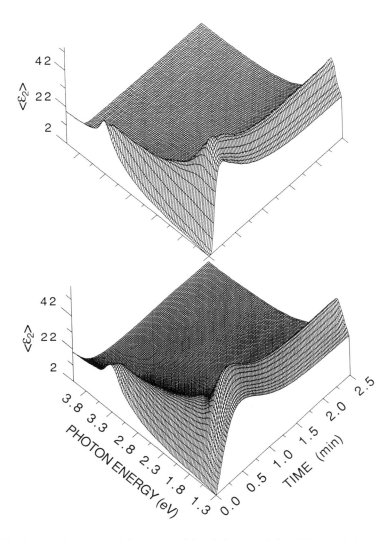

FIG. 13. Imaginary part of the pseudo-dielectric function deduced from real-time measurements during aluminum film preparation by high-rate (280 Å/min) evaporation onto a room-temperature silicon wafer covered with a 71 Å thermally grown oxide layer (top). A total of 121 spectra, each consisting of 83 data points, were collected over a deposition time of 2.5 minutes. Simulated spectra appear at the bottom for the hypothetical optical model in which an aluminum film grows perfectly uniformly on a substrate of the same structure, assuming a thickness-independent aluminum dielectric function characteristic of bulk material. After Ref. 16.

A comparison of the experimental and simulated results in Figs. 12 and 13 finds that they are outwardly very similar. However, a careful examination of the grid lines on both the $\langle\varepsilon_1\rangle$ and $\langle\varepsilon_2\rangle$ spectra near 1.5 eV in the first 30 s of growth indicates that the (200) parallel-band features in the experimental data develop more slowly than those in the simulation. As noted earlier, this effect cannot be attributed to a chemical interaction between the growing film and the substrate. Later in this section, it will be shown to be due to a thickness dependence of the aluminum optical properties. Although this can result from a nucleation and coalescence sequence (and interpreted using an EMT), the effect in the vicinity of the (200) parallel-band feature will be shown (in Subsection C) to be due to a significantly altered dielectric function for the particles in the nucleation regime in comparison to those of the bulk film.

The first step in a formal analysis of the results of Figs. 12 and 13 (top surfaces) is to compare the optical properties of the final, opaque film with the best available data in the literature. As a reference standard, it is reasonable to select the dielectric function determined by Shiles *et al.* over a wide photon energy range, based on reflectance measurements of aluminum films prepared and maintained in ultrahigh vacuum (*24,28*). From 0.7 to 2.5 eV, these results closely match those of Mathewson and Myers, obtained by *in situ* SE measurements of similarly prepared aluminum films (*55*). The reference dielectric function has been used as input in a linear regression analysis (*23*) of the final experimental ($\langle\varepsilon_1\rangle, \langle\varepsilon_2\rangle$) spectra of Figs. 12 and 13. By employing the Bruggeman EMT in a three-medium model for the near-surface of the aluminum film [i.e., (ambient)/(surface roughness layer)/(bulk film)], the analysis led to the following photon energy-independent morphological parameters: (i) bulk film density deficit relative to the reference material, 0.001 ± 0.005; (ii) surface roughness layer thickness, 10.5 ± 0.5 Å; and (iii) relative surface roughness layer density deficit, 0.51 ± 0.02. The first result indicates that the void volume fraction in the bulk layer of the final film is identical to that of the film measured to obtain the reference standard. Thus, both films most likely exhibit a density close to that of single-crystal aluminum.

Figure 14 (points) presents the bulk dielectric function obtained from the final experimental ($\langle\varepsilon_1\rangle, \langle\varepsilon_2\rangle$) spectra of Figs. 12 and 13, after analytically removing (by a mathematical inversion method) the 10 Å surface roughness layer deduced in the linear regression analysis. The dielectric function of this layer is computed from the Bruggeman EMT as a 0.5/0.5 mixture of bulk (underlying) aluminum and void. The reference dielectric function of Shiles *et al.* (*28*) is plotted as the solid line in Fig. 14 for comparison. Overall, there is reasonable agreement between the two dielectric functions. The (200)

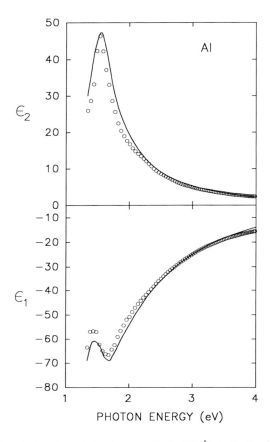

FIG. 14. Dielectric function of aluminum in thick (700 Å), bulk film form, deduced from real-time measurements during growth by high-rate (280 Å/min) evaporation onto room-temperature SiO_2/Si (open circles). The results were obtained from the final (2.5 min) pseudodielectric function of the deposition of Figs. 12 and 13, after correcting for a 10 Å roughness layer. The dielectric function of Shiles *et al.* (*28*) is reproduced for comparison (solid lines). After Ref. 16.

parallel-band feature is somewhat sharper for the film prepared in the real-time study, however, and this may be attributed to a larger aluminum grain size in accordance with Eq. (5).

In the second analysis step, Eqs. (1)–(3) can be employed to fit the bulk dielectric function $\varepsilon_b = \varepsilon_{1b} + i\varepsilon_{2b}$ of the final aluminum film in the real time experiment (points in Fig. 14). The goal is to apply linear regression analysis to deduce the photon energy–independent parameters that characterize the film (e.g., τ_j and U_K); however, the accessible information is limited by the spectral range of the multichannel ellipsometer. This is particularly true for the intraband and (111) parallel-band transitions. The intraband contribution to ε_{2b} is small compared to the interband contribution, as can be seen in Fig. 1. Furthermore, since $\omega\tau_D \gg 1$ over the experimental spectral range, ε_{1b} is also nearly independent of τ_D. For these two reasons, the intraband relaxation time for the bulk film must be fixed in the linear regression analysis at the literature value, $\tau_D = 10.6 \times 10^{-15}$ s (24). In addition, the (111) parallel-band term in Eq. (3) is very weak above 1.3 eV. Thus, the literature value of $|U_{111}| = 0.25$ eV is used, and τ_{111} is equated to τ_{200}. With these constraints and with $\hbar\Omega_p$ set at 12.7 eV (based on a calculation of the optical mass (18)), a linear regression analysis of ε_b in the neighborhood of the (200) parallel-band transitions yields $|U_{200}| = 0.73 \pm 0.01$ eV and $\tau_{200} = (4.1 \pm 0.7) \times 10^{-15}$ s. These values are in excellent agreement with those obtained by Mathewson and Myers (55), who found $|U_{200}| = 0.74$ eV and $\tau_{200} = 3.8 \times 10^{-15}$ s. In summary, because of the dominance of the (200) parallel-band transitions above 1.3 eV (see Fig. 1), $|U_{200}|$ and τ_{200} can be determined with comparatively narrow confidence limits in fits to the bulk aluminum film dielectric functions, deduced from real-time measurements with acquisition times as short as 0.3 s. This capability will be exploited in the next section.

In the third analysis step, an attempt is made to understand the evolution of the experimental ($\langle\varepsilon_1\rangle, \langle\varepsilon_2\rangle$) solely from a morphological standpoint, using linear regression analysis and simple EMTs to extract photon energy–independent morphological parameters such as layer thicknesses and void volume fractions from the real-time SE data (23). Such an approach has been successful in explaining the differences in the optical functions of gold films reported by different researchers (11). Optical models employing one and two layers for the aluminum film are utilized. A one-layer model consists of a single layer of aluminum supported by the SiO_2/c-Si substrate, whereas the two-layer model consists of a surface roughness layer atop a bulk layer, both supported by the substrate. The morphology of thin films often evolve from isolated particles (one-layer model) in the nucleation stage to coalesced structures, and end with a surface roughness layer atop a uniform bulk layer (two-layer model). In the linear regression analysis, dielectric functions of all density-deficient layers are calculated from the bulk film dielectric function and the void volume fractions in the layer by applying a simple EMT, such as that of Bruggeman, Maxwell-Garnett, or Lorentz–Lorenz (22). For the bulk film dielectric function, the analytical expression of Eqs. (1)–(3) has been employed using the combination of fixed and best fit parameters of the previous paragraph.

The overall outcome of this analysis for the aluminum film evaporated at high rate is shown in Fig. 15. The analysis has been performed in three regimes, applying the simple Bruggeman EMT throughout. For $t \leqslant 6$ s, a one-layer model is used with two free parameters, the void volume fraction in the layer and the layer thickness. As shown in Fig. 15, the void volume fraction during this period drops rapidly to zero as the film thickness increases to about 50 Å. This behavior is expected for the initial nucleation and coalescence stages of growth. For $7.5 \leqslant t \leqslant 30$ s, the film is semi-transparent, and a two-layer model is used. In this case, the two free parameters are the layer thicknesses; the void volume fractions for the surface and bulk layers are set at 0.5 and at a best fit value obtained for the opaque film (at $t \sim 1$ min), respectively. In this regime, the bulk thickness is nearly a linear function of time, and an average deposition rate can be extracted. The roughness layer on the other hand, decreases somewhat from 7–8 Å to 3 Å. For $t > 60$ s, the film is opaque and the substrate can be omitted from the two-layer model. In this regime, near-constant surface roughness and bulk void fraction values of 3 ± 3 Å and -0.01 ± 0.03 are deduced.

The difference between results for the analysis of the final opaque film of Fig. 15 (3 ± 3 Å, -0.01 ± 0.03) and that of Fig. 14 (10.5 ± 0.5 Å, 0.00 ± 0.005), the latter using the reference dielectric function of Shiles et al. (28), suggests minor inadequacies in the analytical formulas (Eqs. (1)–(3)) that are unknown at present. These are reflected in higher values of the unbiased estimator of the mean square deviation between experimental and best fit $(\tan \psi, \cos \Delta)$ spectra, observed when linear regression analysis is performed with the analytical formula for the bulk dielectric function ($\sigma \sim 0.02$–0.025), in comparison to when it is performed with the experimental reference dielectric function ($\sigma \sim 0.012$). Such larger σ values are also reflected in the larger confidence limits that do not allow one to determine the roughness thickness to within 6 Å (see Fig. 15, at far right).

The uncertainties in the analytical form for the dielectric function of bulk aluminum (equivalent in magnitude to the optical effect of a few angstroms of surface roughness) are relatively minor compared with the problems in fitting the initial stages of film growth in Fig. 15. Here the σ value increases by a factor of three to $\sigma \sim 0.07$ before returning to a stable value of 0.02–0.025. This behavior is quite reproducible, observed in similar analyses of evaporated and sputtered films, irrespective of deposition rate. Even when other simple EMTs (i.e., the Maxwell-Garnett and Lorentz–Lorenz theories) are applied in place of the Bruggeman EMT, within the same morphological framework, the fits in this regime cannot be improved significantly. In Subsection C, a basic assumption of this approach, namely that the bulk

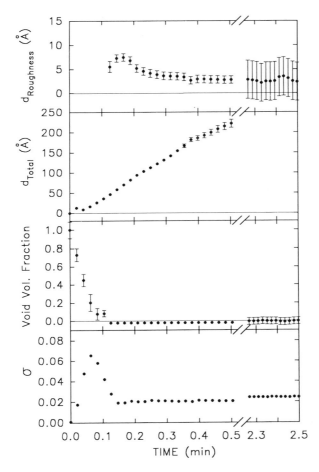

FIG. 15. Interpretation of the evolution of the optical properties using a simple morphological model for the aluminum film of Figs. 12 and 13 prepared by high-rate evaporation. Linear regression analysis was used to deduce the following photon-energy independent parameters (from top to bottom): surface roughness layer thickness ($t \geq 0.12$ min); total film thickness (surface roughness layer plus underlying bulk layer); and void fraction in the nucleating layer ($t < 0.12$ min) or underlying layer ($t \geq 0.12$ min). The bottom panel shows σ, the unbiased estimator of the mean square deviation for the fit. The Bruggerman effective medium theory was used in this analysis, along with an aluminum dielectric function generated analytically from Eqs. (1)–(3), using fixed bulk film values for parameterization. In the first 0.1 min, a one-layer model for the aluminum film is used to simulate the nucleation/coalescence process; for longer times, a two-layer model is used. Error bars represent the 90% confidence limits. After Ref. 16.

film optical functions can be utilized in the isolated particle stage, will be relaxed with much improved results. Furthermore, in Subsection B it will be shown that subtle changes in the aluminum dielectric function not evident in the analyses of Fig. 15 also occur as a function of film thickness after coalescence. In spite of this latter effect, the bulk dielectric function is close enough so that the morphological approach does provide a good estimate of the film thickness and deposition rate after complete coalescence ($7.5 \leqslant t \leqslant 30\,\text{s}$, in Fig. 15).

A preferable replacement for the third analysis step, discussed in the previous paragraphs, is one whose goal is to extract the dielectric function of the aluminum film independently at each thickness by a mathematical inversion method. The advantage of this approach (when it is possible) is that the dielectric functions can be compared for different thicknesses and then modeled separately. Thus, one also avoids the difficulty of having to determine the thicknesses simultaneously with other free parameters. The disadvantage is that only a one-layer, isotropic model for the film can be employed. As will be shown in Subsection C, this does not present a problem in the isolated particle stage of growth.

The mathematical inversion method to extract the dielectric function may not be possible, however, in the most general situation (46). In the experiment, two photon energy–dependent parameters, (ψ, Δ) [or $(\langle \varepsilon_1 \rangle, \langle \varepsilon_2 \rangle)$], are obtained; yet in the analysis, there are the two photon energy–dependent unknowns, the real and imaginary parts of the dielectric function $(\varepsilon_1, \varepsilon_2)$, but also one photon energy–independent unknown, the thickness d. For the aluminum film growth experiments in which the SiO_2/Si substrate shows the strong optical structure of the silicon E'_0–E_1 transitions (see Figs. 12 and 13 at $t = 0$), the inversion procedure developed by Arwin and Aspnes for static films can be applied (56). In this procedure, a one-layer isotropic model for the film is assumed, and a trial value of d is chosen. With this trial value, the (ψ, Δ) [or $(\langle \varepsilon_1 \rangle, \langle \varepsilon_2 \rangle)$] spectra can be inverted to obtain a trial dielectric function $(\varepsilon_1, \varepsilon_2)$ for the aluminum film. The trial dielectric function is then inspected for structure in the vicinity of the E'_0–E_1 transitions of the substrate. If the initial thickness choice was incorrect, then substrate-related features will appear there as artifacts. Thus, by eliminating such artifacts that appear both in ε_1 and ε_2, the correct thickness and effective dielectric function will be obtained.

To illustrate the sensitivity of the substrate artifact minimization procedure, Fig. 16 shows trial dielectric functions calculated from a single pair of real-time (ψ, Δ) spectra collected in $0.32\,\text{s}$ during aluminum film growth at low rate. (This deposition will be discussed in detail in Subsection C.) Different choices of the trial thickness from 22 to 30 Å have been used in the

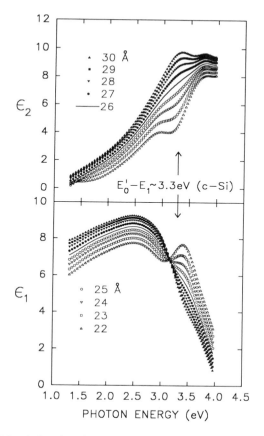

FIG. 16. Trial dielectric functions for an aluminum particle film determined in an isotropic, one-layer analysis of real-time ellipsometric spectra collected at 36 s during preparation onto a SiO_2/c-Si substrates by low-rate evaporation (43 Å/min). The different guesses for the film thickness allows mathematical inversion of (ψ, Δ) [or $(\langle \varepsilon_1 \rangle, \langle \varepsilon_2 \rangle)$] to obtain these results. The correct thickness (26 Å) is the one that eliminates artifacts in the inverted dielectric function that arise from structure in the substrate dielectric function. After Ref. 16.

calculation, and the resulting trial dielectric functions have been smoothed for clarity. Figure 16 reveals that a thickness choice of 26 ± 0.5 Å minimizes substrate artifacts and leads to the correct effective dielectric function for the film at this point in the growth process. The fact that the artifacts in both ε_1 and ε_2 are nearly eliminated suggests that the assumption of a one-layer model is close to reality. Because the generalized Maxwell-Garnett EMT predicts uniaxial behavior for films consisting of isolated particles, it is also very important to assess the validity of the restrictive assumption of isotropy, employed to extract the effective dielectric functions using the artifact minimization criterion. Such an assessment is nontrivial and will be treated in Subsection C.

Figure 17 shows dielectric functions for the high-rate aluminum deposition at selected times during the growth process, using the artifact minimization criterion to establish the correct thickness. This procedure works well to a thickness of ~ 125 Å. Above this thickness, the substrate feature is too weak to provide a sensitive criterion because of the strong absorbance of the overlying aluminum. In Fig. 17, the ε_2 spectrum of the 23 Å thick film (open

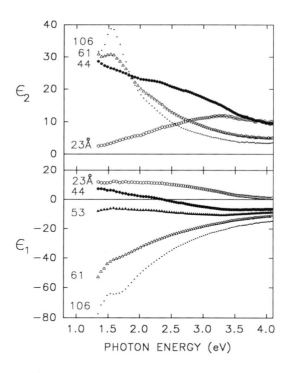

FIG. 17. Selected dielectric functions obtained from spectroscopic ellipsometry measurements (see Figs. 12 and 13) acquired during preparation of an aluminum film by high-rate evaporation. These results were obtained using the isotropic, one-layer analysis technique of Fig. 16. The imaginary part of the dielectric function for the 53 Å thick film is omitted for clarity (see Fig. 21). After Ref. 16.

circles) exhibits a broad resonance absorption band centered at 3.4 eV (similar to the 26 Å film of Fig. 16), and the corresponding ε_1 spectrum exhibits the associated dispersive behavior. These results are characteristic of dipolar plasmon–polariton excitations and are consistent with a thin aluminum film of isolated but interacting particles. As growth continues, the absorption band increases in magnitude and shifts to lower energies, as indicated by the spectra for the 44 Å thick film. The origin of this behavior is similar to that for $\varepsilon_\parallel$ in Figs. 7 and 9, namely, an increase in the strength of the dipole interactions among the particles.

For thicknesses in Fig. 17 between 44 and 53 Å, an abrupt decrease in ε_1 from positive to negative values is observed at the lowest photon energies, indicating that the percolation threshold has been crossed. The thickness range of 45–50 Å observed here for the percolation threshold lies within the range of values reported earlier for evaporated aluminum films (57, 58). For clarity, only the ε_1 spectrum for the 53 Å film is provided in Fig. 17. (Note that the corresponding ε_2 spectrum appears in Fig. 21.) Although metallic optical characteristics have developed at a thickness of 53 Å, as indicated by the negative value of ε_1 at the lowest photon energies, there is still no evidence of the (200) parallel-band feature (see also Fig. 21). Figure 9 suggests that if (i) the effective ε_2 value near 1.5 eV is significant and (ii) the bulk dielectric functions are appropriate for the aluminum component of the film, then the (200) parallel-band feature should be clearly visible, irrespective of the film morphology. The fact that it is not visible in ε_2 for the 44 and 53 Å thick films indicates that the bulk dielectric functions cannot describe either the aluminum particles in the nucleation stage or the partially coalesced film above the percolation threshold. As can be seen in Fig. 17, however, the (200) parallel-band feature develops rather abruptly in ε_2 for film thicknesses between 53 and 61 Å, and dominates the low-energy part of the spectrum at 106 Å. The changes in the visibility of the (200) parallel-band feature over this thickness range cannot be attributed to the changes in morphology as modeled in Fig. 15. The results of Fig. 4 suggest that the effect can be attributed to an increase in relaxation time with increasing thickness between 53 and 106 Å and will be modeled as such in the next two sections.

In summary, the dielectric functions in Fig. 17 present a clear, but qualitative, view of the optical properties of aluminum films, and in this mode the real-time SE capability is expected to be applicable to any metallic films. The artifact minimization analysis provides the thickness ranges associated with the different morphological stages of the film. For aluminum films evaporated at high rate, these include the discontinuous growth, or particle, regime ($d < 45$ Å), the particle contact regime and percolation threshold ($45 < d < 50$ Å), and the coalescence and continuous film growth regime ($d > 50$ Å). In order to provide quantitative details, further analysis focusing on the (200) parallel-band feature has been undertaken for the high-rate aluminum film above the percolation threshold in the continuous film growth regime ($d > 50$ Å). This analysis will be described in the next section. As noted earlier, for the films evaporated at lower rate, the particle regime extends over a larger range of thicknesses, typically up to 60 Å. For such films, the gradual coalescence process, during which particles and larger coalesced structures coexist, makes analysis beyond the percolation threshold more difficult. As a result, analysis of these films in the continuous growth regime will be avoided.

B. Analysis of the Optical Functions of Continuous Films

In analyzing the dielectric functions of the high-rate aluminum film (280 Å/min) in the continuous growth regime, the artifact minimization procedure is not effective for thicknesses greater than 125 Å. Thus, one must return to the linear regression analysis approach that was used for the morphological analysis depicted in Fig. 15. The results for the two thicker films in Fig. 17 (61 and 106 Å) suggest that there may be gradual changes in the sharpness of the (200) parallel-band feature in the continuous film regime, and these changes may provide information on the crystalline structure of the film through Eqs. (4) or (5). To incorporate this possibility, Eqs. (1)–(4) are combined to describe the optical properties of the aluminum component in the Bruggeman EMT, and linear regression analysis is performed using the film thickness, void volume fraction, and a common value of the electron mean free path (defined by Eq. (4)) as the variable parameters. The use of a variable mean free path allows for the possibility of a thickness-dependent ($\varepsilon_1, \varepsilon_2$) for the aluminum component in the EMT and distinguishes this approach from the morphological one of Fig. 15, as described in the previous section. In this refined analysis, the mean free path is the parameter of greatest interest.

In many other features, the analysis of this section parallels that of Fig. 15. First, the same values of 4.1×10^{-15} s are used for the bulk (111) and (200) parallel-band relaxation times in Eq. (4), and the bulk intraband relaxation time from the literature is used (10.6×10^{-15} s). Second, the best fit value of $|U_{200}| = 0.73$ eV is used in Eq. (3), along with the literature values of $|U_{111}|$ (Eq. (3)) and Ω_p (Eq. (2)) (24). Thus, in this analysis, it is assumed that changes in the band structure or electron density with thickness are negligible in the continuous growth regime. Furthermore, in order to maintain consistency with the morphological analysis of Fig. 15 for the bulk aluminum film in the opaque film limit, a fixed roughness layer of 3 Å is assumed. The exact choice of this thickness value, over the possible range from 0 to 10 Å, is not critical to the analysis and has negligible influence on the parameter of interest, the mean free path.

In order to include the mean free path as a variable parameter that defines the thickness dependence of the dielectric function, it is assumed that all relaxation times in Eq. (4) are related through the single parameter $\lambda \equiv \lambda_D = \lambda_{111} = \lambda_{200}$, and the electron velocities v_j are all equal to the Fermi velocity. As discussed in connection with Figs. 3–5, the fits to determine λ will be affected only by the (200) parallel-band contribution for $\lambda > 50$ Å. As λ decreases below 50 Å, however, the Drude contribution becomes more influential. Under no conditions will the fitted parameters be influenced by the (111) parallel-band contribution over the 1.3 to 4.0 eV spectral range.

Figure 18 shows a typical fit (solid line) to the experimental ε_2 spectrum

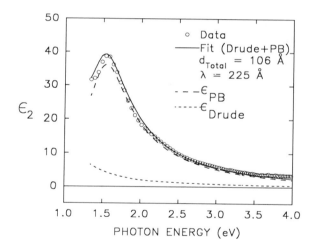

FIG. 18. Imaginary part of the dielectric function for the aluminum film of Fig. 17 at a thickness of 106 Å, obtained from real-time measurements during preparation by high-rate evaporation (points). Linear regression analysis and the Bruggerman effective medium theory were applied to fit these data using free parameters of void volume fraction, thickness, and electron mean free path (solid line). The latter two values are included in the legend. The best-fit dielectric function is decomposed into its intraband (Drude) and interband (parallel-band: PB) components (broken lines). The mean free path in this case is characteristic of the (200) parallel-band electronic transitions. After Ref. 16.

(points) for the high-rate evaporated film at 106 Å. The decomposition of the best-fit ε_2 spectrum into its intraband and interband contributions (broken lines) emphasizes the dominance of the latter. In order to obtain the results in this figure, the experimental (ψ, Δ) spectra and their fits were both processed to extract the bulk layer dielectric function, using the fixed surface roughness and best-fit bulk layer thicknesses. The bulk layer thickness extracted in the linear regression analysis, 103 Å, is close to the 106 Å value obtained by artifact minimization, the difference being attributed to the incorporation of the 3 Å surface roughness layer in the linear regression analysis. The larger width and smaller peak magnitude of the (200) parallel-band feature at 106 Å in Fig. 18, in comparison to the result at 700 Å in Fig. 14, together suggest a reduced relaxation time in accordance with the behavior in Fig. 4. This can be accounted for with a best fit mean free path of 225 Å, leading to a 25% reduction in the parallel-band relaxation time.

Figure 19 shows the common mean free path (open triangles, $d > 50$ Å) versus the total thickness, as determined in the linear regression analysis procedure throughout the continuous film growth stage for the aluminum film prepared by high-rate evaporation. The parallel-band relaxation time (filled triangles, $d > 50$ Å), calculated from Eq. (4), is also included in Fig. 19. Other data appear that will be discussed in the next section (triangles for $d < 50$ Å for the high-rate film, and circles for $d < 60$ Å for the low-rate film). A very interesting effect is observed in Fig. 19, namely, the increase in λ by a factor of nine, from 10 to 90 Å as d increases from 55 to 60 Å. This behavior is reflected in the rapid appearance of the (200) parallel-band feature for film thicknesses between 53 and 61 Å, as can be seen by comparing Figs. 17 and 21. Tentatively, this abrupt change can be attributed to the formation of single-crystalline grains that extend throughout the thickness of the film, from a highly defective or fine-grained nanocrystalline structure that exists in the nucleation and early coalescence processes. There is a more gradual increase in λ for $d > 60$ Å in Fig. 19 that may be caused by the reduction in grain boundary scattering that accompanies an increase in average grain size with increasing thickness. The latter regime will be considered in greater detail first.

In making a quantitative assessment of grain boundary scattering for $d > 60$ Å, Eq. (5) can be adopted in an attempt to describe the variation in the interband relaxation time in Fig. 19. In doing this, four parameters are fixed. First, the bulk relaxation time is set at 4.1×10^{-15} s, as determined by fitting the dielectric function of the final 700 Å film in Fig. 14. Second, the electron velocity is taken to be the Fermi velocity, $v_F = 2.02 \times 10^8$ cm/s. Third, the grain boundary reflection coefficient $\mathfrak{R}$ is set at 0.15, the value determined for free electrons (30). Finally, the mean free path λ_g in Eq. (5)

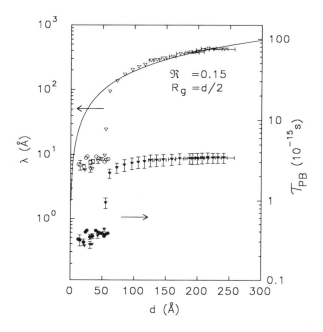

FIG. 19. Electron mean free path (open symbols; from Eq. (4)) and (200) parallel-band relaxation time (solid symbols) as a function of film thickness d deduced from real time observations during the growth of aluminum films at low and high rates (circles: 43 Å/min; triangles: 280 Å/min). The triangles for $d > 50$ Å were determined by fitting (ψ, Δ) spectra with a continuous film model and the Bruggerman effective medium theory. All other results were determined by fitting effective dielectric functions using a particle film model and the generalized Maxwell-Garnett theory. The solid line is a fit to λ for $d > 60$ Å using the grain growth model of Eq. (5). For a grain radius of $d/2$, the solid line corresponds to a grain boundary reflection coefficient of 0.15. Error bars are the 90% confidence limits. After Ref. 16.

is equated to the grain radius, R_g, which in turn is chosen to be $d/2$. This simple choice is based on the approximate equality between grain size and thickness noted in early studies (59). For this latter relationship to occur, dominant grains must increase in size with increasing thickness at the expense of their neighbors; however, this is expected to lead to a complex statistical distribution in the grain size. Such a complication is not included in the simple model.

Figure 20 (points) shows the experimental values of τ_{PB}^{-1} from Fig. 19 plotted versus $1/d$ for $d > 60$ Å. The error bars in Fig. 20 are the 90% confidence limits in the linear regression analysis, which establish the

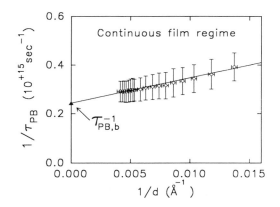

FIG. 20. The (200) parallel-band electron relaxation rate from Eq. (4) plotted versus reciprocal thickness in the continuous film regime ($d > 60$ Å), as deduced from real-time observations during aluminum film growth by high-rate evaporation. The solid line represents a prediction based on Eq. (5) with a mean free path in the grain of $\lambda_g = d/2$, a grain boundary reflection coefficient of $\Re = 0.15$, and a Fermi velocity of $v_F = 2.02 \times 10^8$ cm/s. After Ref. 16.

absolute uncertainty in τ_{PB}^{-1} and d due to parameter correlations. The relative uncertainties in τ_{PB}^{-1} appear to be much better, since the trend in the width of the parallel-band transition can be identified in the dielectric functions directly. Also included in Fig. 20 is the prediction from Eq. (5), using the preceding four parameter choices. This prediction as it applies to the effective λ value of Eq. (4) is also included in Fig. 19 (solid line). Very good agreement between the experimental results and the simple theory is obtained, suggesting that the hypothesis regarding grain growth is valid, at least to first order. It is possible, however, that the agreement between the slopes of the calculated and experimental results in Fig. 20 is fortuitous, since the former relies on the three parameters v_F, $\Re$, and $\lambda_g = R_g = d/2$, and the choices for v_F and $\Re$ are expected to be valid only for free electrons. For example, the same calculated result can be obtained with $\Re = 0.08$ and $\lambda_g = R_g = d/4$. Thus, although a grain growth model appears to be correct, the exact relationship between grain size and thickness is not identified unambiguously from the fits to the dielectric function.

Next, the abrupt transition in λ observed in Fig. 19 for $55 < d < 60$ Å requires a closer look, as well. In particular, the structure of the film prior to the transition is of great interest. Figure 21 includes the ε_2 spectrum of the high-rate aluminum film just prior to the transition. (The associated ε_1 spectrum appears in Fig. 17.) As noted earlier in this section, the film at this

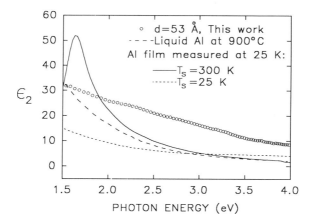

FIG. 21. Imaginary parts of the dielectric functions for three different pure aluminum materials in which the (200) parallel-band featuare is absent: a 53 Å thick aluminum film prepared by evaporation and measured during coalescence by real-time spectroscopic ellipsometry (SE) (open circles); an opaque aluminum film prepared by evaporation onto a substrate held at 25 K and measured *in situ* by SE (short-broken line, after Ref. 61); and bulk liquid aluminum at 900°C (long-broken line, after Ref. 60). For comparison, results are also presented for an opaque aluminum film prepared by evaporation onto a room-temperature substrate and measured *in situ* at 25 K (solid line, after Ref. 61). After Ref. 16.

point in the growth process exhibits metallic behavior, yet no evidence of the (200) parallel-band transition. In fact, the overall features of the dielectric function can be reasonably well fitted using a mean free path of ~ 10 Å in Eqs. (1)–(4), which reduces the relaxation time of electrons participating in both interband and intraband transitions (for comparison, see simulations in Fig. 5). In fact, the dielectric function cannot be fitted unless both modifications are included. If the (200) parallel-band relaxation time is fixed at the bulk value, the absence of a visible feature at 1.5 eV cannot be explained; if the intraband relaxation time is fixed at the bulk value, the relatively strong ε_2 spectrum and the photon energy–independent ε_1 spectrum cannot be explained.

It is of interest to compare the ε_2 spectrum for the 53 Å film to previous experimental results on pure aluminum systems for which the absence of the (200) parallel-band feature has also been reported. These include liquid aluminum measured at 900°C (*60*), and evaporated aluminum quenched onto substrates at 25 K and measured *in situ* by spectroscopic ellipsometry (*61*). Both results appear in Fig. 21 as the broken lines. Also included for contrast is the spectrum for an aluminum film deposited at room temperature and measured at 25 K (solid line) (*61*). In this latter case, a very clear (200) parallel-band feature is observed.

First, one might speculate that the absence of the (200) parallel-band feature for the 53 Å film results from a "liquid" state that remains for a few seconds during the coalescence of ~ 50 Å particles. In fact, the liquid metal state has been observed in ultrathin particle films, well below bulk melting temperatures, because of a particle size–dependent reduction in the melting point (*62,63*). This explanation can be tested by fitting the ε_2 spectrum for the 53 Å film using one intraband term alone (see Eq. (2)), a form that has provided a reasonable description of the measured optical properties of liquid aluminum shown in Fig. 21 (*60*). The fact that such fitting has been unsuccessful and that the shapes of the ε_2 spectra for the 53 Å film and the liquid are significantly different, suggest that the original speculation is not correct. A conclusion based on such arguments, however, needs to be tempered in light of very recent measurements of liquid aluminum that differ significantly from the original report (*64*).

Evidence against liquid or even fully amorphous states for both the 53 Å and quenched films of Fig. 21 comes from electron diffraction measurements suggesting that films prepared under these conditions exhibit a nanocrystalline grain structure (*65,66*). In fact, the similar shapes of the ε_2 spectra for the ultrathin and quenched films in Fig. 21 also suggest a similar nanocrystalline structure for the two materials. The difference in magnitude between the two ε_2 spectra apparently results from a much higher volume fraction of voids in the quenched film. In concluding, one can only speculate that the optical properties measured here are characteristic of a film that is far from equilibrium, existing in the narrow regime between coalescence and bulk film growth. Additional discussion of such points will appear in the next section, where the optical properties of isolated particle films will be discussed.

C. Analysis of the Optical Functions of Particle Films

In Fig. 16, a method for determining the dielectric function and thickness d of aluminum films with $d < 125$ Å has been presented, and representative results for the high-rate deposition were presented in Fig. 17. This approach is ideal for interpreting the optical properties in the isolated particle regime, except for the potential problem noted earlier. As shown by Eqs. (6) and (7), the films in the particle regime are predicted to be uniaxial, in general. A sample consisting of a substrate-supported uniaxial film with its optic axis perpendicular to the substrate surface exhibits a diagonal Jones matrix in the p–s reference frame (46). Thus, the experimental ellipsometry spectra in (ψ, Δ) can be acquired using the same methods as in the case of an isotropic sample, but the spectra must be analyzed according to the techniques of an anisotropic stratified structure.

One characteristic of uniaxial anisotropy in a thin film is the inability of a single thin-film dielectric function to characterize (ψ, Δ) at different angles of incidence. Figure 22 shows results collected on an aluminum film obtained in a deposition on SiO_2/Si that was terminated under real-time control in the particle regime. The film was maintained under a N_2 purge as the windows to the chamber were removed, and spectra were collected through the open ports at different angles of incidence from 65° to 75°. The artifact minimization procedure of Fig. 16 was applied to the spectra collected at 70°, leading to an effective dielectric function having the imaginary part shown in the lower panel of Fig. 22 (56). When this effective dielectric function was employed in a linear regression analysis fit to the (ψ, Δ) spectra at other angles of incidence, the resulting best fits led to thicknesses within ± 1 Å of the original 23 Å. In fact, these best fits (broken lines in Fig. 22) match the experimental (ψ, Δ) spectra so closely that anisotropy cannot be detected. The physical origin behind this interesting observation will be discussed later in this section.

Previous results have suggested that, for weakly anisotropic films, the deduced dielectric functions obtained assuming isotropy are most representative of $\varepsilon_\parallel$ (67). As a result, Eq. (6) is employed, upon substitution of Eqs. (1)–(4), to model by linear regression analysis the effective dielectric functions of thin aluminum films in the isolated particle regime. In this analysis, three free parameters are used, $F_\parallel$, Q, and λ, with λ defining ε_p in Eq. (6). Because a simplified model for the film morphology is used that consists of identical spheroidal particles with symmetry axes perpendicular to the substrate (see Fig. 6), the deduced morphological parameters need further confirmation. However, it will be shown that this simplified model does not invalidate the determination of the mean free path λ.

In implementing this analysis, it is tacitly assumed that the electronic features influencing the dielectric function, other than relaxation time, remain the same in aluminum bulk and particle forms. For selected analyses, however, the plasma frequency Ω_p has been used as a free parameter, but bulk values are deduced within the confidence limits. Thus, $\hbar\Omega_p$ is fixed at 12.7 eV in the analyses of the particle films. It also turns out that the best fit values of λ are so small (see Figs. 19 and 28), and the apparent parallel-band transitions so severely broadened, that the spectra are relatively insensitive to variations in $|U_{200}|$. Thus, this parameter is also fixed—at 0.73 eV. In the

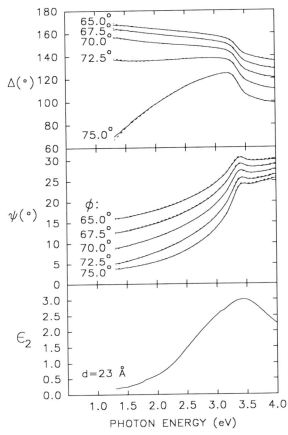

FIG. 22. Experimental (ψ, Δ) spectra for a static 23 Å film prepared by evaporation and measured *in situ* by spectroscopic ellipsometry at different angles of incidence from 65° to 75° (solid lines). The fits to the data (broken lines) were obtained by linear regression analysis using a single effective dielectric function for the film. This dielectric function was deduced from the (ψ, Δ) spectra at a 70° angle of incidence using the artifact minimization procedure of Fig. 16, and the imaginary part appears in the bottom panel. The excellent fits to the data show that the film is isotropic.

overall analysis, the intended role of the interband λ is to broaden a set of transitions that are common to the bulk material and particles; however, it is difficult to rule out the possibility that a finite λ is actually simulating a new set of electronic transitions that exist in particles because of changes in the band densities of states. In this latter case, λ still provides useful qualitative information on the loss of the critical-point structure in the joint density of states that derives from the parallel bands.

Figure 23 shows selected experimental effective dielectric functions from artifact minimization analysis of (ψ, Δ) spectra collected during the particle stage of growth for an evaporated aluminum film (points). As noted earlier, a film with a low deposition rate of 43 Å/min was chosen for analysis in

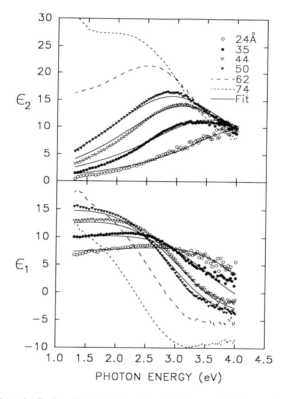

FIG. 23. Selected effective dielectric functions in the particle film regime, obtained from real-time observations of aluminum growth by low-rate evaporation (43 Å/min). The isotropic, one-layer analysis procedure of Fig. 16 was used to extract the effective dielectric functions and thicknesses from spectra in (ψ, Δ). The solid lines represent three-parameter, linear regression analysis fits of the dielectric functions to the generalized Maxwell-Garnett effective medium theory of Eq. (6). The free parameters include the aluminum volume fraction Q, the interaction parameter $F_\parallel$, and a common value of the mean free path λ, which defines the dielectric functions of the particles. After Ref. 16.

order to extend the isolated particle regime to thicknesses of 50–60 Å. The broken lines in Fig. 23 represent selected dielectric functions for $d > 60$ Å, as the particles begin to make contact. In the regime from 60 to 75 Å, the optical spectra are very complicated, and the films appear to consist of smaller particles and larger coalesced structures, yet the percolation threshold has not been reached. The solid lines in Fig. 23 are the best three-parameter linear regression analysis fits to the effective dielectric functions of the particle films as described in the previous paragraphs. Results for the three free parameters and derived quantities are included in Figs. 24–28.

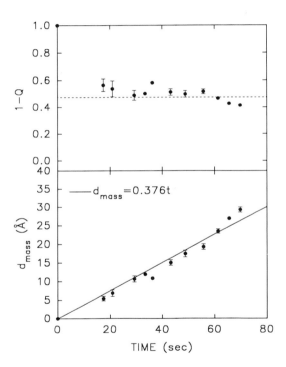

FIG. 24. Void volume fraction plotted versus time (upper panel) in the particle regime for thin-film aluminum prepared by low-rate evaporation. The lower panel shows the mass thickness, defined by $d_{mass} = Qd$, where d is the film thickness and Q is the aluminum volume fraction. This quantity is proportional to the rate at which aluminum atoms are incorporated into the film. These results were calculated by combining the data from the artifact minimization analysis of Fig. 16 with those from the linear regression analyses of the effective dielectric functions of Fig. 23. In the latter analyses, the generalized Maxwell-Garnett effective medium theory of Eq. (6) was used.

First, the two morphological parameters will be discussed briefly, and the physical mechanism behind the observed isotropy will be addressed (Figs. 24–27). Then, attention will be focused on the mean free path determination (Fig. 28).

Figure 24 (upper panel) shows values for the void volume fraction $1 - Q$ plotted versus time. (Because of the time-consuming nature of the effective dielectric function determination as shown in Fig. 16, not all (ψ, Δ) spectra from the real-time data set were subjected to the analysis.) The very weak time dependence of the void volume fraction suggests that the particles are flattened and cover a large fraction of the substrate area. This behavior is very similar to that observed for a-Si:H growth on SiO_2/c-Si, as described in the previous chapter. If the particles are in fact spheroidal, according to the model of Fig. 6, then the results of Fig. 24 suggest that the spacing between particles increases almost linearly with film thickness, as in a statistical aggregation process. The lower panel in Fig. 24 includes the time dependence of the mass thickness, defined by $d_{mass} = Qd$, which is proportional to the rate at which aluminum atoms are incorporated into the film. A reasonably good linear trend in the d_{mass} values is observed, exhibiting deviations of less 3 Å from the values predicted by the best-fit linear accumulation rate. Because Q is nearly constant at 0.5, the physical thickness d also increases linearly, but at a rate greater by a factor of two.

Figure 25 shows the interaction parameter $F_{iso} \equiv F_{\parallel}$ plotted as a function of Q. Also included on this plot is the prediction of the simple (isotropic) Maxwell-Garnett theory, namely $F_{iso} = (1 - Q)/3$. Although the interaction parameter measured experimentally decreases with increasing Q and d, the interaction between the particles is stronger than that predicted from the Lorentz local-field calculation. Resorting instead to the generalized Maxwell-Garnett theory to describe the interaction parameter, an estimated axial ratio γ can be determined by solving Eq. (8), after substituting the experimental values of $F_{\parallel}$ and Q from Fig. 25. Two solution branches exist, and the physically realistic, oblate branch ($\gamma > 1$) is selected and plotted versus thickness in Fig. 26. Different choices of the parameter ξ correspond to different assumptions concerning the arrangement of particles, e.g., $\xi = 1$ corresponds to a square array (8) and $\xi = 1.4$ has been calculated for an "amorphous" distribution (7). In earlier studies of gold particle films, ξ has been used as a fitting parameter (9). In Figure 26, the larger values of ξ give the physically reasonable result that γ gradually increases in thickness and saturates prior to coalescence. This saturation value of 2.5–3 is in reasonable agreement with $\gamma = 4$, estimated from direct electron microscopy of aluminum particle films on SiO_2/c-Si substrates (66).

Next, the values of Q and γ, deduced from Eq. (8) with different choices

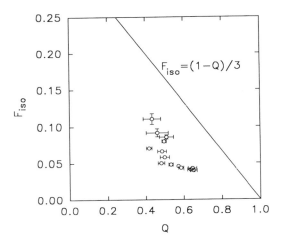

FIG. 25. Interaction parameter $F_{iso} \equiv F_\parallel$ plotted versus the solid material volume fraction Q in the particle regime for thin-film aluminum prepared by low-rate evaporation. These results were deduced in linear regression analyses of the effective dielectric functions of Fig. 23, using the generalized Maxwell-Garnett effective medium theory (EMT) of Eq. (6). The solid line is based on the predictions of the simple Maxwell-Garnett EMT, assuming that the particles interact via the Lorentz local field.

of ξ, can be substituted into Eq. (9) to determine $F_\perp$. Then, a measure of the degree of anisotropy, $\delta_A \equiv |F_\parallel - (F_\perp - Q)|$, can be calculated. The results for different values of ξ appear as a function of thickness in Fig. 27. The results show that for $1.0 \leq \xi \leq 1.4$, the anisotropy is very small and nearly independent of thickness to $d = 40$ Å. This result appears to explain the experimental observation of isotropy in Fig. 22. The reason for this behavior is that the anisotropy due to particle shape is effectively canceled by the anisotropy in the dipole interaction among particles (see Eq. (8)). It is reasonable to choose the solution $\xi = 1$ in Figs. 26 and 27, corresponding to particles on a square grid, because for this choice the predicted isotropy extends to the boundary of the coalescence regime.

The physically reasonable results, including the increase and saturation of the axial ratio with thickness, as well as the prediction of isotropy for a range of interparticle geometries, argues for the validity of the model embodied in Eqs. (6)–(9), even though complications due to particle size or shape distributions are not included. One possible reason for this will be discussed shortly. In spite of the favorable outcome, the deduced mor-

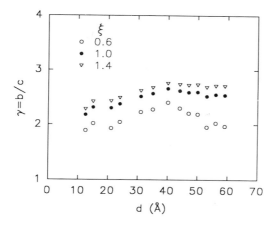

FIG. 26. Spheroidal particle axial ratio plotted as a function of film thickness in the particle regime for aluminum prepared by evaporation at low rate. These results represent the oblate solution ($\gamma > 1$) to Eq. (8), calculated upon substituting $F_\parallel$ and Q from Fig. 25. Different values of the parameter ξ are assumed in Eq. (8), representing different geometrical distributions of particles on the substrate. The choice $\xi = 1$ corresponds to particles located on a square grid. After Ref. 16.

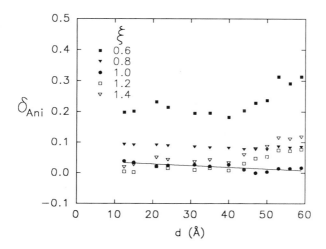

FIG. 27. Degree of anisotropy, defined by $\delta_A = |F_\parallel - (F_\perp - Q)|$, plotted as a function of film thickness in the particle regime for aluminum prepared by evaporation at low rate. These results were calculated using $F_\parallel$ and Q from Fig. 25 and $F_\perp$ from Eq. (9). Different values of the parameter ξ are assumed, representing different geometrical distributions of particles. The solid line is a guide to the eye for the $\xi = 1$ results (solid circles), corresponding to particles located on a square grid. After Ref. 16.

phological parameters are not completely free of possible problems. It may be noted that the values of $1 - Q$ in Fig. 24 lie just above 0.48, which is the minimum void volume fraction for non-contacting, spheroidal particles located on a square grid. In fact, for film thicknesses above 40 Å, the void volume fraction drops below 0.48, yet the dielectric functions still show evidence of isolated particles. This suggests that the particles may be hemispheroidal or disk-shaped, or possibly exhibit a closer packing than a square grid. These geometries would permit a smaller minimum void volume fraction. The different particle shapes would require a reformulation of Eqs. (8) and (9), however.

In the last part of this section, the presentation will concentrate on the common mean free path λ, obtained along with $F_\parallel$ and Q from the best fits to the effective dielectric functions in Fig. 23. It should be recalled that the common value of λ reduces the relaxation time of electrons participating in both intraband and interband transitions through Eq. (4) and (in contrast to $F_\parallel$ and Q) provides insights into the electronic properties of the particles themselves. Figure 28 includes results for λ versus thickness d for the

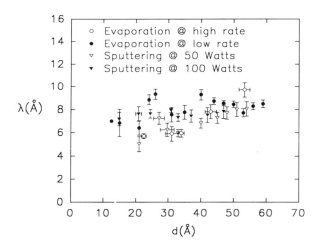

FIG. 28. Common value of the mean free path λ plotted versus film thickness characterizing the evolution of electronic properties in the particle regime for two evaporated and two sputtered aluminum films. These results were deduced in linear regression analyses of effective dielectric functions such as those in Fig. 23, using the generalized Maxwell-Garnett effective medium theory of Eq. (6). The best-fit values of λ, which reduce the electron relaxation times from their bulk values, are influenced equally by the free-electron and (200) parallel-band transitons. After Ref. 17.

low-rate aluminum film of Figs. 23–27 (solid circles), the high-rate aluminum film (open circles), and for films prepared by sputtering (triangles). The values for the low- and high-rate evaporated films in the particle regime are included in Fig. 19 (circles and triangles, respectively), for comparison with the corresponding results in the continuous film regime. Two important observations can be made from Figs. 19 and 28.

First, it is noted from Fig. 19 that λ (as well as τ_{PB} and τ_D) for both low- and high-rate aluminum films in the particle regime connect smoothly with the value for the 53 Å thick, high-rate film that is above the transition to continuous film growth. It should be emphasized that the generalized Maxwell-Garnett EMT is used in the analysis of the effective dielectric function of the films in the particle regime, whereas the Bruggeman EMT is used in the continuous film regime. In the former case, λ is determined through the width of the resonance absorption band in the effective dielectric function (see Fig. 10), whereas in the latter, λ is determined by the overall features in the continuous film dielectric function. The continuity in λ across the coalescence transition suggests that the electronic structure of the aluminum particles persists even after the particles make contact. The few-second delay between the percolation threshold (45–50 Å) and large-scale crystallization (55–60 Å) in the high-rate film appears to occur as remaining void structures collapse.

Thus, the continuity in λ across the transition provides support for the overall modeling approach. For example, one might propose that the widths of the resonance absorption bands in the effective dielectric functions of the particle films are controlled, not by λ, but by a particle shape distribution, an effect that has been neglected in the morphological analysis. This would lead to true values of λ in the particle regime significantly larger than those plotted in Fig. 19. Thus, a large drop in λ upon coalescence would occur, which is a counterintuitive result. Thus, it is concluded that the width of the resonance band is in fact controlled by the imaginary part of the dielectric function of the particles through the parameter λ.

The second interesting observation from Figs. 19 and 28 is that, although there is some scatter, λ falls within the narrow range of 7.5 ± 2 Å, remarkably independent of film thickness d (when the confidence limits on each value are considered). This same range is observed for both low- and high-rate evaporated particle films, as well as for the particle films prepared by sputtering. This result implies that λ is also independent of particle size $R \sim c$, as well, since R is expected to increase linearly with d. Therefore, the electron mean free path cannot be limited by scattering at the particle surfaces. Such a model would predict that $\lambda \sim R$ (see discussion of Eq. (4)), and thus λ should increase from ~ 6 Å for the thinnest film in Fig. 28 to 30 Å for the thickest film. Since this trend is absent, it is evident that scattering sites internal to the particles control the mean free path of the electrons. One can also infer that the quantum size effects on the free electrons that lead to discrete transitions described by the particle-in-the-cube formalism will not be observable, even if the particle size distribution is sufficiently narrow.

The important influence of internal defects on the optical properties of metal particles has been suggested in previous studies. Kreibig first observed that the resonance absorption bands in silver particle films prepared at low temperature were much broader than those in films prepared at higher temperature, and the latter were in reasonable agreement with Eq. (4) when $\lambda \sim R$ *(10)*. The observations were attributed to defects in the silver particles prepared at low temperatures.

The results of Fig. 28 for the different films indicate that a similar internal structure and intrinsic dielectric function for the metal particles is obtained irrespective of the particle size and method of preparation. This in turn suggests that the structure of the particles has reached a common equilibrium state, in spite of the fact that the energies of the incident aluminum atoms in the sputtering process are significantly higher than those in evaporation. In contrast, if one were to propose that the width of the resonance band in the particle optical properties is controlled by the particle shape distribution, rather than by λ, then it would be difficult to imagine how similar distributions would be present in all films. In fact, the average morphological parameters themselves are different for these films. Specifically, the particles in the thinnest films prepared by sputtering ($d = 10$–$20\,\text{Å}$) exhibit a larger axial ratio than those in the evaporated films of similar thickness (see Fig. 26). This behavior is consistent with the higher-energy incident atoms in the sputtering process. Finally, it should also be emphasized that the size-independent particle dielectric function observed here effectively eliminates one potential source of error in the analysis that results from neglecting the particle size distribution.

In summary, it can be concluded from the results in Figs. 19 and 28 that a common intrinsic dielectric function describes aluminum particles, irrespective of size. The same dielectric function also describes partially coalesced aluminum films evaporated at a high rate. To emphasize this, Fig. 29 shows this dielectric function ε_p, corresponding to aluminum particles with $\lambda = 8.5\,\text{Å}$, compared to that for bulk film aluminum. Both dielectric functions have been calculated from Eqs. (1)–(4). The dielectric function with $\lambda = 8.5\,\text{Å}$ shows features similar to that for the 53 Å film in Figs. 17 and 21, as would be expected based on the continuity of λ between the two regimes in Fig. 19 as noted earlier. The similar features include an ε_1 spectrum that depends weakly on photon energy, an enhanced ε_2 spectrum relative to that

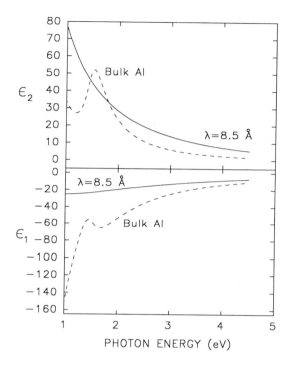

FIG. 29. Characteristic dielectric function for substrate-supported aluminum particles under all conditions of preparation (both sputtering and evaporation) as well as for partially coalesced continuous films prepared by high-rate evaporation (solid lines). These results are obtained from Eqs. (1)–(4), with $\lambda = 8.5$ Å in Eq. (4). The dielectric function of aluminum in bulk form is also included for comparison (broken lines).

of the bulk, and no evidence of the (200) parallel-band transitions. One might suggest that the atomic-scale structure of the particles is also similar to that of the partially coalesced film. Thus, the particles most likely consist of small nanocrystallite clusters connected by an intergranular phase that scatters electrons, or single nanocrystals with a high density of internal defects that serve the same role. A nanocrystalline structure for small aluminum particles has also been suggested by transmission electron diffraction studies (66).

Before concluding, it is important to contrast the recently developed real-time SE approach with that used in earlier investigations of films consisting of isolated particles. The most advanced studies of particle films have usually relied on *ex situ* normal-incidence transmittance spectra, $T(h\nu)$

(anticipating that measurements performed at non-normal incidence would be influenced by anisotropy in the films). *Ex situ* electron microscopy has also been used to obtain morphological information and, as a result, to reduce the number of free parameters required to simulate $T(hv)$. Because of the relatively low atomic number of aluminum, edge-on electron microscopy, often used to provide the particle shape and thickness directly, tends to be difficult and inaccurate. Furthermore, the high reactivity of aluminum implies that the particle size observed by electron microscopy does not correspond to the true metal particle size because of post-oxidation. Thus, even when direct microscopy is successful, fitting parameters that describe the morphology, e.g., the thickness of the oxide coating on the particles, must still be used in the simulations of $T(hv)$. In addition, such studies generally have relied on untested assumptions concerning the intrinsic dielectric functions of the metal particles. For example, in one study of aluminum particles, only an intraband contribution was used (*68*). In another study, both intraband and interband contributions were used, and the former was modified according to the classical size effect of Eq. (4) with $\lambda_D = R$, the particle radius (*69*). In these studies, simulated and experimental spectra have generally agreed to no better than 10% in T over the full spectral range, and one source of the inadequacy may be incorrect intrinsic dielectric functions for the metal particles.

Figure 30 shows the magnitude of the problem that may arise if the choice of the dielectric function of the aluminum particles is incorrect. In this figure, the effective dielectric function of the $d = 50$ Å film of Fig. 23 is shown, along with its best three-parameter fit, obtained with $Q = 0.58$, $F_\parallel = 0.044$, and $\lambda = 8.5$ Å (solid line). Also included are the two effective dielectric functions from Fig. 10, calculated assuming (i) that the dielectric function for the aluminum particles is identical to the bulk film dielectric function (dotted line), and (ii) that the common mean free path for both intraband and interband contributions is limited by scattering at particle surfaces, leading to $\lambda = d/2 = 25$ Å (broken line). As noted in the discussion of Fig. 10, the values of $F_\parallel$ and Q have been adjusted to match the peak magnitude and position of the resonance absorption band in the experimental ε_2 spectrum. It is clear, based on the arguments presented in this section, that these assumptions are not correct. As a result, the bands in the two simulated spectra are much narrower than the observed band. These results show that incorrect assumptions concerning the dielectric function of the particles may also lead to distortions in the best-fit morphological parameters used in the analysis.

In contrast to an approach based on *ex situ* transmittance measurements of individual films prepared to different thicknesses, the real-time SE

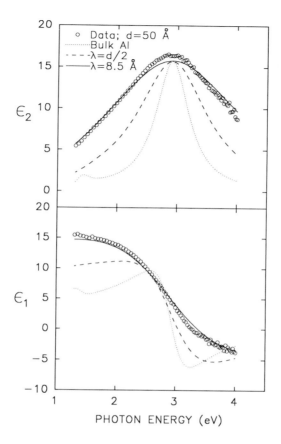

FIG. 30. Effective dielectric function for a 50 Å thick aluminum film prepared by low-rate evaporation reproduced from Fig. 23 (open circles). The solid lines represent a three-parameter, linear regression analysis fit of the effective dielectric function to the generalized Maxwell-Garnett effective medium theory of Eq. (6), yielding a common value of the mean free path, $\lambda = 8.5$ Å, that defines the dielectric functions of the particles. Also shown are effective dielectric functions of Fig. 10, calculated from Eqs. (1)–(4) using other assumptions regarding the dielectric function of the particles. For the broken line result, λ is fixed at $d/2 = 25$ Å, and for the dotted line result, the bulk dielectric function is chosen (i.e., $\lambda \to \infty$). After Ref. 16.

approach provides both real and imaginary parts of the effective dielectric function and the thickness d of a single film as a continuous function of time. The latter capability is made possible by a self-consistent analysis that determines the correct thickness, based on the criterion that substrate-related artifacts disappear in the effective dielectric function of the aluminum particle film (56). The effective dielectric function is then fitted by linear regression analysis, using free parameters that describe the morphology and the electronic properties of the particles. In order to assess the quality of these fits in comparison with previous attempts, the transmittance spectra $T(h\nu)$, calculated from effective dielectric functions for the low-rate aluminum film (see Fig. 23), are presented as a function of photon energy in Fig. 31 (points). Simulated $T(h\nu)$ calculated from the best fits to the effective dielectric functions are also included (solid lines). For simplicity these

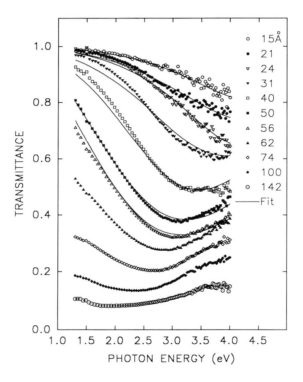

FIG. 31. Evolution of the transmittance spectra with thickness for thin-film aluminum prepared by low-rate evaporation (points), as calculated from the experimental effective dielectric functions (see Fig. 23) assuming a free-standing film. The solid lines correspond to the transmittance spectra calculated from the generalized Maxwell-Garnett effective medium theory using free parameters $(Q, F_\parallel, \lambda)$ that provide the best fit to the full complex effective dielectric function. After Ref. 16.

calculations were performed assuming a free-standing aluminum particle film. Agreement between the "experimental" and best-fit simulated spectra in Fig. 31 is obtained to within $\sim 2\%$ in T over the full spectral range from 1.5 to 4.0 eV. This agreement represents a significant improvement over previous such attempts.

It should also be noted that the fits in Fig. 31 go beyond any previous attempts in that they not only match the transmittance spectra, but also reproduce the full complex optical response. If the dielectric function of the particles themselves were to be fixed by setting a common value of the mean free path at 7.5 Å for the full series of spectra in Fig. 31, excellent fits can be achieved (in the absence of any other characterization) with only two morphology-related free parameters for each spectrum. In fact, the validity of this fixed dielectric function is further established by its ability to characterize suitably-prepared coalescing films, as well. Such continuous films do not require modeling by the generalized Maxwell-Garnett theory. The time dependence of the best-fit morphological parameters, the metal volume fraction Q and the axial ratio γ, exhibit trends with thickness that are reasonable and in fact predict the isotropy that is observed experimentally for the particle films. Thus, if the particles are assumed to be oblate ($\gamma > 1$) and γ is established from Q by the constraint that the effective dielectric function be isotropic, then the close fits in Fig. 31 can be achieved with just one free parameter for each spectrum.

V. Summary

In this chapter, a detailed case study has been presented in which the evolution of the optical functions of aluminum thin films has been investigated by real-time spectroscopic ellipsometry (SE) as a continuous function of thickness throughout the nucleation, coalescence and bulk film growth regimes. This investigation is the first of its kind and has been made possible by the development and perfection of the multichannel ellipsometer described in greater detail in the previous chapter of this volume. Parallel detection with the multichannel ellipsometer provides acquisition times as short as 0.32 s for ellipsometric spectra (ψ, Δ) over the photon energy range from 1.3 to 4.0 eV. This capability yields monolayer resolution when SE is performed in real time during aluminum film growth at rates as high as 400 Å/min.

Aluminum has been chosen for this case study of metal deposition because it is one of the most technologically important metals for which a theoretical framework has been developed to relate the dielectric function to the band structure. Extensive experimental optical studies of aluminum thin films have supported this framework. In particular, the bulk film optical response of aluminum can be described using an analytical expression that includes intraband (or free-electron) and interband contributions. The interband contribution is dominated by electronic transitions between nearly parallel bands that occur in planes parallel to the square (200) Brillouin zone faces. The (200) parallel-band transitions give rise to pronounced features in the real and imaginary parts of the dielectric function near 1.5 eV. The widths of these features provide qualitative information on the relaxation time of the electrons excited into the upper band.

As a result, the dielectric function of aluminum can be expressed in terms of a few photon-energy–independent parameters so that the evolution of the electronic characteristics of the film can be determined in the different growth stages. In this investigation, the possible effects of electron scattering by surfaces, subsurface defects, and grain boundaries have been included in both the intraband and interband contributions to the dielectric function of aluminum in the different growth stages. This appears to be an additional key element that distinguishes this case study from the numerous previous attempts at understanding the optical properties of metal films in the isolated particle stage.

In this study, a detailed analysis of the discontinuous regime, when the film consists of isolated particles, has been performed on films prepared by evaporation and sputtering at relatively low deposition rates (~ 40 Å/min). At the low rates, the thickness range over which isolated particles can be observed increases. On the other hand, an analysis of the continuous film growth regime has been performed on an aluminum film prepared at a rate higher by a factor of seven (280 Å/min). At the high rate, the morphological transition separating the particle and continuous film growth regimes is much more abrupt than that observed at low rates. In this way, one avoids the difficulty of modeling films that consist of a mixture of both particles and larger coalesced structures.

In the particle growth regime for aluminum films prepared by low-rate evaporation, the effective dielectric functions and thicknesses (typically, $d < 60$ Å) have been extracted from the real-time (ψ, Δ) spectra using a one-layer, isotropic model for the aluminum film. In this procedure the correct thickness choice is identified by trial and error as the one that avoids artifacts in the deduced dielectric function of the aluminum film. These artifacts arise from the features in the dielectric function of the substrate, in this case a thermally oxidized silicon wafer. The assumption of isotropy has been justified on the basis of (ψ, Δ) measurements performed on a static particle film at different angles of incidence. The effective dielectric functions deduced by the artifact minimization procedure are then fitted by linear regression analysis using a generalized Maxwell-Garnett effective medium theory (EMT) and a simplified morphological model. The model consists of identical, spheroidal particles distributed on the substrate surface, and dipole interactions among the particles and their images in the substrate are included in the EMT. In general, such a model predicts a uniaxial anisotropic film with the optic axis normal to the substrate. The experimentally observed isotropy, however, suggests that one can restrict the interpretation to a single principal component of the effective dielectric function, the one parallel to the substrate plane ($\varepsilon_\parallel$).

The linear regression analysis fits involve three free parameters. The first two parameters characterize the morphology, namely the aluminum volume fraction Q in the particle film, and an interaction parameter $F_\parallel$ that depends on the particle shape, distribution, and Q. The third parameter is a common value of the electron mean free path λ. This parameter serves to reduce the intraband and interband electron relaxation times and takes into account possible electron scattering at defects, surfaces, and grain boundaries. The mean free path defines the particle dielectric function and provides insights into the electronic processes within the particles. The best-fit free parameters are found to provide excellent agreement between the simulated and experimental effective dielectric functions throughout the particle regime.

Although the evolution of the best-fit values of Q with thickness is very similar to that measured for other thin-film systems by real-time SE, the values near coalescence appear to be larger than predicted on the basis of spheroids located on a square grid. This may be an indication that the particles are disk-shaped or hemispherical, rather than spheroidal. In spite of this possible inconsistency with the model, the best fit values of Q and $F_\parallel$, when analyzed within the framework of the generalized EMT, predict particles that flatten as a function of thickness, reaching an approximate axial ratio of 2.5. This prediction is in reasonable agreement with previous direct measurements by electron microscopy. The best-fit Q and $F_\parallel$ also predict near-isotropy for the film over the full range of thickness ($d < 60$ Å) when the particles are distributed on a square array, and this prediction is also in good agreement with observations. In fact, if isotropy is used as a constraint in the generalized EMT, $F_\parallel$ becomes fixed by Q, and good fits to the effective dielectric functions can be obtained with only two free parameters, Q and λ.

All aluminum particles in the discontinuous film regime, whether prepared by evaporation or sputtering, irrespective of rate, exhibit a nearly constant best-fit value of the electron mean free path, $\lambda = 7.5 \pm 2$ Å. Thus, a common dielectric function intrinsic to the particles is observed, independent of their size and shape. As a result, one potential source of error in the optical modeling that results from neglecting particle size and shape distributions is eliminated. For λ values of 7.5 Å, the dielectric functions of the particles are influenced equally by the intraband and interband contributions. Thus, it is important that both contributions be included (and modified by a reduced electron relaxation time) in any analysis of the optical properties of aluminum particle films. The fact that the mean free path is independent of particle size demonstrates that electron scattering at crystalline defects or nanocrystalline boundaries within the particles, rather than at particles surfaces, determines the particle dielectric function.

Another important observation is that, for an aluminum film prepared at high rate showing abrupt coalescence characteristics, the mean free path λ is continuous across the percolation threshold at $d = 45$–50 Å and remains low for a few seconds into the coalescence regime. It should be stressed that because the volume fraction of aluminum in the partially coalesced film is quite high, particles are not present, and a different optical model based on the Bruggeman EMT must be applied to extract λ. Thus, the continuity of λ suggests that the overall optical analyses in both particle and coalescence regimes are correct and that the analysis in the particle regime is not affected by errors, such as those that may result from neglecting distributions in particle shape or size. The continuity in λ, in fact, leads to the reasonable conclusion that the same defective or nanocrystalline-grain structure of the particle also remains in the partially coalesced films for a brief time.

The continuity of λ across the percolation threshold implies that a similar dielectric function is obtained for the partially coalesced film and the aluminum particles. Thus, if the dielectric functions of the particles in the discontinuous aluminum films are constrained to be equal to that of the partially coalesced film, then excellent fits to the effective dielectric functions of aluminum particle films can be obtained over a large range of thickness with only one free parameter, Q. This one free parameter can provide agreement between transmittance spectra deduced from experimental and best-fit effective dielectric functions that differ by no more than 2%. This represents a significant improvement over previous attempts to model transmittance spectra.

For the high-rate aluminum film above the percolation threshold ($d = 45$–50 Å) in the continuous film regime, there is another transition in the optical functions at a thickness of $d = 55$–60 Å. The transition is characterized by nearly an order of magnitude increase in the mean free path, resulting in relaxation times approaching the bulk values. The transition is evident even in the raw ellipsometric data, as the (200) parallel-band feature, invisible at lower thicknesses, makes a rapid appearance. This behavior may be explained by the conversion of the defective or nanocrystalline internal structure into one in which high-quality single-crystalline grains extend through the thickness of the film. For thicknesses above the transition, the mean free path increases more gradually with thickness, exactly as would be expected for electron scattering at the boundaries of grains that increase in size linearly with film thickness.

Future directions will involve additional investigations along these lines to determine whether the overall features described here for aluminum are also observed in other metallic systems. As was done here, the effective optical functions of the metal thin film must be expressed in terms of photon-energy–independent parameters that characterize the morphology and electronic properties. Then linear regression analysis can be employed to extract values for the parameters in fits to spectroscopic ellipsometry data derived from real-time measurements throughout the growth process. The parameters must be checked for consistency with the predictions of simple models of nucleation, coalescence, and bulk film growth and also for consistency with other available direct structural measurements, as well as independent measurements designed to detect film anisotropy.

Acknowledgments

This research was supported by the National Science Foundation under Grant Nos. DMR-8957159, DMR-8901031, and DMR-9217169. This study would not be possible without the financial assistance provided by Philips-DuPont Optical Company and the DuPont Corporation. The authors gratefully acknowledge Dr. G. H. Johnson of DuPont Corporation for his support, his involvement in this research, and his constant encouragement. The authors are also indebted to their collaborators at Penn State, in the research areas of metal film growth and real-time spectroscopic ellipsometry, Profs. R. Messier, L. Pilione, and K. Vedam.

References

1. M. Faraday, *Philos. Trans. R. Soc. Lond.* **147**, 145 (1857).
2. J. C. Maxwell-Garnett, *Philos. Trans. R. Soc. Lond.* **A203**, 385 (1904); **A205**, 237 (1905).
3. R. S. Sennett and G. D. Scott, *J. Opt. Soc. Am.* **40**, 203 (1950).

4. S. Yamaguchi, *J. Phys. Soc. Jpn.* **15**, 1577 (1960); **17**, 184 (1962).
5. A. Meessen, *J. Phys. (Paris)* **33**, 371 (1972).
6. T. Yamaguchi, S. Yoshida, and A. Kinbara, *J. Opt. Soc. Am.* **64**, 1563 (1974).
7. D. Bedeaux and J. Vlieger, *Physica (Utrecht)* **73**, 287 (1974).
8. T. Yamaguchi, H. Takahashi, and A. Sudoh, *J. Opt. Soc. Am.* **68**, 1039 (1978).
9. S. Norrman, T. Andersson, C. G. Granqvist, and O. Hunderi, *Phys. Rev. B* **18**, 674 (1978).
10. U. Kreibig, *Z. Phys. B* **31**, 39 (1978).
11. D. E. Aspnes, E. Kinsbron, and D. D. Bacon, *Phys. Rev. B* **21**, 3290 (1980).
12. D. E. Aspnes, Chapter 15 *in* "Optical Properties of Solids: New Developments" (B. O. Seraphin, ed.), North-Holland, Amsterdam, 1976.
13. Y.-T. Kim, R. W. Collins, and K. Vedam, *Surf. Sci.* **233**, 341 (1990).
14. I. An, Y. M. Li, H. V. Nguyen, and R. W. Collins, *Rev. Sci. Instrum.* **63**, 3842 (1992).
15. H. V. Nguyen, I. An, and R. W. Collins, *Phys. Rev. Lett.* **68**, 994 (1992).
16. H. V. Nguyen, I. An, and R. W. Collins, *Phys. Rev. B* **47**, 3947 (1993).
17. H. V. Nguyen and R. W. Collins, *J. Opt. Soc. Am. A* **10**, 515 (1993).
18. N. W. Ashcroft and K. Sturm, *Phys. Rev.* **3**, 1898 (1971).
19. U. Kreibig, *J. Phys. F* **4**, 999 (1974).
20. G. A. Niklasson, D. E. Aspnes, and H. G. Craighead, *Phys. Rev. B* **33**, 5363 (1986).
21. G. A. Niklasson, C. G. Granqvist, and O. Hunderi, *Appl. Opt.* **20**, 26 (1981).
22. D. E. Aspnes, *Thin Solid Films* **89**, 249 (1982).
23. D. E. Aspnes, *Proc. Soc. Photo-opt. Instrum. Eng.* **276**, 188 (1981).
24. For a review, see D. Y. Smith, E. Shiles, and M. Inokuti, *in* "Handbook of Optical Constants of Solids" (E. D. Palik, ed.), Academic Press, Orlando, Florida, 1985, p. 369; D. Y. Smith and B. Segall, *Phys. Rev. B* **34**, 5191 (1986).
25. N. W. Ashcroft, *Philos. Mag.* **8**, 2055 (1963).
26. L. W. Bos and D. W. Lynch, *Phys. Rev. Lett.* **25**, 156 (1970).
27. F. Wooten, "Optical Properties of Solids," Academic Press, New York, 1972.
28. E. Shiles, T. Sasaki, M. Inokuti, and D. Y. Smith, *Phys. Rev. B* **22**, 1612 (1980).
29. U. Kreibig and C. von Fragstein, *Z. Phys.* **224**, 307 (1969).
30. A. F. Mayadas and M. Schatzkes, *Phys. Rev. B* **1**, 1382 (1970).
31. G. F. Feng and R. Zallen, *Phys. Rev. B* **40**, 1064 (1989).
32. H. V. Nguyen and R. W. Collins, *Phys. Rev. B* **47**, 1911 (1993).
33. D. M. Wood and N. W. Ashcroft, *Phys. Rev. B* **25**, 6255 (1982).
34. For a review, see U. Kreibig and L. Genzel, *Surf. Sci.* **156**, 678 (1985).
35. C. A. Neugebauer, Chapter 8 *in* "Handbook of Thin Film Technology" (L. I. Maissel and R. Glang, eds.), McGraw-Hill, New York, 1970.
36. J. R. Blanco, P. J. McMarr, K. Vedam, and J. M. Bennett, *in* "Multiple Scattering of Waves in Random Media and Random Rough Surfaces" (V. V. Varadan and V. K. Varadan, eds.), The Pennsylvania State University, University Park, 1987, p. 679.
37. R. H. Doremus, *J. Appl. Phys.* **37**, 2775 (1966).
38. P. N. Sen, C. Scala, and M. H. Cohen, *Geophysics* **46**, 781 (1981).
39. D. E. Aspnes, *Phys. Rev. B* **33**, 677 (1986).
40. B. T. Sullivan and R. R. Parsons, *J. Vac. Sci. Technol. A* **5**, 3399 (1987).
41. T. Yamaguchi, S. Yoshida, and A. Kinbara, *Thin Solid Films* **21**, 173 (1974).
42. J. A. Osborn, *Phys. Rev.* **67**, 351 (1945).
43. E. C. Stoner, *Philos. Mag.* **36**, 803 (1945).
44. D. N. Jarrett and L. Ward, *J. Phys. D* **9**, 1515 (1976).
45. See, for example, U. Kreibig, B. Schmitz, and H. D. Breuer, *Phys. Rev. B* **36**, 5027 (1987).

46. R. M. A. Azzam and N. M. Bashara, "Ellipsometry and Polarized Light," North-Holland, Amsterdam, 1977.
47. N. V. Nguyen, B. S. Pudliner, I. An, and R. W. Collins, *J. Opt. Soc. Am. A* **8**, 919 (1991).
48. J. H. Halford, F. K. Chin, and J. E. Norman, *J. Opt. Soc. Am.* **63**, 786 (1973).
49. D. E. Aspnes and A. A. Studna, *Phys. Rev. B* **27**, 985 (1983).
50. I. H. Malitson, *J. Opt. Soc. Am.* **55**, 1205 (1965).
51. J. R. Black, *IEEE Reliability Phys. Conf.* **15**, 257 (1977).
52. Y. E. Strausser and K. S. Majumder, *J. Vac. Sci. Technol.* **15**, 238 (1978).
53. R. J. Blattner and A. J. Braundmeier, *J. Vac. Sci. Technol.* **20**, 320 (1982).
54. R. Brendel and R. Hezel, *J. Appl. Phys.* **71**, 4377 (1992).
55. A. G. Mathewson and H. P. Myers, *Phys. Scr.* **4**, 291 (1971); *J. Phys. F* **2**, 403 (1972).
56. H. Arwin and D. E. Aspnes, *Thin Solid Films* **113**, 101 (1984).
57. Y. Yagil and G. Deutscher, *Thin Solid Films* **152**, 465 (1987).
58. S. Kar, R. Varghese, and S. Bhattacharya, *J. Vac. Sci. Technol. A* **1**, 1420 (1983).
59. R. Meservey and P. M. Tedrow, *J. Appl. Phys.* **42**, 51 (1971).
60. J. C. Miller, *Philos. Mag.* **20**, 1115 (1969).
61. L. G. Bernland, O. Hunderi, and H. P. Myers, *Phys. Rev. Lett.* **31**, 363 (1973).
62. M. Takagi, *J. Phys. Soc. Jpn.* **9**, 359 (1954).
63. P. Buffat and J.-P. Borel, *Phys. Rev. A* **13**, 2287 (1976).
64. S. Krishnan and P. C. Nordine. *Phys. Rev. B* **47**, 11780 (1993).
65. W. Buckel, *Z. Phys.* **138**, 136 (1954); H. Bulow and W. Buckel, *Z. Phys.* **145**, 141 (1956).
66. S. Roberts and P. J. Dobson, *Thin Solid Films* **135**, 137 (1986).
67. M. L. Jones, H. H. Soonpaa, and B. S. Rao, *J. Opt. Soc. Am.* **64**, 1591 (1974).
68. E. Dobierzewska-Mozrzymas, A. Radosz, and P. Bieganski, *Appl. Opt.* **24**, 727 (1985).
69. C. G. Granqvist and O. Hunderi, *J. Appl. Phys.* **51**, 1751 (1980).

Optical Characterization of Inhomogeneous Transparent Films on Transparent Substrates by Spectroscopic Ellipsometry

P. Chindaudom and K. Vedam

Materials Research Laboratory and Department of Physics
Pennsylvania State University, University Park, Pennsylvania

I. Introduction 191
II. Inhomogeneity in Thin Films 193
III. Experimental 196
   A. Spectroscopic Ellipsometry 196
   B. RAE with a Compensator 198
   C. Detector Nonlinearity 202
   D. Data Analysis 205
   E. Measurements on Vitreous Silica as a Test of Accuracy
      of our Modified SE System 207
IV. Experimental Results on Transparent Films 215
   A, Particulars of Films Studied 215
   B. Preliminary Results on MgO and LaF_3 Films 216
   C. Results on Transparent Films on Vitreous Silica Substrates . . 220
      1. Films with a Micro-rough Surface Overlayer 221
      2. Smooth Films on Top of an Interface between Substrate
         and Film 227
      3. Films That Can Be Modeled as Composed of Three-Layers . . 230
V. Discussion 231
   A. General 231
   B. Effect of Surface Roughness 233
   C. Optical Functions of the Dielectric Film Materials 234
      1. Comparison with Previous Results 237
VI. Summary and Conclusion 243
   References 244

I. Introduction

It is well known that the quality of the films prepared by the standard thermal or electron beam evaporation techniques are strongly dependent on the various parameters of the preparatory conditions. In fact, even if such parameters are not varied, the characteristics of the films prepared in subsequent runs of the same preparatory system are rarely reproducible to the degree of perfection required in modern high-tech materials and systems. The main reason for this state of affairs is the inhomogeneity introduced in these films during film growth. Here the term inhomogeneity covers all the chemical, morphological, and micro- and nanostructural features of the films. The inhomogeneities in thin films have also played a major role in our inability to determine the true optical functions $[n(\lambda) + ik(\lambda)]$ of many materials that are available only in thin-film forms. Hence, if we are to prepare reproducibly high-quality films with the desired characteristics, or if we are interested in fundamental properties such as the optical function of the film material, then we must first understand the origin or the cause of the inhomogeneity in the films. Only then can we meaningfully develop suitable methods to overcome or circumvent this problem of inhomogeneity. This is addressed in Section II of this article. When once the film is prepared, it must be characterized nondestructively and non-invasively, and if possible in real time, so that the film can be modified *in situ* as desired as it is grown.

Of the various techniques currently available for such studies, only optical techniques satisfy the criteria just mentioned. Of the available optical techniques, spectrophotometry which measures the reflectance and the transmittance as a function of wavelength has been extensively used in the past for this purpose. However, closer examination of these studies reveal that most of them invoke the tacit assumption that the film is homogeneous before analyzing the measured data. In the recent past there have been a few spectrophotometric studies (1–4) wherein some efforts were made to detect and characterize the distribution of inhomogeneities in the thin films. But even these studies suffer from a serious drawback in that they had to make an arbitrary assumption that the inhomogeneity varies linearly across the thickness of the film, before they could analyze the spectrophotometric data. A few moments' reflection will immediately reveal that such an assumption is not justified in many cases where the inhomogeneity arises just from the fact that the surface of the film is not perfectly smooth. In such cases the film can be considered as made up of a perfect film covered with an overlayer of micro-rough surface. Obviously the assumption of linear variation of inhomogeneity in the film is not valid in such cases. Here it is relevant to point out a recent round-robin study conducted by the Optical Society of America (3), which unequivocally demonstrated that even for nominally identical films, the evaluated values of the optical constants can be quite different, depending on the model assumed for the films for the analysis of the spectrophotometric data.

An optical technique that is capable of solving this problem is spectroscopic ellipsometry (SE), by virtue of its ability to determine two independent parameters—namely, the relative changes in the amplitude and phase of the light on reflection from the film at each wavelength—as against the spectrophotometric technique, which measures just the changes in intensity (or the amplitude) alone as a function of wavelength. However, till now SE was not used for such studies on optically transparent films because of two inherent difficulties encountered in such SE measurements: (i) the near zero (or 180°) relative phase change in the ellipsometric parameter Δ introduced by the transparent samples, and (ii) the very low reflectance of these samples. De Nijs and van Silfhout (5) have recently shown that the experimental errors involved in measuring the ellipsometric parameters Δ and Ψ are proportional to $(\sin \Delta)^{-1}$ and $(\sin 2\Psi)^{-1}$. However, as will be shown later in this article, the spectral variation of Δ contains information on the inhomogeneity in the film, and hence it is imperative that Δ be measured accurately, even though it may be close to 0° (or 180°). Both these problem areas have been successfully overcome by the present authors, with the result that we can now detect the presence of inhomogeneity in the film and also depth-profile the film nondestructively in order to determine the distribution of the inhomogeneity across the thickness of the film. In addition, it has also been possible to determine the optical function of the film material at the same time. Section III of this article describes the various modifications and corrective procedures that had to be incorporated into the SE system to obtain reliable and meaningful SE data on transparent films on transparent substrates, and also the special data analyis procedures that had to be adopted for these cases. Section IV describes the results of measurements on a bulk vitreous silica sample in order to test the reliability and accuracy of the modified SE system. The final results obtained on a number of fluoride and oxide optical coatings are then presented and discussed in Section V. Section VI presents the summary and conclusion.

II. Inhomogeneity in Thin Films

In general, inhomogeneities in thin films deposited from the vapor phase arise from the following:
   (i) microstructure or porosity in the film:
      (a) columnar structure,
      (b) orientation of columns,
      (c) "form birefringence" due to columnar structure, and
      (d) nanostructure within the columns;
   (ii) chemical inhomogeneity;

(iii) phase or crystallographic inhomogeneity;
(iv) stress in the film and hence the stress birefringence;
(v) adsorption of water in the micropores in the film; and
(vi) the surface micro-roughness.

Of these six contributors to the total inhomogeneity of the film, microstructure plays the dominant role in the physical (including the optical) properties exhibited by the film. The reason for this becomes obvious from Fig. 1, which is a schematic diagram of the general physical structure of a thin film prepared under low-mobility conditions. Figure 1 shows the scanning electron micrograph, transmission electron micrographs and field ion micrograph of a sputtered amorphous-Ge film. Almost similar features have been obtained with metallic (6) as well as dielectric (ceramic) materials (7, 8), provided the normalized substrate temperature T_s/T_m is <0.3. Here T_s represents the temperature (in kelvins) of the substrate during the deposition of the film, and T_m the melting point of the bulk film material. In this connection it is relevant to draw attention to the structure zone model (SZM) originally proposed by Movchan and Demchishin (6) in 1969 to explain the origin of microstructure of the film in terms of a single deposition parameter. Since then this model has been extended by a number

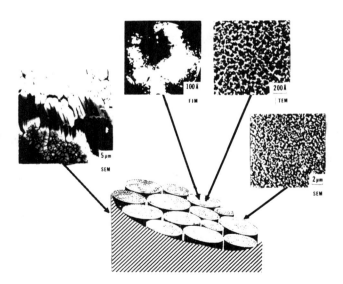

FIG. 1. Schematic representation of the general physical structure of a thin film prepared under low-mobility conditions. The scanning electron, transmission electron, and field ion micrographs are for sputterd amorphous-Ge films. After Messier et al. (10).

of workers (9–12) to include additional process parameters such as the residual gas pressure during sputter deposition of the thin film, and the bombardment energy of ionized or otherwise accelerated particles. Figure 2 shows the most recent generalized SZM model of thin films proposed by Guenther (13), wherein the total particle energy is represented by its thermal equivalent T_s and the melting point of the film material T_m represents the influence of the activation energy for surface self-diffusion. As a consequence of our understanding of the origin of the microstructure of the film, considerable effort (14–17) has been devoted to find ways and means to eliminate or at least minimize the microstructure. For example, the development of the various plasma-assisted techniques, such as ion-assisted deposition (IAD) (18), ionized cluster beam (ICB) (19), and reactive low-voltage ion plating (RLVIP) (20,21) techniques, can be attributed to this goal, and this has led to the production of nearly perfect films with minimal microstructure (13, 22, 23).

All the dielectric films that will be discussed in this article were deposited in high vacuum on heated vitreous silica substrates (at ~225°C) by the electron beam evaporation technique. For such films it is obvious from Fig. 2 that the important parameter T_s/T_m (the normalized substrate temperature) is <0.3, and hence the films will have porous columnar structure. Since

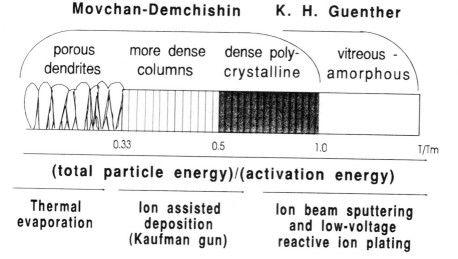

FIG. 2. Extended and generalized structure zone model of thin growth. The total particle energy is represented by its thermal equivalent T_s, and the melting point of the film material T_m reparesents the influence of the activation energy. After Guenther (12).

the film growth begins from the nucleation at the high points of the substrate, and taking into account the shadowing of the low areas, the columnar microstructures of these films can be represented by the schematic diagram (*11*) shown in Fig. 3. It shows that in general these films can be modeled as composed of three-layers, i.e., a substrate/film interface layer, with some voids, below a nearly void-free middle layer and a micro-rough surface overlayer. Such a model will be used to analyze the SE experimental data.

III. Experimental

A. Spectroscopic Ellipsometry

As mentioned in the introduction, the only technique that can nondestructively detect and characterize the distribution of the inhomogeneities in a transparent film, and also determine the optical function of the film material at the same time, is spectroscopic ellipsometry. Of late, SE has become a popular technique, and numerous articles (*24–33*) have appeared in the literature describing the principles, the experimental details, and the error correction and data analysis procedures. Hence, in what follows, only particulars relevant to the study of transparent films on transparent substrates will be described.

A schematic diagram of a rotating analyzer ellipsometer (RAE) is shown

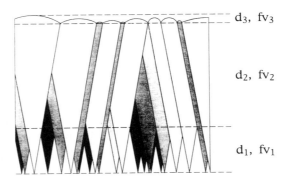

FIG. 3. Model of cross-section of film showing nucleation and growth features.

in Fig. 4, and Fig. 5 shows the RAE in the Jones matrix representation of its various optical components. A monochromatic light beam is first linearly polarized, and then allowed to be incident on the sample under study at the chosen angle of incidence. The reflected light, which in general will be elliptically polarized, is then passed through an analyzer rotating at a constant frequency ω. The sinusoidal signal measured by the detector at any instant is given by (26,27)

$$I = I_0(1 + \alpha \cos 2\omega t + \beta \sin 2\omega t), \tag{1}$$

where I_0 is the average intensity of the signal, and α and β are the normalized Fourier coefficients. Traditionally the ellipsometric measurements are expressed in terms of the ellipsometric parameters Δ and Ψ, which are given by the complex ratio of the Fresnel reflection coefficients:

$$\rho = r_p/r_s = \tan \Psi e^{i\Delta}$$

$$= \frac{(\cot Q - ia)}{(1 + ia \cot Q)} \tan P. \tag{2}$$

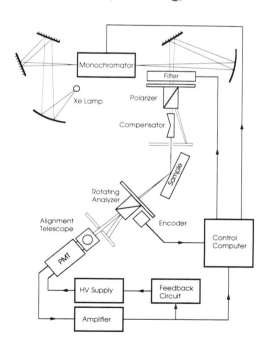

FIG. 4. Schematic diagram of a rotating analyzer ellipsometer (RAE) system.

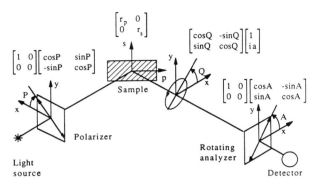

FIG. 5. The corresponding Jones matrix representation for the optical components of the RAE shown in Fig. 4.

In Eq. (2), r_p and r_s refer to the Fresnel reflection coefficients of the parallel and perpendicular polarization components of the reflected light; P is the azimuth angle of the polarizer; and Q and a are parameters characterizing the elliptically polarized light incident on the rotating analyzer. As has been shown by Aspnes (26, 27), a, the minor/major-axis ratio (with $|a| \leq 1$), and the azimuth angle Q of its major axis (with $|Q| \leq \pi/2$) can be described in terms of the experimentally determined parameters α and β, evaluated from Eq. (1) as

$$Q = \frac{1}{2}\tan^{-1}\left(\frac{\alpha}{\beta}\right) + \frac{\pi}{2}u(-\alpha)\,\text{sgn}(\beta), \tag{3a}$$

$$a = \frac{\pm\sqrt{1-\xi^2}}{1+\xi}, \tag{3b}$$

where $u(x) = 0$ for $x < 0$ and 1 for $x \geq 0$, $\text{sgn}(x) = -1$ for $x < 0$ and $+1$ for $x \geq 0$, and $\xi = \sqrt{\alpha^2 + \beta^2} \geq 0$.

B. RAE WITH A COMPENSATOR

De Nijs and van Silfhout (5) have shown that the experimental errors in the measurement of Δ and Ψ by RAE are proportional to $1/\sin\Delta$. Thus, in ellipsometric measurements of a transparent surface or a transparent thin film on transparent substrate, the experimental errors can be unacceptably large because the values of Δ for these samples are near 0° or 180° and hence

$\sin \Delta \approx 0$. Very recently the present authors have developed (34) a new variation of the preceding SE method by using an achromatic compensator to artificially shift the near 0° (or 180°) phase change to near 90°, in order to satisfy the optimum condition for rotating analyzer ellipsometry (27, 35), and thus were able to successfully characterize a thin film of LaF_3 on a vitreous silica substrate (36).

The basic design of the three-reflection type achromatic compensator (37–40) used in this study is shown in Fig. 6. This was chosen over the more well-known Babinet–Soleil compensator, first because it provides an $\sim 90°$ phase shift automatically throughout the spectral range of our measurement, i.e., between 300 and 700 nm, as shown by the value of Δ for the achromatic compensator in Fig. 7. Secondly, since this is made of vitreous silica, it does not suffer from the optical activity problems associated with the α-quartz used in the Rochon prisms and the Babinet–Soleil compensator. Since detailed descriptions of this compensator are available in the literature (37–40), it will not be discussed any further here. The achromatic compensator was mounted (i.e., allowed to rest) on a removable base so that the unit could be quickly and kinematically installed in the ellipsometer system (or removed from it) without altering the system alignment. It was noticed that any pressure from the mounting on the compensator could hinder the performance of this achromatic compensator because of the stress birefringence caused by even the slightest mechanical stress applied directly to the body of the achromatic compensator. The achromatic compensator was initially aligned in the straight-through position by adjusting the micropositioner of the compensator mounting so that the light was not deviated. Then, the fine tuning of the compensator alignment was accomplished by comparing the light intensities of the two corresponding analyzer positions between two consecutive optical cycles with the help of an oscilloscope.

First, the analyzer arm of the ellipsometer is swung to the straight-through position, the achromatic compensator inserted in the polarizer arm,

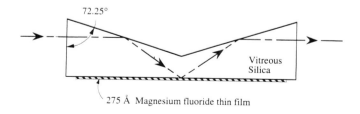

FIG. 6. Achromatic three-reflection quarter wave system. After King and Downs (39).

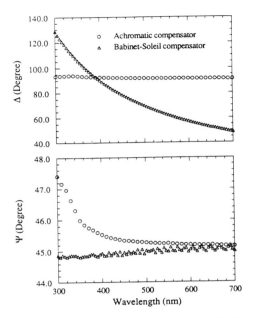

FIG. 7. Ellipsometric parameters Δ and Ψ of the Babinet–Soleil compensator and the achromatic compensator as a function of wavelength, measured in the PCA configuration.

aligned as shown schematically in Fig. 8a, and the ellipsometric parameters Δ_1 and Ψ_1 are measured over the spectral range of interest in the straight-through polarizer–compensator–analyzer (PCA) configuration. From Eq. (2) the ellipsometric equation can then be written as

$$\rho_1 = \frac{r_{1p}}{r_{1s}} = \frac{E_{1p}}{E_{1s}/E_s} \frac{E_p}{,} \tag{4a}$$

$$= \tan \Psi_1 e^{i\Delta_1} = \frac{(\cot Q_1 - ia_1)}{(1 + ia_1 \cot Q_1)} \tan P, \tag{4b}$$

where E_{1p} and E_{1s} are, respectively, the p and s components of the light transmitted by the compensator. This allows the calibration of the compensator and also provides a complete description of the state of polarization of the light striking the sample surface at every wavelength in the subsequent

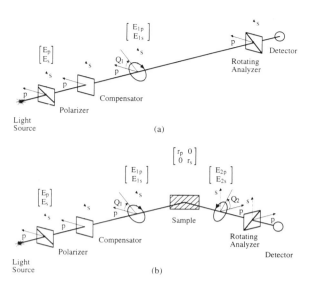

FIG. 8. Schematic diagrams showing the ellipsometer arrangements for the measurement with a compensator: (a) PCA configuration and (b) PCSA configuration.

experiment. In other words, this measurement can be treated as the *calibration of the effective light source.*

Next, the sample is inserted and aligned at the desired angle of incidence in the usual polarizer–compensator–sample–analyzer (PCSA) ellipsometric configuration, as in Fig. 8b. Then the newly determined ellipsometric parameters Δ_2 and Ψ_2 over the same wavelength range provide two spectra containing the lumped information on the optical properties of the compensator and the sample. As before, E_{2p} and E_{2s} are, respectively, the p and s components of the light reflected by the sample. Then Eqs. (4a) and (4b) can be rewritten as

$$\rho_2 = \frac{r_{2p}}{r_{2s}} = \frac{E_{2p}/E_p}{E_{2s}/E_s}, \tag{4c}$$

$$= \tan \Psi_2 e^{i\Delta_2} = \frac{(\cot Q_2 - ia_2)}{(1 + ia_2 \cot Q_2)} \tan P. \tag{4d}$$

Since the 90° phase shift of the compensator adds to the near-zero Δ of the sample, the measured Δ_2 is also near 90°. Further, since both the foregoing measurements were made under optimum conditions *(27,35)* for Δ (i.e., Δ near 90°), the errors in both measurements were minimal.

From the definitions, the p and s component reflection coefficients of the sample alone are $r_p = E_{2p}/E_{1p}$ and $r_s = E_{2s}/E_{1s}$, respectively. Then the desired true Δ and Ψ of the sample at any desired wavelength can be evaluated by dividing Eq. (4c) by Eq. (4a) and Eq. (4d) by Eq. (4b):

$$\rho = \frac{r_p}{r_s} = \tan\Psi e^{i\Delta} = \frac{E_{2p}/E_{1p}}{E_{2s}/E_{1s}} = \frac{\tan\Psi_2}{\tan\Psi_1} e^{i(\Delta_2 - \Delta_1)}$$

$$= \frac{(\cot Q_2 - ia_2)(1 + ia_1 \cot Q_1)}{(\cot Q_1 - ia_1)(1 + ia_2 \cot Q_2)}, \quad (5)$$

where ρ is the complex reflectance ratio of the sample. It is seen that ρ, is independent of the settings of the achromatic compensator and the polarizer. Therefore, exact calibration for the precise knowledge of the achromatic compensator and the polarizer azimuth angles is not as critical as in the conventional compensator method. Further, it is evident that this technique is also insensitive to the errors due to the compensator imperfections, provided that the compensator is not disturbed between the two consecutive measurements described in the last two paragraphs.

Figure 7 compares the results of straight-through PCA measurements on a Babinet–Soleil compensator at a fixed setting and the achromatic compensator employed in this study. The Δ spectra show that the retardation of the Babinet–Soleil compensator varies from 50° to 130° for a fixed setting, whereas the achromatic compensator produces an almost constant near-90° phase shift over the wavelength range from 280 to 700 nm without any further adjustment. Consequently, for automated spectroscopic ellipsometry, the advantage of using the achromatic compensator is obvious. The structures in both Ψ spectra suggest that there are wavelength-dependent errors in both compensators, since Ψ should nearly be constant for an ideal compensator. Aspnes (41) has discussed the wavelength-dependent errors due to compensator imperfections. It is evident that these errors can cause an error in Ψ of the experimental sample ($\delta\Psi > 1°$), if not corrected properly. On the other hand, the measurement technique described in this section effectively compensates for such errors at each measured wavelength.

C. Detector Nonlinearity

As mentioned in the introduction, the second major obstacle facing SE measurements on transparent substrates and films is the low reflectance of the experimental samples. In other words, the low signal-to-noise ratio in these measurements demands that we examine the source of every possible error in the system and institute appropriate error correction procedures.

Many of these errors in the RAE system have been examined in detail, and appropriate methods have been developed to overcome or minimize these errors (5, 24–32). But this approach becomes more difficult as the errors are amplified rapidly when Δ approaches 0° or 180°. As discussed in Section III.B, the use of an achromatic compensator in the SE system artifically shifts the Δ values to ~90°, and hence the preceding corrective procedures can be employed with good results.

Accurate calibration and appropriate corrections of errors due to any imperfection in the various optical elements are necessary prerequisites for any high-precision measurements. The effects of component optical activity of the polarizer and analyzer elements in the rotating-analyzer ellipsometer, as well as the appropriate calibration procedures, have been discussed in detail by Aspnes (26, 27) and hence will not be repeated here. In view of the low reflectivity of transparent samples (<5%), other problems such as the ambient light and the nonlinearity of the detector are also critical (32, 41, 42).

The theoretical irradiance at the detector, given by Eq. (1), contains both dc and ac components with respect to the analyzer position. In practice, the light intensity is converted to an electrical signal by a signal-processing circuit with some filtering to reduce noise. Thus, while the static and dynamic intensity levels would be detected with the same sensitivity by an ideal detector, in practice the general expression for the measured intensity can be represented as (27, 32)

$$I = I_0[1 + \eta(\alpha' \cos 2A + \beta' \sin 2A)], \tag{6}$$

where the parameter η is the correction term to take into account the transfer characteristics of the signal processing circuit. η is related to the ratio of the gain for the dc to that for the ac signals and in general is not equal to unity. α' and β' are the measured normalized Fourier coefficients. Thus, α and β in Eq. (1), corresponding to the case of an ideal detector, should be substituted by $\eta\alpha'$ and $\eta\beta'$, respectively, for use with real systems.

Equation (6) was first introduced by Aspnes (27) assuming a linear detector. Aspnes also suggested a calibration procedure, known as residual calibration, to determine simultaneously a number of calibration parameters including η. De Nijs and co-workers (5) have recently shown that the residual method is inaccurate when Δ of the sample is near zero, and they suggested an alternative method called the phase calibration procedure. Nevertheless, both methods require a number of measurements to be performed at each wavelength, and thus become inconvenient if η is wavelength-dependent. Indeed, we have found that η varies significantly, especially at both ends of the spectral range where the light-source intensity is low.

We have measured η directly as a function of photomultiplier voltage, as follows. The light beam was allowed to pass through the polarizer set at $P = 45°$ in the straight-through position. Following the normal procedure, the polarization state of the light passing through the polarizer is characterized by the rotating analyzer to obtain the Fourier coefficients α and β. Neglecting the optical activity of the polarizer and analyzer, from Eqs. (2) and (3), the mode parameter a for the linearly polarized light must be zero (i.e., $\sqrt{\alpha^2 + \beta^2} = 1$), while Q defines the azimuth angle of the polarizer. Finally, using $\alpha = \eta\alpha'$ and $\beta = \eta\beta'$, η can be evaluated through the expression

$$\eta = \frac{1}{\sqrt{\alpha'^2 + \beta'^2}}. \tag{7}$$

The average intensity of light reaching the photomultiplier tube (PMT) was varied by adjusting a variable neutral-density filter positioned before the polarizer perpendicular to the light beam. With the feedback loop active, the PMT voltage was allowed to vary from 600 to 1,000 V, corresponding to the variation of the average intensity caused by the neutral density filter. The values of η calculated by Eq. (7) are plotted as a function of PMT voltage in Fig. 9. This plot suggests a linear relationship between η and the PMT voltage. These data were fitted to a linear equation and the values of the fitting parameters were stored in the computer, and hence the correct η can be calculated for any PMT voltage corresponding to different wavelengths or sample characteristics. Such calculated η values for particular PMT voltages can be used in the evaluation of

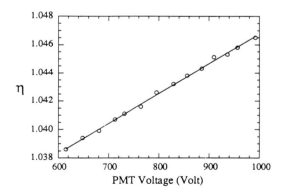

FIG. 9. The η parameter as a function of PMT voltage.

Δ and Ψ at any wavelength. It should be mentioned here that, since accurate data for α' and β' are required in the calculation of η values, the calibration procedure for η should be preceded by a background noise correction. It has been observed that without the proper correction for background noise, the η plot curves upwards at high PMT voltages.

D. Data Analysis

The SE data obtained on a number of transparent film samples were then individually analyzed following the usual procedure (29, 32, 33) using linear regression analysis (LRA) to simultaneously depth-profile the film for the distribution of voids and determine the optical function of the film material. In brief, the procedure involves constructing realistic model based on Fig. 3 for the film under study and fitting the calculated [$\Delta(\lambda)$, $\Psi(\lambda)$] spectra of the model to the corresponding experimental [$\Delta(\lambda)$, $\Psi(\lambda)$] spectra. Starting with the simplest structure, such as a homogeneous single-layer model (SLM), the complexity of the model is gradually increased as needed to take into account the inhomogeneity of the film by subdividing the film into multilayers with varying fractions of void (air) content. The selection of the final model was based on the *simultaneous* fulfillment of the following five criteria: (i) the model must be physically realistic; (ii) must have a low value of σ, the unbiased estimator of the goodness of fit; (iii) there are reasonably low values of the 90% confidence limits of the variable parameters; (iv) there is good agreement between the calculated and the observed (Δ, Ψ)λ spectra over the entire spectral region studied; and (v) there are acceptably low values of the cross-correlation coefficients between the evaluated parameters. The confidence limits not only give some insight as to how well the chosen model fits the data, but also provide information as to whether the experimental data are really determining the wavelength-independent parameter values, or whether too many parameters have been used. Thus, LRA ensures that too many parameters or strongly correlated parameters result in drastic increases in the confidence limits. Thus, the confidence limits provide a built-in check against the tendency to add parameters indiscriminately in order to reduce the value of the unbiased estimator of the goodness of fit σ. Here

$$\sigma = \frac{1}{\sqrt{m-p-1}} \left[\sum_{i=1}^{m} \left\{ (\Delta_{i,\exp} - \Delta_{i,\text{cal}})^2 + (\Psi_{i,\exp} - \Psi_{i,\text{cal}})^2 \right\} \right]^{1/2}. \quad (8)$$

Here $\Delta_{i,\exp}$, $\Psi_{i,\exp}$ are the measured ellipsometric data: $\Delta_{i,\text{cal}}$, $\Psi_{i,\text{cal}}$ are the calculated values of the particular model selected; m is the number of independent measurements corresponding to the different wavelengths (>80) at which the measurements were made; and p is the number of the unknown parameters. The unknown parameters for these films are, for example, the thickness of each sublayer, the volume fraction of incorporated voids in each sublayer, and the optical function of the film material. The optical function of these transparent film materials can be described by the simplest equation, such as the well-known Sellmeier dispersion equation,

$$n^2(\lambda) = A + B\lambda^2/(\lambda^2 - \lambda_0^2), \qquad (9)$$

where λ is the wavelength in nanometers, λ_0 is the wavelength of the effective oscillator, B the oscillator strength and A the contribution of the far ultraviolet terms. A, B and λ_0 are the unknown parameters to be determined in the modelling process. Bruggeman's (43) effective medium theory was used to calculate the effective refractive index of the composite layer of film material and voids (or air). Malitson's (44) data on the dispersion of the refractive index of vitreous silica were used for the substrate in the preceding computations.

Before describing the results, it is necessary to emphasize the importance of using the proper error function in analyzing the SE data obtained on transparent samples or transparent films on transparent substrates using the linear regression analysis technique (29,32). The proper form for evaluating the unbiased estimator σ for modeling the SE data has been investigated by many workers. Modeling with either $\sigma(\cos\Delta, \tan\Psi)$ or $\sigma(\alpha, \beta)$ was recommended by Kim and Vedam (45) in the cases of semiconductor and metals. These workers did not consider transparent samples. Very recently the present authors have examined this problem (36) for the case of LaF_3 film on vitreous silica substrate in some detail, and hence their discussion will not be repeated here. But in what follows we will just summarize their conclusions and list the advantages of using $\sigma(\Delta, \Psi)$ over either $\sigma(\cos\Delta, \tan\Psi)$ or $\sigma(\alpha, \beta)$ in the case of transparent films on transparent substrates.

Since Δ for transparent materials is expected to be near $0°$ or $180°$ over the entire spectral range for transparent materials, $\cos\Delta$ will be approximately ± 1.0 and becomes insensitive to the small variation of Δ. Consequently, the use of ($\cos\Delta$ and $\tan\Psi$) instead of (Δ and Ψ) in computations using error function equations similar to Eq. (8) puts great weight on $\tan\Psi$ and essentially ignores $\cos\Delta$. Thus, in real films, which are invariably inhomogeneous, information about the microstructural inhomogeneities is often lost in such computations based on $\sigma(\cos\Delta, \tan\Psi)$. Hence, in the case of transparent materials, the proper function for the unbiased estimator σ that should be used is Eq. (8), which is based on $\sigma(\Delta, \Psi)$. Furthermore, when $\sigma(\Delta, \Psi)$ is used, the computations yield the final results of σ in degrees, and thus provide a clue as to when we should stop increasing the complexity of the model. For example, when σ is of the same order of magnitude as the experimental errors in Δ and Ψ, the evaluated parameters are also of the same order of magnitude as the experimental errors in Δ and Ψ. Then there is no justification for increasing the complexity of the model by adding more parameters just to decrease the value of σ. Some of these features will become evident when the results on vitreous silica and LaF$_3$ film are discussed in Sections III.E and IV.B, respectively.

E. Measurement on Vitreous Silica as a Test of Accuracy of Our Modified System

Before undertaking any SE measurements on dielectric films, it was felt to be desirable first to carry out measurements on a transparent solid of known optical function, say, vitreous silica, to check the accuracy, reliability, and limitations of our SE system and of the data analysis procedures. Vitreous silica was selected as the test material (34) because accurate data on its optical properties over a wide spectral range are available in the literature (44). Since vitreous silica is optically transparent, it is important to insure that reflection from the back surface of the sample is not picked up by the detector. A large piece of vitreous silica sample, 3.0 cm diameter by 0.8 cm thick, was cut and polished with the backside in the shape of a wedge to eliminate the reflection from the backside of the sample. A low-energy He–Ne laser beam coincident with the incident light beam of the ellipsometer was very helpful in locating the reflected beams from the back and the front faces and in precisely aligning the sample on the ellipsometer.

The vitreous silica sample was measured at a 70° angle of incidence by the standard RAE method both with and without the use of the achromatic compensator as described in Section III. The SE data obtained are plotted as a function of wavelength in Fig. 10. The exact ellipsometric parameters were also calculated by using Eq. (5) and the reference data on the spectral variation of the refractive index of vitreous SiO$_2$ determined by Malitson (44). For comparison, such calculated Ψ values are also plotted as solid curves in Fig. 10. Since the vitreous silica sample is transparent in the wavelength region studied ($k < 10^{-6}$), the Δ values over this wavelength range are expected to be zero ($< 10^{-4}$ deg). It is immediately evident from Fig. 10 that the precision of near-zero Δ over the entire wavelength range is improved dramatically by using the compensator, whereas the precision of Ψ remains the same. The root-mean-square (rms) deviation of Ψ is less than 0.02°. These results indicate that Ψ can be determined by RAE with similar

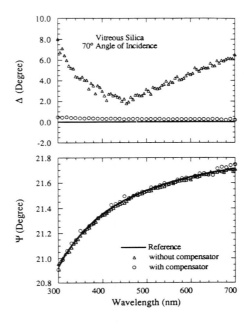

FIG. 10. Ellipsometric parameters (Δ, Ψ) of vitreous silica measured with and without the achromatic compensator. After Chindaudom and Vedam (34).

precision with or without the use of a compensator; on the other hand, measurements on Δ by RAE without a compensator will yield erroneous results. In fact, it is seen from Fig. 10 that the measured values of Δ are greater than 2.5° without the use of the compensator, as opposed to less than 0.5° when the achromatic compensator is used.

The effective optical constants $\langle n, k \rangle$ of vitreous silica can be evaluated from the measured SE data (Fig. 10) using direct inversion of (Δ and Ψ) *assuming* that the experimental sample is truly homogeneous and smooth. We have used the term "effective optical constants" because these direct inversion calculations were made without any corrections for the possible presence of micro-rough surface on the sample. Figure 11 is a plot of the evaluated $\langle n \rangle$ and $\langle k \rangle$ of vitreous silica as a function of wavelength. The solid curve is the plot of the reference refractive index of vitreous silica (44). It can be seen that the values of $\langle n \rangle$ determined by RAE with and without the compensator are comparable, and furthermore, their rms deviations from the reference data on n are <0.001. On the other hand, the large error in Δ in the measurements without the compensator (see Fig. 10) results in a large error in $\langle k \rangle$, as shown in Fig. 11. Use of the compensator reduces the value of Δ to $\sim 0.5°$, which yields $\langle k \rangle$ values that are closer to zero, so that at 500 nm, k is reduced from values greater than 0.04 to values less than 0.004. Moreover, the spectral curve of $\langle k \rangle$ evaluated from the data obtained with the compensator is flatter, indicating that the experimental errors have been minimized considerably by using the compensator in the SE measurements. It should be reiterated here that these $\langle k \rangle$ values, which are still much too high for vitreous silica in the visible region of the spectrum, are not the true k values, but rather the "effective k" of the *sample*. The nonzero Δ, and hence the nonzero $\langle k \rangle$ values evaluated can be attributed to the effects of the presence of either (i) some inhomogeneities in the vitreous silica sample or (ii) a micro-rough surface layer on the vitreous silica *sample*. We can rule out the first possibility since the samples used were Suprasil-1 grade manufactured by Amersil, Inc., and according to the manufacturer's specifications the samples are optically homogeneous and free of strain,

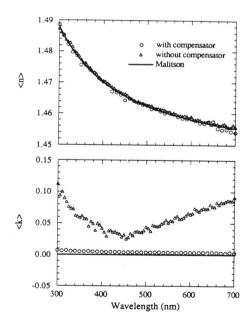

FIG. 11. The effective optical constants of vitreous silica evaluated from the ellipsometric parameters (Δ, Ψ) given in Fig. 10 with and without the achromatic compensator. Solid lines are data calculated from the dispersion equation of vitreous silica by Malitson (*44*). After Chindaudom and Vedam (*34*).

striae, bubbles and inclusions. On the other hand, the presence of a micro-rough surface layer on the vitreous silica sample is more likely, since even with our current state of the art only rarely we are able to produce atomically smooth surfaces of glasses (46) and at the same time the phase measurements or Δ in SE is sensitive to even nano-scale micro-roughness on the surface of the sample. Hence, this problem was tackled by the standard method of determining the optical function $n(\lambda)$ of a transparent material by SE. Following the usual procedure (29, 32, 33) the experimentally observed SE data on vitreous silica (of course measured using the achromatic compensator) were analyzed by linear regression analysis (LRA) using appropriate models, to determine *simultaneously* the optical function of vitreous silica and the parameters of the micro-rough surface layer. As discussed earlier, the optical function $n(\lambda)$ of vitreous silica in the visible region of spectrum can be expressed by the Sellmeier dispersion equation, Eq. (9). Since the values of Δ are small, the unbiased estimator σ in the form of $\sigma(\Delta, \Psi)$ was used instead of the usual $\sigma(\cos \Delta, \tan \Psi)$ (see Section III.D) during the regression analysis computations. Figure 12, which is essentially the same as Fig. 10 but plotted in an enlarged scale, shows a comparison of the measured SE data and the calculated values using the best-fit model. The model includes a micro-rough surface layer on the vitreous silica sample that can be treated as an interface layer composed of SiO_2 and air. The optical constants of the interface were evaluated by using Bruggeman effective medium theory (43). Figure 13 shows the best-fit model for our vitreous silica sample with its interface. Here it should be mentioned that the LRA revealed a strong correlation between the thickness and the volume fraction of the void (or air) in the interface layer. This is not surprising when it is realized that Δ changes only from 0.5° to 0.2°, over the spectral region of 300 to 700 nm, and furthermore the observed Δ spectrum is structureless. Hence, we fixed the volume fraction of the voids as 50%, and then the LRA results revealed that the thickness of the interface layer was 0.94 ± 0.03 nm. Here it is relevant to point out that Jellison and Sales (35) also had to resort to a similar procedure of fixing the volume fraction of voids in the micro-rough surface layer before they could determine its thickness, for the same reasons, in their very recent studies on the optical functions of transparent glasses. It is seen from Figs. 12 and 13 that with the incorporation of a 0.94 nm thick micro-rough surface layer with an effective 50/50 composition of SiO_2 and air, the observed $[\Delta(\lambda), \Psi(\lambda)]$ spectra are in complete agreement with the corresponding theoretically calculated spectra, and there is no need to assume that our sample of vitreous silica is optically absorbing in the visible region of spectrum. In other words, the observed nonzero Δ spectrum is an artifact of the surface micro-roughness. In such a case we can

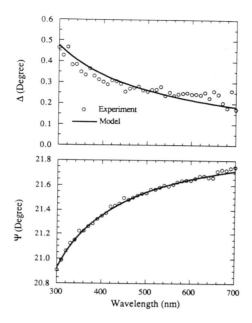

FIG. 12. Comparison of the observed ellipsometric parameters (Δ, Ψ) of vitreous silica with the values from the best-fit model shown in Fig. 13. After Chindaudom and Vedam (34).

mathematically subtract the effect of the micro-roughness (i.e., treat $\Delta(\lambda)$ as zero at all wavelengths in the observed spectral region), and then by direct inversion of the $\Psi(\lambda)$ spectrum we can obtain the true optical function $n(\lambda)$ of vitreous silica. An alternative method of obtaining the optical function $n(\lambda)$ of vitreous silica is to use the Sellmeier dispersion relation given by Eq. (9) with the numerical values for the three variable parameters evaluated by

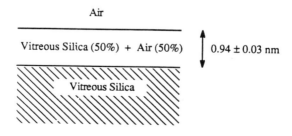

FIG. 13. Model of the vitreous silica sample showing the micro-rough surface layer.

LRA of the SE data as $A = 1.000$, $B = 1.0984 \pm 0.0023$, $\lambda_0 = 92.58 \pm 1.93$ nm. Such computations were carried out, and the final results on the numerical values of n at various wavelengths are presented in Table I, along with the best values of n of vitreous silica reported in the literature (44).

It is seen from Figs. 10 and 12 that the rms deviations of the SE data measured with the compensator, as against the theoretically computed values, are $0.029°$ in Δ and $0.015°$ in Ψ over the spectral range 300–700 nm. Thus, the accuracy of measurement in Δ of transparent materials (measured with an achromatic compensator), estimated from the measurements of the vitreous silica test sample, is about $0.03°$ over the entire spectral range studied. Jellison and Sales (35) have also obtained the same order of accuracy with their two-channel spectroscopic polarization modulation ellipsometry (47) on vitreous silica. Typically, the accuracy of measurements on Δ by RAE have been quoted as a few hundredths of a degree when Δ is larger than $15°$. However, it has been shown here for the first time that, using an achromatic compensator, the accuracy of measurement in Δ by RAE can be as high as $0.03°$ even when Δ is close to $0°$ over the spectral range 300–700 nm.

Some remarks regarding the proper error function to be used for the data analysis are relevant here. Figure 14 shows the SE data for the vitreous silica sample and the best-fit models obtained by the minimization of $\sigma(\Delta, \Psi)$ (solid curve) and $\sigma(\cos \Delta, \tan \Psi)$ (dashed curve). Clearly, $\sigma(\Delta, \Psi)$ is seen to result in much better fit in Δ than $\sigma(\cos \Delta, \tan \Psi)$. With the help of the compensator, the precision in measurement of Δ is now comparable to that of Ψ, even when Δ is close to 0 or $180°$. Therefore, fitting by favoring Ψ over Δ is no longer necessary. In addition, since Δ is much more sensitive

TABLE I
COMPARISON OF THE REFRACTIVE INDEX OF VITREOUS SILICA AS DETERMINED BY SE IN THIS STUDY WITH THE BEST REPORTED VALUES IN THE LITERATURE (44)

Wavelength (nm)	$n_{lit.}$ (Malitson)	n (this study)	δn ($n_{lit.} - n$)
300	1.4878	1.4879	−0.0001
350	1.4769	1.4768	0.0001
400	1.4701	1.4699	0.0002
450	1.4656	1.4652	0.0004
500	1.4623	1.4620	0.0003
600	1.4580	1.4578	0.0002
700	1.4553	1.4553	0.0000

to the surface layer than Ψ, and $\cos\Delta$ is insensitive to the change of Δ when Δ is near zero, $\sigma(\Delta, \Psi)$ is more suitable for modeling the SE data of transparent samples. During the LRA analysis a value of $0.035°$ was obtained for the unbiased estimator $\sigma(\Delta, \Psi)$, whereas the value evaluated for $\sigma(\cos\Delta, \tan\Psi)$ was 0.00036 from the same modeling. Note that $\sigma(\Delta,\Psi)$ is comparable to the estimated error in Δ in the measurements. Since the lowest value of σ is determined by the errors in the experiment as well as the model employed for the analysis, $\sigma(\Delta, \Psi)$ provides some insights on the error in the modeling when it reaches the limit of the errors in the measurement. On the other hand, even though the evaluated value of $\sigma(\cos\Delta, \tan\Psi)$ as 0.00036 appears to be extraordinarily good, for reasons explained earlier in this article and in the earlier article (36) on the optical function of LaF_3 film, such a low value of σ obtained by ignoring $\cos\Delta$ in the computations is meaningless.

Figure 15 shows the plot of $\langle n, k \rangle$ of Fig. 11, but in an enlarged version of $\langle k \rangle$ as evaluated from the SE data and the best-fit model as a function of wavelength using $\sigma(\Delta, \Psi)$. The rms deviations are less than 0.0006 in $\langle n \rangle$

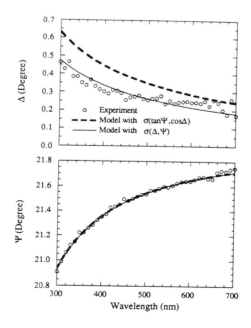

FIG. 14. Comparison of the fits for the vitreous silica sample using $\sigma(\Delta, \Psi)$ and $\sigma(\cos\Delta, \tan\Psi)$.

and less than 0.0004 in $\langle k \rangle$. Thus, the accuracies in n and k of transparent materials as measured by RAE with an achromatic compensator are estimated for our system as 0.001. It should be mentioned that the presence of a micro-rough layer of 1 nm or less in the model results in small changes in the values less than 0.001° in Ψ and less than 5×10^{-5} in $\langle n \rangle$ and are not detectable by SE; nevertheless, the changes in Δ and $\langle k \rangle$ are appreciable and detectable by SE using a compensator. In other words, the value of $\langle n \rangle$ of vitreous silica evaluated in this study is actually its true refractive index.

Since we are interested in n, the "true refractive index" of vitreous silica in bulk form, and not in the effective refractive index $\langle n \rangle$, i.e., not in the values of n influenced by the presence of a micro-rough surface, Table I presents the numerical values of the refractive index of vitreous silica at a few selected wavelengths as determined in this study, along with the corresponding values reported by Malitson (44), which are considered to be the most accurate values reported in the literature. It is seen that the maximum deviation between the two sets of values is only 0.0004, and hence we can safely conclude that we can now determine the refractive index of

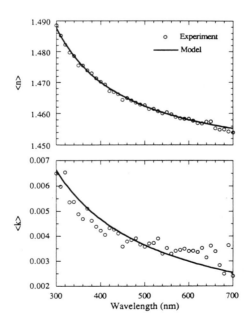

FIG. 15. Effective refractive index and extinction coefficient of the vitreous silica sample with surface micro-roughness. After Chindaudom and Vedam (34).

transparent materials by SE definitely to within 0.0005.

In summary these results on vitreous silica clearly demonstrate that the ellipsometric parameters, Δ and Ψ, and hence the refractive index n and the extinction coefficient k (provided $k \geqslant 0.001$) of transparent samples can be obtained as accurately as those of metals and semiconductors, by RAE with the help of an achromatic compensator. In other words, the central difficulty in studying transparent samples by RAE has been overcome, and the well-known precision and accuracy inherent in the SE technique can be utilized fully. This in turn enables SE to become the primary nondestructive optical technique available in determining the micro-structure and surface roughness of transparent films, and thus SE can detect and characterize the inhomogeneity in thin films and also simultaneously determine the true optical functions of the film materials.

IV. Experimental Results on Transparent Films

A. Particulars of Films Studied

Before presenting the experimental results, a few remarks on the preparatory history of the experimental samples, as well as some specific details of the SE measurements on them, are appropriate here. Optical-quality thin films of AlF_3, CeF_3, HfF_4, LaF_3, ScF_3, YF_3, Al_2O_3, HfO_2, Sc_2O_3, ThO_2, Y_2O_3, and ZrO_2 were deposited on heated ($\sim 225°C$) vitreous silica substrates (3 cm diameter and 0.6 cm thick) in high vacuum by the electron-beam evaporation technique at low rates (0.3–0.7 nm/s). The substrates were rotated during deposition to improve uniformity and minimize optical anisotropy of the films. The thicknesses of these films were in the range of 130–450 nm. The back sides of the substrates were ground as a wedge at a small angle, in order to eliminate the complicating effect of reflection from the back side of the substrate during ellipsometric measurements. The samples were stored in a desiccator for a long period (two years) before the SE measurements were carried out. However, there were no visual indications of water vapor adsorption, such as haze or cloudy patches, on the samples, despite the lengthy holding time.

All the SE measurements were carried out over the spectral range 300–700 nm by a rotating analyzer ellipsometer, at an angle of incidence of 75° in the case of fluorides, and 70° in the case of the oxide coatings. The actual details of the alignment, calibration and the various error-correction procedures were discussed earlier in this article in Section III. The preceding angles of incidence were chosen so that most of the Ψ values are $>20°$, and hence the experimental errors (*3, 5*) that are proportional to $(1/\sin 2\Psi \sin \Delta)$ can be minimized. Since an achromatic compensator was employed in these measurements, the measured values of the "effective Δ" were around 90°, even though the true values of Δ are around 0° (or 180°), and hence we were able to obtain reproducibility and reliable values of Δ.

The analyses of these SE data were carried out by LRA as described earlier in Section III.D with the help of realistic models. As described in Section II, following Movchan and Demchishin (*6*) and Yang et al. (*11*), an optical model depicting the microstructure of these dielectric films can in general be described by (i) a substrate/film interface larger with some voids, (ii) a central layer that is a denser (or even void-free) region representing the region where the columns have coalesced, and (iii) an outermost film/air interface layer representing the surface micro-roughness. Thus, the model parameters to be determined for the most general case are the following: (i) d_1 and f_{v1}, the thickness and volume fraction of the voids in the film/substrate interface layer; (ii) d_2 and f_{v2}, the thickness and the volume fraction of the voids in the central region; (iii) d_3 and f_{v3}, the thickness and the volume fraction of the voids in the surface layer; and (iv) $n(\lambda)$, the refractive index of the film material and its dispersion with wavelength.

As far as possible, all of the model parameters in the trial models for the various films were adjusted simultaneously by LRA for the best value of each parameter. However, it was found that in some cases, the correlations between the model parameters were too strong, causing unrealistically large 90% confidence limits for these unknown parameters. Therefore, in such cases the few parameters that strongly influence the values of the other parameters and their 90% confidence limits, were kept constant in order to obtain a realistic model. For example, in the modeling of the relatively weak dispersive film materials studied here, it was necessary to fix the parameter A of the dispersion equation(8) as unity in the case of fluoride optical coatings.

B. Preliminary Results on MgO and LaF$_3$ Films

The first example to be discussed in this review is the case of an MgO film on a vitreous silica substrate studied by SE—the very first simultaneous determination of the refractive index, its dispersion, and the depth profile of an inhomogeneous MgO thin film by SE without the use of a compensator, by Vedam and Kim (*48*). Since the refractive index of MgO is much higher than that of the vitreous silica substrate, the reflectance was high enough that precision measurements of Ψ could be carried out at a large number

(~ 300) of wavelengths distributed in the spectral range 275–825 nm at a 60° angle of incidence. By performing LRA analysis of such data using $\sigma(\alpha, \beta)$ instead of $\sigma(\Delta, \Psi)$ of Eq. (8), these workers were able to obtain good agreement between the observed and calculated values of the optical dielectric constant, as shown in Fig. 16. At the same time they were also able to obtain the values of the optical function of MgO. Table II presents their values of the refractive index of MgO for a few selected wavelengths, along with the best values of the refractive index reported in the literature (49). It is seen that the two values are in excellent agreement with each other to within 0.001. In retrospect we now feel that the success of this work was due to (i) the high-precision measurement of $\Psi(\lambda)$ and (ii) the very large number of wavelengths (~ 300) at which measurements were carried out.

The next example to be discussed is the case of our early measurements (36) on LaF_3 at an angle of incidence of 70°, which (i) dramatically demonstrates the failure of the homogeneous single-layer model as against the two-layer inhomogeneous model, and (ii) brings out the importance of using $\sigma(\Delta, \Psi)$ instead of $\sigma(\cos \Delta, \tan \Psi)$ for the data analysis. Figure 17a

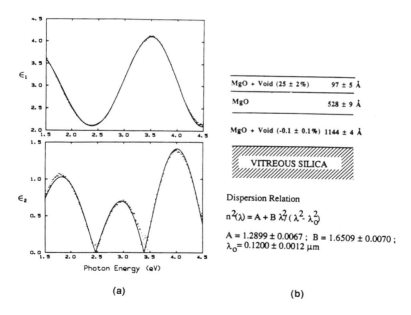

FIG. 16. (a) Observed and calculated ellipsometric [$\Delta(\lambda)$, $\Psi(\lambda)$] spectra for the MgO film on vitreous silica substrate. Angle of incidence 60°. (b) Best-fit model of the MgO film on vitreous silica. Vedam and Kim (48).

shows the comparison of the observed Δ and Ψ spectra with the corresponding calculated spectra using the single-layer model (SLM) and the double-layer model (DLM) presented in Fig. 17b for the film. Figure 17b also shows the value of σ and the various wavelength-independent model parameters, along with their 90% confidence limits. Figure 17b clearly indicates that the LaF_3 film is inhomogeneous and that this inhomogeneity arises from the micro-roughness of the 9.0 ± 0.9 nm thick surface layer of the film. The presence of the micro-rough surface on the film is also corroborated by the increased surface scattering compared to that of the substrate, as observed by the manufacturer of the film with a laser scatterometer. It also can be seen in Fig. 17a that the addition of an inhomogeneous surface layer has a stronger influence on the change of the Δ spectrum, which indicates that the phase parameter Δ is much more sensitive to the structural inhomogeneity of the film than Ψ. Thus, for the study of inhomogeneous films, it is important to be able to measure Δ accurately.

The reason for our earlier remarks in Section III.D on the choice of (Δ and Ψ) instead of (cos and tanΨ) in Eq (8), for the data analyses becomes obvious by comparing Fig. 17a with Fig. 18, where (cos Δ, tan Ψ) was used for the modeling instead of (Δ, Ψ). Since Δ is expected to be near $0°$ or $180°$ over the entire spectral range for transparent films on transparent sub-

TABLE II
COMPARISON OF THE REFRACTIVE INDEX OF MgO AS DETERMINED BY SE (48) WITH THE BEST REPORTED VALUES IN THE LITERATURE (49)

λ(nm)	n(lit)	n(calc)	Δn
250	1.8490	1.8534	−0.0044
300	1.8028	1.8042	−0.0014
350	1.7779	1.7778	0.0001
400	1.7627	1.7618	0.0009
450	1.7527	1.7513	0.0014
500	1.7458	1.7440	0.0018
550	1.7408	1.7388	0.0020
600	1.7371	1.7348	0.0023
650	1.7342	1.7318	0.0024
700	1.7319	1.7294	0.0025
750	1.7301	1.7275	0.0026
800	1.7286	1.7259	0.0027
850	1.7274	1.7246	0.0028
900	1.7264	1.7236	0.0028

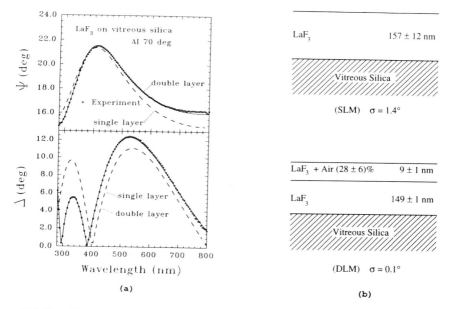

FIG. 17. (a) Observed and calculated ellipsometric [$\Delta(\lambda)$, $\Psi(\lambda)$] spectra for the LaF$_3$ film on vitreous silica. Angle of incidence 70°. (b) Best-fit single-layer model (SLM) and double-layer model (DLM) of the LaF$_3$ film on vitreous silica substrate. After Chindaudom and Vedam (36).

strates, cos Δ will be around $\pm$ 1.0 and become insensitive to the small variation of Δ. Consequently, the use of (cos Δ, tan Ψ) puts enormous weight on tan Ψ and essentially ignores cos Δ in the computations. Therefore, the details on the micro-structural inhomogeneities in the film, which are contained in cos Δ spectrum, are often lost in such computations. Further, in Fig. 18, if one had plotted cos Δ in the same scale as tan Ψ, the disagreement between the calculated and the observed data will appear to be negligible. Furthermore, it is seen that the tan Ψ curve does not even differentiate between the single-layer and two-layer models, implying thereby that if one relies on the Ψ spectrum alone, the presence of inhomogeneity in the film may not even be noticed.

There is another advantage of using (Δ and Ψ) spectra, in that comparison of the σ value in degrees with the experimental error limits can be used as an indicator of the goodness of the calculated model. For example, from Fig. 17a considering the experimental errors in Δ and Ψ are about $\pm$ 0.03°, the evaluated final value of σ as 0.13° indicates that the final values of the evaluated parameters are already of the same order of magnitude as the actual experimental errors. This implies that there is no justification to

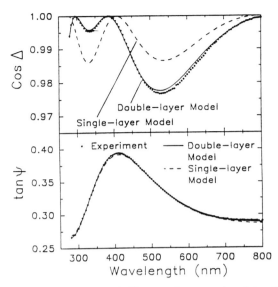

FIG. 18. Observed and calculated ellipsometric spectra ($\cos \Delta$, $\tan \Psi$) for the LaF_3 film on vitreous silica showing the single-layer and double-layer model fittings using $\sigma(\cos \Delta, \tan \Psi)$. After Chindaudom and Vedam (36).

increase the complexity of the model any further, as, for example, by the addition of a microrough (substrate/film) interface layer in order to decrease the value of σ. On the other hand, the value of σ as 0.001 obtained by using ($\cos \Delta$ and $\tan \Psi$) for the data analysis appears to be extraordinarily good, but for reasons explained in the last paragraph, such a low value of σ obtained by ignoring $\cos \Delta$ in the computations is meaningless.

C. Results on Transparent Films on Vitreous Silica Substrates

After the completion of the preceding preliminary studies on LaF_3, more exhaustive procedures for the calibration and corrections for the various errors in the SE system were incorporated before the present series of measurements was undertaken. Since the results obtained were reliable and reproducible compared to our earlier studies on LaF_3, the SE measurements on LaF_3 were repeated at a 75° angle of incidence, and the revised results are included in the following discussion.

Tables III and IV present the final results of LRA respectively for all the fluoride and oxide optical coatings studied in this work. The values of σ and the model parameters evaluated by LRA, such as (a) the Sellmeier oscillator parameters for the film material, (b) the thicknesses d_1, d_2 and d_3 of the sublayers of the film, and (c) the void fractions f_{v1} and f_{v3} in these sublayers, along with their 90% confidence limits for these films, are given in Tables III and IV. Instead of presenting all the corresponding tables of the cross-correlation parameters of the 12 films reported here, the data on the most general and complicated film studied—namely, Y_2O_3 films alone—are given in Table V. Some of these results have already appeared in print as in a conference proceedings (50) and in (51).

The detailed analyses of the SE data obtained on all 12 dielectric films on vitreous silica substrates reveal that *all* the films are indeed inhomogeneous, without exception. Further, based on their internal micro-structure it is found that all these films can be grouped into two categories—namely, films that can be modeled as composed of either two layers or three layers. The two-layer model category can be further subdivided into (i) films with a micro-rough surface layer on top of a void-free layer, and (ii) smooth films on top of a substrate/film interface layer with voids. The three-layer model corresponds to a central void-free film with a substrate/film interface layer underneath the central layer and also a micro-rough surface overlayer.

1. Films with a Micro-rough Surface Overlayer

CeF_3, LaF_3, YF_3, Al_2O_3, HfO_2, Sc_2O_3, ThO_2 and ZrO_2 films on vitreous silica fall into this category. These films can be modeled as composed of

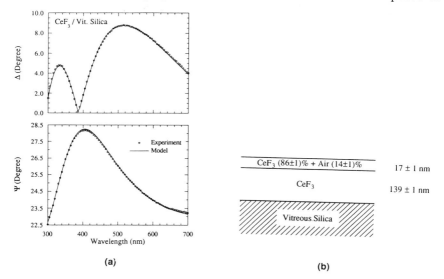

FIG. 19. (a) Observed and calculated ellipsometric $[\Delta(\lambda), \Psi(\lambda)]$ spectra for the CeF_3 film on vitreous silica. Angle of incidence 75°. (b) Best-fit model of the CeF_3 film on vitreous silica. After Chindaudom and Vedam (51).

TABLE III
VALUES OF THE MODEL PARAMETERS FOR THE FLUORIDE OPTICAL COATINGS, AS EVALUATED BY THE LEAST SQUARES REGRESSION ANALYSIS OF THE SE DATA

	AlF_3	CeF_3	HfF_4	LaF_3	ScF_3	YF_3
σ(deg)	0.07	0.13	0.07	0.07	0.05	0.06
A	1.00	1.00	1.00	1.00	1.00	1.00
B	0.904 ± 0.002	1.558 ± 0.003	1.401 ± 0.002	1.509 ± 0.002	1.093 ± 0.001	1.448 ± 0.013
λ_0(nm)	69.4 ± 1.8	106.1 ± 1.4	91.2 ± 1.3	88.5 ± 1.4	105.3 ± 1.6	72.4 ± 6.2
d_3(nm)	—	16.7 ± 1.1	—	13.9 ± 0.7	15.1 ± 0.1	6.1 ± 0.4
						98.7 ± 15.8
d_2(nm)	14.2 ± 3.6	138.8 ± 1.1	32.8 ± 4.0	144.9 ± 0.7	115.1 ± 1.8	66.3 ± 17.4
d_1(nm)	183.9 ± 3.6	—	136.9 ± 4.0	—	55.7 ± 4.4	—
f_{v3}	—	0.140 ± 0.009	—	0.176 ± 0.009	0.500	0.500
f_{v1}	0.080 ± 0.017	—	0.010 ± 0.001	—	0.045 ± 0.003	0.023 ± 0.004

TABLE IV
VALUES OF THE MODEL PARAMETERS FOR THE OXIDE OPTICAL COATINGS, AS EVALUATED BY THE LEAST SQUARES REGRESSION ANALYSIS OF THE SE DATA

	Al_2O_3	HfO_2	Sc_2O_3	ThO_2	Y_2O_3	ZrO_2
σ(deg)	0.08	0.27	0.45	0.32	0.60	2.12
A	1.25	2.07 ± 0.28	1.81 ± 0.28	1.00	1.68 ± 0.54	1.00
B	1.34 ± 0.01	1.96 ± 0.28	1.67 ± 0.27	2.28 ± 0.01	1.73 ± 0.52	2.86 ± 0.02
λ_0(nm)	124.0 ± 0.5	162.2 ± 7.7	168.0 ± 9.0	137.6 ± 1.0	141.8 ± 15.6	142.3 ± 1.0
d_3(nm)	1.64 ± 0.03	11.6 ± 1.5	7.8 ± 0.2	10.0 ± 1.2	18.5 ± 1.4	19.2 ± 1.7
		74.8 ± 0.8	83.1 ± 4.9	85.8 ± 1.9	135.7 ± 6.2	394.3 ± 4.1
					55.1 ± 1.5	
d_2(nm)	167.8 ± 0.1	50.0 ± 1.6	50.1 ± 7.0	55.6 ± 3.6	182.0 ± 2.5	52.8 ± 2.0
d_1(nm)	—	—	—	—	47.0 ± 1.7	—
f_{v3}	0.500	0.262 ± 0.029	0.574 ± 0.234	0.415 ± 0.086	0.380 ± 0.032	0.302 ± 0.022
		0.057 ± 0.002	0.029 ± 0.002	0.036 ± 0.002	0.085 ± 0.006	0.088 ± 0.009
					0.046 ± 0.003	
f_{v2}	—	—	—	—	—	—
f_{v1}	—	—	—	—	0.071 ± 0.004	—

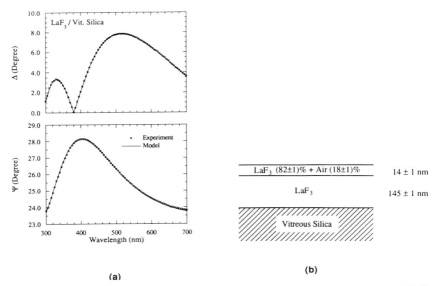

FIG. 20. (a) Observed and calculated ellipsometric [$\Delta(\lambda)$, $\Psi(\lambda)$] spectra for the LaF$_3$ film on vitreous silica. Angle of incidence 75°. (b) Best-fit model of the LaF$_3$ film on vitreous silica. After Chindaudom and Vedam (51).

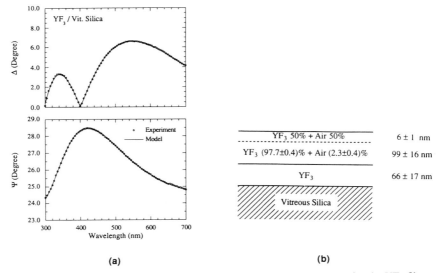

FIG. 21. (a) Observed and calculated ellipsometric [$\Delta(\lambda)$, $\Psi(\lambda)$] spectra for the YF$_3$ film on vitreous silica. Angle of incidence 75°. (b) Best-fit model of the YF$_3$ film on vitreous silica. After Chindaudom and Vedam (51).

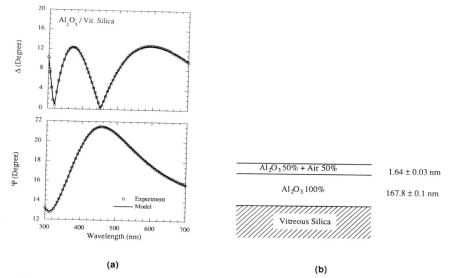

Fig. 22. (a) Observed and calculated ellipsometric [$\Delta(\lambda), \Psi(\lambda)$] spectra for the Al_2O_3 film on vitreous silica. Angle of incidence 70°. (b) Best-fit model of the Al_2O_3 film on vitreous silica.

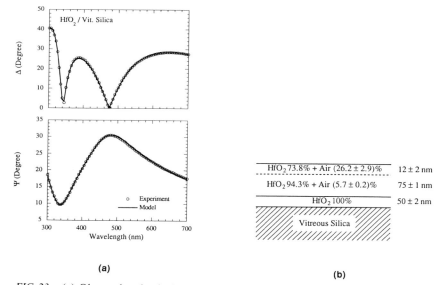

FIG. 23. (a) Observed and calculated ellipsometric [$\Delta(\lambda), \Psi(\lambda)$] spectra for the HfO_2 film on vitreous silica. Angle of incidence 70°. (b) Best-fit model of the HfO_2 film on vitreous silica.

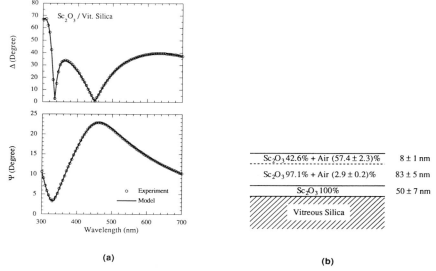

FIG. 24. (a) Observed and calculated ellipsometric [$\Delta(\lambda)$, $\Psi(\lambda)$] spectra for the Sc_2O_3 film on vitreous silica. Angle of incidence 70°. (b) Best-fit model of the Sc_2O_3 film on vitreous silica.

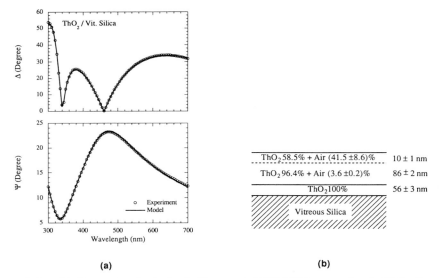

FIG. 25. (a) Observed and calculated ellipsometric [$\Delta(\lambda)$, $\Psi(\lambda)$] spectra for the ThO_2 film on vitreous silica. Angle of incidence 70°. (b) Best-fit model of the ThO_2 film on vitreous silica.

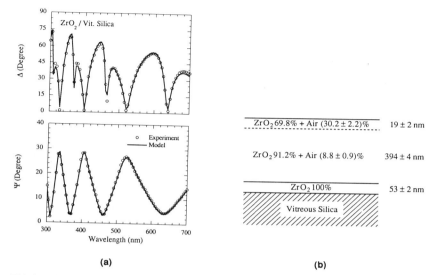

FIG. 26. (a) Observed and calculated ellipsometric [$\Delta(\lambda), \Psi(\lambda)$] spectra for the ZrO_2 film on vitreous silica. Angle of incidence 75°. (b) Best-fit model of the ZrO_2 film on vitreous silica.

two-layers, i.e., these films have a micro-rough surface overlayer on top of a $\cong$ void-free layer. Figures 19a through 26a show the agreement between the experimentally observed [$\Delta(\lambda), \Psi(\lambda)$] spectra and the corresponding calculated spectra, using the models shown in Figs. 19b through 26b, respectively, for the cases of CeF_3, LaF_3, YF_3, Al_2O_3, HfO_2, Sc_2O_3, ThO_2 and ZrO_2 films on vitreous silica. In these films it appears that the film growth initiated from a very large number of nuclei, and hence there does not appear to be any voided interface layer between the substrate and the film. However, during the evolution of the film growth, because of severe competition between the growing columns, a few survive, and their dome-shaped growing surfaces exhibit the characteristics of a rough surface.

2. Smooth Films on Top of an Interface between Substrate and Film

These films can be modeled as composed of two layers, i.e., a substrate/film interface layer with voids, below a denser void-free smooth layer. AlF_3, and HfF_4 films on vitreous silica substrates fall under this category. Figures 27a and 28a show the agreement between the experimentally observed [$\Delta(\lambda), \Psi(\lambda)$] spectra and the corresponding calculated spectra, using the models shown in Figs. 27b and 28b, respectively, for the cases of AlF_3 and HfF_4 films on vitreous silica. Figures 27b and 28b also list the numerical

TABLE V
Correlation Matrix of the Model Parameters of Y_2O_3 Film on Vitreous Silica Substrate

		f_{v1}	d_1	d_2	f_{v3}	d_3	f_{v4}	d_4	f_{v5}	d_5	B	λ_0
1	f_{v1}	1.00	−0.03	−0.03	0.10	0.31	0.01	−0.42	−0.43	0.39	0.78	−0.56
2	d_1	−0.03	1.00	−0.64	0.20	−0.10	0.18	−0.05	0.01	−0.03	0.47	−0.55
3	d_2	−0.03	−0.64	1.00	−0.30	0.08	−0.47	−0.51	−0.45	0.46	−0.31	0.35
4	f_{v3}	0.10	0.20	−0.30	1.00	0.17	0.88	0.20	0.35	−0.19	0.13	0.03
5	d_3	0.31	−0.10	0.08	0.17	1.00	0.06	−0.53	−0.34	0.39	0.28	−0.19
6	f_{v4}	0.01	0.18	−0.47	0.88	0.06	1.00	0.46	0.58	−0.39	0.01	0.07
7	d_4	−0.42	−0.05	−0.51	0.20	−0.53	0.46	1.00	0.91	−0.86	−0.52	0.36
8	f_{v5}	−0.43	0.01	−0.45	0.35	−0.34	0.58	0.91	1.00	−0.95	−0.51	0.35
9	d_5	0.39	−0.03	0.46	−0.19	0.39	−0.39	−0.89	−0.95	1.00	0.46	−0.27
10	B	0.78	0.47	−0.31	0.13	0.28	0.01	−0.52	−0.51	0.46	1.00	−0.79
11	λ_0	−0.56	−0.55	0.35	0.03	−0.19	0.07	0.36	0.35	−0.27	−0.79	1.00

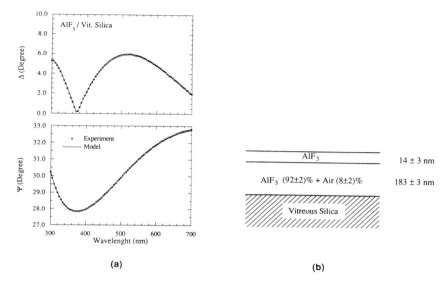

FIG. 27. (a) Observed and calculated ellipsometric [$\Delta(\lambda), \Psi(\lambda)$] spectra for the AlF$_3$ film on vitreous silica. Angle of incidence 70°. (b) Fig. 1a. Best-fit model of the AlF$_3$ film on vitreous silica. After Chindaudom and Vedam (51).

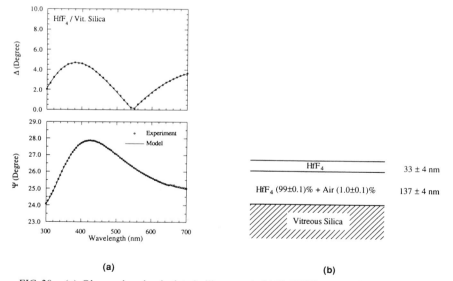

FIG. 28. (a) Observed and calculated ellipsometric [$\Delta(\lambda), \Psi(\lambda)$] spectra for the HfF$_4$ film on vitreous silica. Angle of incidence 70°. (b) Best-fit model of the HfF$_4$ film on vitreous silica. After Chindaudom and Vedam (51).

values of the evaluated model parameters, along with their 90% confidence limits. In these cases the films appear to have grown from a limited number of nuclei that grew outwards as ever-expanding conical columns and that eventually coalesced with each other to form a void-free overlayer, in accordance with the generally accepted model (*10*, *13*) of film growth from the vapor phase.

3. Films That Can Be Modeled as Composed of Three Layers

These films can be modeled as composed of a substrate/film interface layer below a void-free middle layer and a micro-rough surface overlayer. ScF_3 and Y_2O_3 films on vitreous silica substrates belong to this most general category. Figures 29a and 30a show the agreement between the experimentally observed [$\Delta(\lambda)$, $\Psi(\lambda)$] spectra and the corresponding calculated spectra using the models shown in Figs. 29b and 30b, respectively, for the cases of ScF_3 and Y_2O_3 films on vitreous silica.

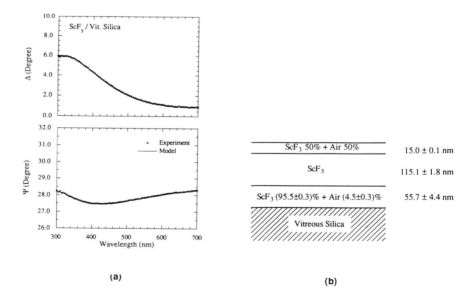

FIG. 29. (a) Observed and calculated ellipsometric [$\Delta(\lambda)$, $\Psi(\lambda)$] spectra for the ScF_3 film on vitreous silica. Angle of incidence 70°. (b) Best-fit model of the ScF_3 film on vitreous silica. After Chindaudom and Vedam (*51*).

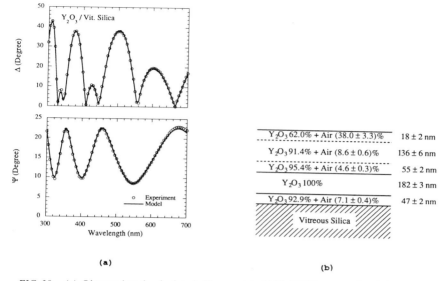

FIG. 30. (a) Observed and calculated ellipsometric [$\Delta(\lambda)$, $\Psi(\lambda)$] spectra for the Y_2O_3 film on vitreous silica. Angle of incidence 75°. (b) Best-fit model of the Y_2O_3 film on vitreous silica.

V. Discussion

A. General

Before we even begin discussing the results, the first question that must be answered is how trustworthy the results are. In other words, does getting a good fit between the observed and the calculated [$\Delta(\lambda)$, $\psi(\lambda)$] spectra provide enough justification to be considered as a proof of the validity of the model? Further, how can one be sure that the complexity of the model has not been increased to get a good fit and to obtain low value for σ? Finally, is it not possible that there may be another model(s) that can give an equally good fit?

It is not possible to give answers to these questions in categorical terms. However, for the following statistically meaningful reasons, it is highly unlikely that there are also other model(s) that will give equally good fits *simultaneously* to both $\Delta(\lambda)$ and $\Psi(\lambda)$ spectra *and also* give the desired model parameters within reasonably good confidence limits without sacrificing acceptable cross-correlation parameters. From Tables III and IV, it is seen that the confidence limits of all the parameters determined by linear regression analysis (LRA) are all fairly low. Similarly, from Table V it is seen that all the cross-correlation parameters are respectably low for even the most complicated case involving 11 unknown parameters. Our own experience with LRA has clearly indicated, a countless number of times during the development of the models, that the moment we are on the wrong track the confidence limits of the evaluated parameters blow up enormously even though the fit may be getting better. Further, the final models arrived at for the different films are physically realistic, as can be deduced from the previous history of the samples and the currently accepted theoretical as well as computer-simulated models of film growth. Furthermore, two independent SE measurements on the same sample (LaF$_3$ film) at two different angles (70° and 75°) yield essentially the same results, even though the first measurement at 70° was not done under the optimum conditions of the second measurements at 75°. Perhaps an even stronger argument can be advanced by the following examples, even though the experiments described were not actually carried out on purely dielectric films. The power of SE to characterize multilayer structures with almost atomic resolution is clearly brought out in Fig. 31, which shows a comparisn between our SE results (52) and cross-section transmission electron microscopy on the same specimen. In this experiment our SE studies and the LRA analysis of the data on the Si$^+$ ion–implanted crystalline silicon was completed *before* the same sample was sent to Oak Ridge National Laboratory to be cut and ion milled for cross-section transmission electron microscopy (XTEM). Since

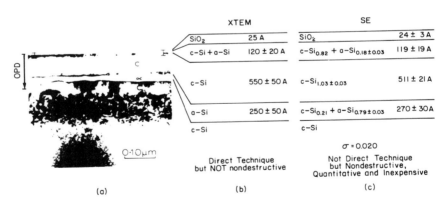

FIG. 31. Comparison of cross-section transmission electron microscopy (XTEM) and spectroscopic ellipsometry (SE) for depth profiling multilayer structure. (c-Si, crystalline silicon; a-Si, amorphous silicon; OPD, optical penetration depth in SE.) (a) Cross-section TEM micrograph of crystalline silicon sample implanted with ^{28}Si$^+$ ions; (b) schematic diagram of model for the same sample as evaluated from the XTEM shown in (a); (c) schematic diagram of model for the same sample as evaluated by SE (Vedam *et al.*, Ref. 52).

then, similar exact agreements between our SE results and XTEM on oxygen ion–implanted silicon-on-insulator (SOI) structures (53) have been obtained. Similarly, very recently, our SE results on nitrogen ion–implanted SOI structures (54) have also shown excellent agreement with the results obtained by XTEM as well as Rutherford back scattering (RBS) studies.

B. Effects of Surface Roughness

From Tables III and IV as well as from Figs. 19b through 30b, it is evident that surface microroughness is the most common inhomogeneity present in the thin films reported in this article. In fact, in only two out of 12 films studied are the films found to be smooth (i.e., for AlF_3 and HfF_4 films, d_3 is zero). From Figs. 10, 12 and 13, it is seen that SE can detect the effective thickness of the rough surface layer to within 0.9 nm, and hence we can conclude that both AlF_3 and HfF_4 films are almost atomically smooth. In all the other cases, the films have rough surfaces with the effective thickness varying from 1.6 nm for Al_2O_3 film to over 400 nm in the case of ZrO_2. Thus, the growing dome-shaped sufaces, as well as the columnar growth features (see Figs. 2 and 3), exhibit the characteristics of a rough surface in almost all films deposited in high vacuum either by thermal or electron beam evaporation. Again, from Figs. 10 and 11 it is evident that this roughness causes the measured value of Δ to be quite different from zero, and this in turn causes the effective optical constant k to be different from zero. This implies that the roughness causes some light to be scattered, and hence to be lost in transmission and reflectance measurements. Neglect of this loss of light by scattering in the R, T measurements of spectrophotometry will naturally yield erroneous results in the derived optical function of the material. The fact that SE can detect and take into account the presence of such atomic-scale micro-roughness clearly shows the advantage and power of SE in such measurements.

In the cases of YF_3, HfO_2, Sc_2O_3, ThO_2, Y_2O_3 and ZrO_2 films, the roughness layers have been further subdivided into two or three sublayers without sacrificing the reliability of the final results. However, it is seen from the values of f_{v3} given in Tables III and IV that, while the volume fraction of voids in the outermost surface layer varies from 14% to 57% over a small layer of thickness, 1.6 to 19 nm, representing the dome-shaped growing surfaces, the roughness layer(s) underneath have varying volume fraction of voids (2.3% in the case of YF_3 to 8.8% in the case of ZrO_2) and rather large thickness (75 nm for HfO_2 to 394 nm for ZrO_2), revealing the presence of long close-packed columns underneath the dome-shaped growing surfaces.

This is in complete accord with the model shown in Figs. 2 and 3.

C. Optical Functions of the Dielectric Film Materials

The values of the refractive index of these fluorides and oxides have been calculated, using Eq. (9), for selected wavelengths between 300 and 800 nm and are presented respectively in Tables VI and VII. Figures 32 and 33 show the dispersion curves of the fluoride and oxide films and of the vitreous silica substrate. Note that the refractive index of ScF_3 film is very close (within 0.002) to that of the substrate, over the entire wavelength range studied. Because of this very small difference in the refractive indices, the film exhibits minimal interference contrasts in the $[\Delta(\lambda), \Psi(\lambda)]$ spectra (Fig. 29a). For the same reason, the correlations between the model parameters for the ScF_3 film also become very strong if all nine unknown parameters in the three-layer model are allowed to vary. Additionally, the evaluated 90% confidence limits for the model parameters also become unreasonably large. Unfortunately, a thicker ScF_3 film with better interference contrasts was not available. Hence, in this case, during the regression analysis, the paramter A of the Sellmeier oscillator and f_{v3}, the volume fraction of void in the surface roughness layer, were kept constant. Finally, it should be noted that this study on ScF_3 film clearly demonstrates the power of SE in conjunction with an achromatic compensator to simultaneously depth-profile an in-

TABLE VI
Refractive Index of Fluoride Optical Coating Materials as a Function of Wavelength

λ(nm)	AlF_3	CeF_3	HfF_4	LaF_3	ScF_3	YF_3
300	1.398	1.668	1.595	1.629	1.499	1.593
350	1.393	1.648	1.582	1.616	1.484	1.585
400	1.390	1.636	1.574	1.608	1.474	1.580
450	1.388	1.628	1.569	1.603	1.468	1.577
500	1.386	1.622	1.565	1.599	1.464	1.574
550	1.385	1.618	1.562	1.596	1.461	1.573
600	1.384	1.615	1.560	1.594	1.459	1.571
650	1.384	1.613	1.558	1.593	1.457	1.570
700	1.383	1.611	1.557	1.592	1.455	1.569

TABLE VII

REFRACTIVE INDEX OF OXIDE OPTICAL COATING MATERIALS AS A FUNCTION OF WAVELENGTH

λ(nm)	γ-Al$_2$O$_3$	HfO$_2$	Sc$_2$O$_3$	ThO$_2$	Y$_2$O$_3$	ZrO$_2$
260	1.731	2.298	2.275	2.041	2.035	2.255
280	1.712	2.241	2.207	2.001	2.002	2.204
300	1.696	2.200	2.160	1.972	1.977	2.166
320	1.685	2.170	2.125	1.949	1.958	2.137
340	1.675	2.146	2.098	1.930	1.943	2.114
360	1.668	2.128	2.077	1.916	1.931	2.095
380	1.662	2.113	2.060	1.904	1.921	2.080
400	1.656	2.101	2.047	1.894	1.913	2.068
450	1.646	2.079	2.022	1.875	1.898	2.044
500	1.640	2.064	2.005	1.862	1.887	2.028
550	1.635	2.054	1.993	1.853	1.880	2.016
600	1.631	2.046	1.984	1.846	1.874	2.008
700	1.626	2.035	1.972	1.836	1.867	1.996
800	1.623	2.028	1.966	1.830	1.862	1.988

homogeneous film and also determine the refractive index and its dispersion, even when the refractive index of the film differs by as little as ~ 0.002 from that of the substrate.

The crystallinity of the films was also examined by x-ray diffraction. The x-ray diffraction peaks of the CeF$_3$, LaF$_3$, YF$_3$, HfO$_2$, Sc$_2$O$_3$, Y$_2$O$_3$ and ZrO$_2$ films indicated that they are polycrystalline. The other films did not show clear x-ray peaks, indicating that these films are either amorphous or composed of very fine grains. γ-Al$_2$O$_3$, HfO$_2$, Sc$_2$O$_3$, ThO$_2$ and Y$_2$O$_3$ crystallize in a cubic system, and hence these crystals are optically isotropic. But AlF$_3$, α-Al$_2$O$_3$ and ScF$_3$ crystallize in a trigonal system, while CeF$_3$, LaF$_3$, and YF$_3$ crystallize in a hexagonal system, and hence all these six materials are optically uniaxial in the crystalline phase. In other words, these materials possess two refractive indices, n_o and n_e, the ordinary and extraordinary indices. Hence, the values of the refractive index determined for these films are actually the average value n_{ave}, where $n_{ave} = (2n_o + n_e)/3$. On the other hand, our x-ray diffraction study revealed that the ZrO$_2$ film studied by us was monoclinic. Similarly, HfF$_4$ also crystallizes in the monoclinic system, and hence these two materials are optically biaxial. In

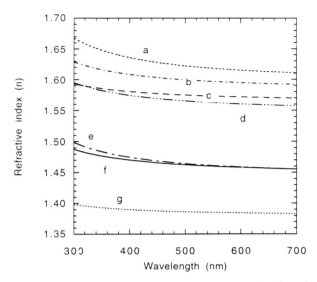

FIG. 32. Summary of the evaluated refractive index as a function of wavelength for the fluoride films. The corresponding data (Malitson, (44) on the dispersion of the vitreous silica substrate are also presented. (a) CeF_3, (b) LaF_3, (c) YF_3, (d) HfT_4, (e) ScF_3, (f) vitreous silica, (g) AlF_3.

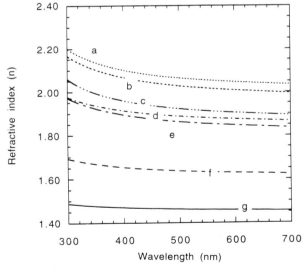

FIG. 33. Summary of the evaluated refractive index as a function of wavelength for the oxide films. The corresponding data (Malitson (44)) on the dispersion of the vitreous silica substrate are also presented. (a) HfO_2, (b) ZrO_2, (c) Sc_2O_3, (d) Y_2O_3, (e) ThO_2, (f) Al_2O_3, (g) vitreous silica.

these cases, the refractive index of the film represents $n_{ave} = (n_1 + n_2 + n_3)/3$, where n_1, n_2, n_3 represent the three principal refractive indices of the materials.

1. Comparison with Previous Results

Our $n(SE)$ data on the refractive indices and their dispersion with wavelength, of seven (CeF_3, LaF_3, Al_2O_3, HfO_2, Sc_2O_3, Y_2O_3, and ZrO_2) films, are compared with the published data on the corresponding films in Figs. 34 through 40. Similar data on the remaining (AlF_3, HfF_4, ScF_3, YF_3 and ThO_2) films are not available in the literature. However, it is appropriate to make a brief mention here of two studies (59,60) on the refractive indices of YF_3 and HfF_4, respectively, at one wavelength. Bezuidenhout and Clarke (59) have recently determined the value of the refractive index of a YF_3 film as 1.53 at 600 nm, assuming the film to be homogeneous, whereas our corresponding value obtained without making any such arbitrary assumption about the homogeneity of the film is much higher, as expected, and is 1.571 at 600 nm, as given in Table VI. HfF_4 crystallizing in the monoclinic system possesses three refractive indices n_1, n_2 and n_3. But only two of three refractive indices of HfF_4 at 589 nm have been determined and are given in the Landolt–Bornstein tables (60) as $n_1 = 1.54$ and $n_2 = 1.58$ for the bulk crystal. From Table VI it is seen that our value of n_{ave} at 600 nm, 1.560, matches the average of these two bulk values.

In all the figures, 34–40, our $n(SE)$ data are shown as bold continuous curves. In the few cases where data on the bulk crystalline samples are

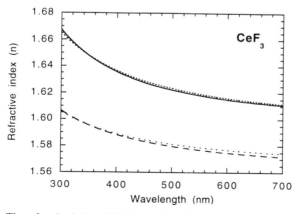

FIG. 34. The refractive index of CeF_3 as a function of wavelength. ———, present work; ----, Smith and Baumeister (55); - - - -, manufacturer; ———, Piegari and Emiliani.

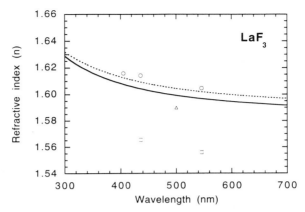

FIG. 35. The refractive index of LaF$_3$ as a function of wavelength. ——, present work; ----, Smith and Baumeister (55); ○ (bulk crystal), Wirick (56); △, Ogura (57); □, King and Downs (58).

available, such data are also plotted in the appropriate figures. Almost all these published data on thin films were determined by first measuring the film reflectance (R) or transmission (T), or both, as a function of wavelength, and then evaluating the parameters of the Sellmeier dispersion equation by a curve-fitting technique. For this it is usually assumed either that the film is homogeneous, or that the inhomogeneity (such as the

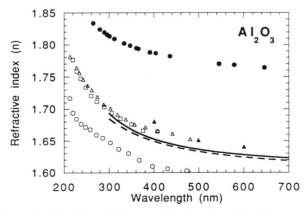

FIG. 36. The refractive index of Al$_2$O$_3$ as a function of wavelength. ——, present work; □, Starke et al. (62), reactive ion plated (IAD) film; △, Starke et al. (62), ion beam sputtered (IAD) film; ○, Starke et al. (62), electron beam evaporated film; ▲ Masetti et al. (63); ———, manufacturer; ●, Malitson (64) (single crystal α-Al$_2$O$_3$).

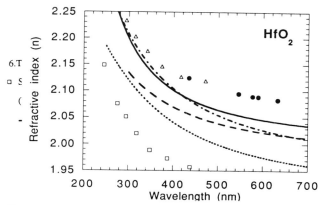

FIG. 37. The refractive index of HfO$_2$ as a function of wavelength. ———, present work; -·-·-, Smith and Baumeister (55); △; Starke et al. (64) (IAD) film—high packing density; □, Starke et al. (64) (IAD) film—low packing density; ----, Borgogno et al. (2); — — —, manufacturer; ●, Aleksandrov et al. (65) (bulk specimen).

distribution of the void fraction) varies linearly with thickness—and as pointed out earlier, both these assumptions are not strictly valid for any of these films. Most of these films were prepared by the usual electron-beam or thermal evaporation techniques. In a few cases the films were prepared by ion-assisted deposition (IAD) techniques such as reactive ion plating

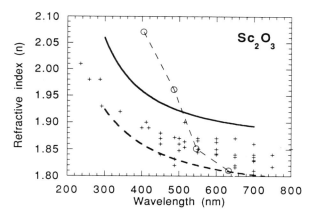

FIG. 38. The refractive index of Sc$_2$O$_3$ as a function of wavelength. ———, present work; +, results of "round robin" experiments (3); ⌀, single-wavelength ellipsometry results (3); — — —, manufacturer.

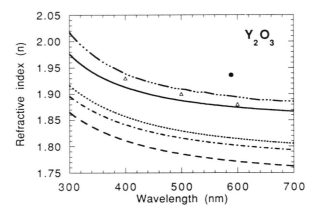

FIG. 39. The refractive index of Y_2O_3 as a function of wavelength. ——, present work; -··-, Smith and Baumeister (55); △, Masetti et al. (63); ----, Borgogno et al. (2); ———, manufacturer; ● (bulk single crystal), Nigara (66); -····-. Scaglione et al. (67) (IAD) film.

(RIP) or ion beam sputtering (IBS). Published data on such IAD samples are indicated in the figures at appropriate places. However, depending on the energy of the "assisting ion" the packing density of the deposited film can vary considerably. Hence, in such cases (see Fig. 37), where data are available on samples prepared with different preparatory conditions of IAD, only values at the two extremes are presented. The manufacturer of our samples had also carried out some measurements on reflectance as a function of wavelength on nominally the same samples as ours and analyzed

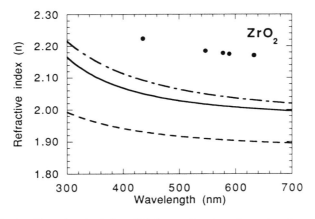

FIG. 40. The refractive index of ZrO_2 as a function of wavelength. ——, present work; -··-, Smith and Baumeister (55); ———, manufacturer; ●, Aleksandrov et al. (65) (bulk).

these data to obtain the optical function $n(\lambda)$ of the films, assuming the films to be homogeneous. These data are also included in the figures as bold dashed curves.

Before discussing these results on the optical functions $n(\lambda)$ of the film materials, it is necessary to make the following comments on the $n(\lambda)$ data on Al_2O_3 and Sc_2O_3 films.

Al_2O_3 films: The starting sample for the deposition of Al_2O_3 film was analytical grade γ-Al_2O_3 grains of 1 μm size. γ-Al_2O_3 is cubic, whereas α-Al_2O_3 is rhombohedral. According to Clausen and Hren (*61*), γ-Al_2O_3 recrystallizes to α-Al_2O_3 on prolonged heating at high temperature. Since an e-beam evaporation procedure was employed to deposit our film in high vacuum, it is most likely that the film will be polycrystalline γ-Al_2O_3 and not α-Al_2O_3, because the sample was not subjected to prolonged heating during or even subsequent to deposition of the film. It is seen from Fig. 36 that the value of n of our Al_2O_3 film is much lower (by about 0.12) than that reported by Malitson (*64*) for bulk single-crystalline α-Al_2O_3. From Fig. 36 it is also seen that the same is the case for every Al_2O_3 film studied thus far, indicating that all these Al_2O_3 films are made up of polycrystalline γ-Al_2O_3, and not of α-Al_2O_3. This is understandable when we consider the fact that all these Al_2O_3 films were deposited from vapor phase using γ-Al_2O_3 as the starting material, and that none of them were subjected to prolonged heating at very high temperatures to convert the γ-Al_2O_3 to α-Al_2O_3 as required (*65*). Hence, it is not surprising to find lower values for the refractive index of Al_2O_3 film, corresponding to that of γ-Al_2O_3.

Sc_2O_3 films: Recently the Optical Society of America chose this film for a case study to perform an unbiased performance test of all existing optical characterization techniques (see Arndt *et al.* (*3*)). Seven renowned laboratories around the world were asked to characterize the optical properties and thicknesses of Sc_2O_3 films prepared in the same deposition run. Six of these seven laboratories employed different methods to analyze the R and T spectra to evaluate the refractive index and thickness of the film. One laboratory employed null ellipsometry with multiple angle-of-incidence measurements. Only three of these laboratories used the linearly varying inhomogeneous film model, while all the others assumed the films to be homogeneous. The results of such a "round-robin" experiment (*3*) are entered in Fig. 38 as data points distributed in the 400–750 nm spectral range. From these results it is evident that the variations in the evaluated refractive index can be as much as 0.08 depending on the method employed and the model assumed for the analysis of the results. Figure 38 also shows our present results for n(SE) on Sc_2O_3 (solid curve) along with the data

obtained by the manufacturer assuming the film to be homogeneous (dashed curve). It is seen that our n(SE) values are much higher than the rest since the effect of voids has been properly taken into account. It may be noticed that the two isolated data points at 405 and 486 nm plotted above n(SE) are the results of null ellipsometry studies in the round-robin experiments. The very strong dispersion, and the fact that these data points are well separated from the others, can be attributed to the higher sensitivity of the ellipsometric measurements to film inhomogeneity. In other words, in the analysis of single-wavelength null ellipsometry data, it is almost impossible to properly take into account the inhomogeneity features, and hence the neglect of the inhomogeneity in the film leads to erroneous results.

From an analysis of the comparative data given in Figs. 34 to 40, the following facts are evident:

(i) In every case our n(SE) values are larger than the values determined by the manufacturer. Even though their films were identical to our samples (they were prepared in the same batch as our specimens), their results are consistently lower than our results—first because they employed the spectro-reflectance (R) data alone for their analyses. In other words, they ignored the phase information contained in the reflected light, and as pointed out earlier in this article, the "phase spectrum" contains the details of the inhomogeneity in the film. Secondly, they made an arbitrary assumption that the films are homogeneous, which again is not true. Treating an inhomogeneous voided film as a homogeneous void-free film will naturally yield an effective value of the refractive index of the film that is lower than its true value.

(ii) In general, our n(SE) values are in good agreement ($\delta n < 0.02$) with those obtained on IAD samples with highest packing density, and they are also higher than the values evaluated from the (R, T) measurements on films prepared by conventional techniques such as electron-beam or thermal evaporation techniques. It is well known that the packing density of the films prepared by IAD techniques can be very high. This implies that in our spectro-ellipsometric method, the presence of voids and their inhomogeneous distribution in the film are *properly* taken into account *before* the optical function is evaluated from the experimentally observed SE data, even though the films were prepared by the electron beam evaporation technique.

(iii) Not *all* IAD films have refractive indices equal to those of the bulk samples. More about this is discussed in the next section.

VI. Summary and Conclusion

It has been shown that by incorporating an achromatic three-reflection quarter wave compensator in a spectroscopic ellipsometer and applying appropriate calibration and error correction procedures, one can successfully detect and delineate the inhomogeneous distribution of voids in transparent films on transparent substrates and provide at the same time all the parameters of the Sellmeier equation describing the dispersion of the refractive index of the film material. Thus, it is no exaggeration to state that the long-standing problem of nondestructive characterization of inhomogeneous transparent films on transparent substrates, which had baffled scientists and engineers for decades, has been overcome. It has been shown that the phase spectrum carries information on inhomogeneities in the film, and thus the success of this non-destructive technique is due to the use of the phase spectrum in addition to the amplitude (or the reflectance) spectrum in the final computations.

Studies on 12 transparent fluoride and oxide optical coatings on vitreous silica substrates reveal that every one of them is inhomogeneous because of either the non-uniform distribution of the void content in the film, or the presence of micro-roughness of the surface—or a combination of both of these inhomogeneities. By properly taking into account these inhomogeneities, one can obtain the true optical function of the film material from the observed spectro-ellipsometric data, as has been shown by good agreement of the $n(\lambda)$ data with those obtained on samples prepared by IAD (ion-assisted deposition) techniques under optimum conditions. However, as can be seen from Fig. 37, not all IAD samples possess optimum characteristics of the bulk film material. If the ion energy is too small, the impinging atoms will not have enough surface mobility to fill in all the voids, and at the same time if the ion energy is too high, then some of the ions will get trapped in the film, and these highly energetic ions can produce damages or strains in the films. In other words, there is a narrow window for the optimum energy of the "assisting" ions in the film deposition process. This is also brought out pictorially in Fig. 2 by Guenther (12) in his revised structure zone model. This means that for every material there are optimum values for the film deposition parameters to obtain reproducibly void- and strain-free, smooth, stoichiometric films. Till now there was no reliable technique that could nondestructively detect and characterize the presence of such voids in transparent optical films. But now with the development of SE with an achromatic compensator this limitation has been overcome. The results of ScF_3 film (*vide* Fig. 32 and Table VI) clearly bring out the power of this technique to simultaneously depth-profile an inhomogeneous film, and also determine the refractive index and its dispersion even if the refractive index of the film differs by as little as 0.002 from that of the substrate. Further, since this technique can easily be adapted for *in situ* operation for monitoring and control of the process parameters during film deposition using IAD, it should be possible to eliminate the columnar structure, the voids, and the surface micro-roughness in the films. Having thus eliminated these three most common and major causes for the inhomogeneity in films as listed at the beginning of Section II, we should soon be able to produce almost perfect transparent optical coatings, even on transparent substrates, in a reproducible fashion.

References

1. J. P. Borgogno, F. Flory, P. Roche, B. Schmitt, G. Albrand, E. Pelletier, and H. A. Macleod, "Refractive index and inhomogeneity of thin films," *Appl. Opt.* **23**, 3567–3571 (1984).
2. J. P. Borgogno, B. Lazarides, and E. Pelletier "Automatic determination of the optical constants of inhomogeneous thin films,"*Appl. Opt.* **21**, 4020–4029 (1982).
3. D. P. Arndt, R. M. A. Azzam, J. M. Bennett, J. P. Borgogno, C. K. Carniglia, W. E. Case, J. A. Dobrowolski, U. J. Gibson, T. T. Hart, F. C. Ho, V. A. Hodgkin, W. P. Klapp, H. A. Macleod, E. Pelletier, M. K. Purvis, D. M. Quinn, D. H. Strome, R. Swenson, P. A. Temple, and T. F. Thonn, "Multiple determination of the optical constants of thin-film coating materials," *Appl. Opt.* **23**, 3571–3596 (1984).
4. A. Piegari and G. Emiliani, "Analysis of inhomogeneous thin films by spectro-photometric measurements," *Thin Solid Films* **171**, 243–250 (1989).
5. J. M. M. De Nijs and A. van Silfhout, "Systematic and random errors in rotating-analyzer ellipsometry," *J. Opt. Soc. Am. A* **5**, 773–781 (1988).
6. B. A. Movchan and A. V. Demchishin, "Study of the structure and properties of thick vacuum condensates of nickel, titanium, tungsten, aluminum oxide and zirconium dioxide," *Phys. Met. Metallogr.* **28**(4), 83–90 (1969).
7. K. H. Guenther, "The influence of the substrate surface on the performance of optical coatings," *Thin Solid Films* **77**, 239 (1981).
8. K. H. Guenther, "Nodular defects in dielectric multilayers and thick single layers," *Appl. Opt.* **20**, 1034–1038 (1981).
9. J. A. Thornton, "Influence of apparatus geometry and deposition conditions on the structure and topography of thick sputtered coatings," *J. Vac. Sci. Technol.* **11**, 666–670 (1974).
10. R. Messier, A. P. Giri, and R. A. Roy, "Revised structure zone model for thin film physical structure," *J. Vac. Sci. Tech.* **A2**, 500–503 (1984).
11. B. Yang, B. L. Walden, R. Messier, and W. B. White, "Computer simulations of the cross-sectional morphology of thin films," *Proc. SPIE* **821**, 68–76 (1987).
12. K. H. Guenther, "Microstructure of vapor-deposited optical coatings," *Appl. Opt.* **23**, 3806–3816 (1984).
13. K. H. Guenther, "Revisiting structure zone models for thin film growth," *Proc. SPIE* **1324**, 2–12 (1990).

14. R. I. Seddon, ed., "Optical Thin Films," *Proc. SPIE* **325**, 1–185 (1982).
15. R. I. Seddon, ed., "Optical Thin Films II: New Developments," *Proc. SPIE* **678**, 1–220 (1986).
16. R. Herrmann, ed., "Optical Thin Films and Applications," *Proc. SPIE* **1270**, 1–305 (1990).
17. R. I. Seddon, ed., "Optical Thin Films III: New Developments," *Proc. SPIE* **1323**, 1–376 (1990).
18. P. J. Martin, "Ion-based methods for optical thin film deposition," *J. Mat. Sci.* **21**, 1–25 (1986).
19. F. K. Urban III and A. I. Bernstein, "Cluster size in ionized cluster beam deposition," *Proc. SPIE* **1323**, 8–18 (1990).
20. J. Edlinger and H. K. Pulker, "Ion currents and energies in reactive low-voltage ion-plating: Preliminary results," *Proc. SPIE* **1323**, 19–28 (1990).
21. K. H. Gunther, "Recent advances in reactive voltage ion plating deposition," *Proc. SPIE* **1323**, 29–38 (1990).
22. F. Flory, "Comparison of different technologies for high quality optical coatings," *Proc. SPIE* **1270**, 172–183 (1990).
23. A. Starke, H. Schink, J. Kolbe, and J. Ebert, "Laser induced damage thresholds and optical constants of ion plated and ion beam spluttered Al_2O_3- and HfO_2-coatings for the ultraviolet," *SPIE* **1270**, 299–304 (1990).
24. R. M. A. Azzam and N. M. Bashara, "Ellipsometry and Polarized Light," North-Holland, Amsterdam, 1977. Chapters 5 and 6, pp. 411–433.
25. D. E. Aspnes, "Optimizing precision of rotating-analyzer ellipsometers," *J. Opt. Soc. Am.* **64**, 639–646 (1974).
26. D. E. Aspnes, "Effects of component optical activity in data reduction and calibration of rotating-analyzer ellipsometers," *J. Opt. Soc. Am.* **64**, 812–819 (1974).
27. D. E. Aspnes and A. A. Studna, "High precision scanning ellipsometer," *Appl. Opt.* **14**, 220–228 (1975).
28. D. E. Aspnes, J. B. Theeten, and F. Hottier, "Investigation of effective-medium models of microscopic surface roughness by spectroscopic ellipsometry," *Phys. Rev. B* **20**, 3292–3302 (1979).
29. D. E. Aspnes, "Microstructural information from optical properties in semiconductor technology," *Proc. SPIE* **276**, 188–195 (1981).
30. D. E. Aspnes, "Studies of surface, thin film and interface properties by automatic spectroscopic ellipsometry," *J. Vac. Sci. Technol.* **18**, 289–295 (1981).
31. D. E. Aspnes and A. A. Studna, "Dielectric functions and optical parameters of Si, Ge, GaP, GaAs, GASb, InAs, InP, and InSb from 1.5 to 6.0 eV," *Phys. Rev.* **B27**, 985–1008 (1983).
32. R. W. Collins, "Automatic rotating element ellipsometer: Calibration, operation and real-time applications," *Rev. Sci. Instrum.* **61**, 2029–2062 (1990).
33. P. J. McMarr, K. Vedam, and J. Narayan, "Spectroscopic ellipsometer: A new tool for nondestructive depth-profiling and characterization of interfaces," *J. Appl. Phys.* **59**, 694–701 (1986).
34. P. Chindaudom and K. Vedam, "Determination of the optical function $n(\lambda)$ of vitreous silica by spectroscopic ellipsometry using an achromatic compensator," *Appl. Opt.* **33**, 2664–2671 (1994).
35. G. E. Jellison and B. C. Sales, "Determination of the optical functions of transparent glasses by using spectroscopic ellipsometry," *Appl. Opt.* **30**, 4310–4315 (1991).
36. P. Chindaudom and K. Vedam, "Determination of the optical constants of an inhomogeneous transparent LaF_3 thin film on a transparent substrate by spectroscopic ellipsometry," *Opt. Lett.* **17**, 538–540 (1992).

37. A. E. Oxley, "On apparatus for the production of circularly polarized light," *Phil. Mag.* **21**, 517–532 (1911).
38. V. A. Kizel, Y. I. Krasilov, and V. N. Shamraev, "Achromatic $\lambda/4$ device," *Opt. Spectroscopy* **17**, 248–249 (1964).
39. R. J. King and M. J. Downs, "Ellipsometry applied to films on dielectric substrates," *Surf. Sci.* **16**, 288–302 (1969).
40. J. E. Bennett and H. E. Bennett, "Achromatic retardation plates," in *"*Handbook of Optics*"* (W. G. Driscoll and W. Vaughan, eds.), McGraw-Hill, New York, 1978, Section 10, pp. 115–124.
41. D. E. Aspnes and A. A. Studna, "Methods for drift stabilization and photomultiplier linearization for photometric ellipsometers and polarimeters," *Rev. Sci. Instr.* **49**, 291–297 (1978).
42. S. H. Russev, "Correction for nonlinearity and polarization-dependent sensitivity in the detection system of rotating analyzer ellipsometers," *Appl. Opt.* **28**, 1504–1507 (1989).
43. D. A. Bruggeman, "Berechnung verschiedener physikalisher Konstanten vor heterogenen Substanzen," *Ann. Phys. (Leipzig)* **24**, 636 (1935).
44. I. H. Malitson, "Interspecimen comparison of the refractive index of fused silica," *Opt. Soc. Am.* **55**, 1205 (1965).
45. S. Y. Kim and K. Vedam, "Proper choice of the error function in modeling spectro-ellipsometric data," *Appl. Opt.* **25**, 2013–2021 (1986).
46. A. A. Tesar, B. A. Fuchs, P. Paul Hed, "Examination of the polished surface character of fused silica," *Appl. Opt.* **31**, 7164–7172 (1992).
47. G. E. Jellison, Jr. and F. A. Modine, "Two-channel polarization modulation ellipsometer," *Appl. Opt.* **29**, 959–974 (1990).
48. K. Vedam and S. Y. Kim, "Simultaneous determination of refractive index, its dispersion and depth-profile of magnesium oxide thin film by spectroscopic ellipsometry," *Appl. Opt.* **28**, 2691–2694 (1989).
49. R. E. Stephans and I. H. Malitson, "Index of refraction of magnesium oxide," *J. Res. Natl. Bur. Stand.* **49**, 249–252 (1952).
50. P. Chindaudom, and K. Vedam, "Studies on inhomogeneous transparent optical coatings on transparent substrates by spectroscopic ellipsometry," *Thin Solid Films* **234**, 439–442 (1993).
51. P. Chindaudom, and K. Vedam, "Characterization of inhomogenous transparent thin films on transparent substrates by spectroscopic ellipsometry: refractive indices $n(\lambda)$ of some fluoride coating materials," *Appl. Opt.* **33**, 2664–2671 (1994).
52. K. Vedam, P. J. McMarr, and J. Narayan, "Nondestructive depth profiling by spectroscopic ellipsometry," *Appl. Phys. Letters* **47**, 339–341 (1985).
53. J. Narayan, S. Y. Kim, K. Vedam, and R. Manukonda, "Formation and nondestructive characterization of ion implanted silicon-on-insulator layers," *Appl. Phys. Letters* **51**, 343–345 (1987).
54. T. Lohner, W. Skorupa, M. Fried, K. Vedam, N. Nguyen, R. Grotzschel, H. Bartsch, and J. Gyulai, "Comparative study of the effect of annealing of nitrogen implanted silicon-on-insulator structures by spectroscopic ellipsometry, cross-sectional transmission electron microscopy, and Rutherford backscattering spectroscopy," *Mat. Sci. Eng.* **B12**, 177–184 (1992).
55. D. Smith and P. Baumeister, "Refractive index of some oxide and fluoride coating materials," *Appl. Opt.* **18**, 111–115 (1979).
56. M. P. Wirick, "The near ultraviolet optical constants of lanthanum fluoride," *Appl. Opt.* **5**, 1966–1967 (1966).

57. S. Ogura. Cited by Targrove et al., *Appl. Opt.* **26**, 3733 (1987).
58. R. J. King and M. J. Downs, "Ellipsometry applied to films on dielectric substrates," *Surf. Sci.* **16**, 288–302 (1969).
59. D. F. Bezuidenhout and K. D. Clarke, "The optical properties of YF_3 films," *Thin Solid Films* **155**, 17–30 (1987).
60. Landolt and Bornstein Tables, Band II Teil 8, p. 153 (1962).
61. E. M. Clausen, Jr. and J. J. Hren, "The gamma to alpha transformation in thin film alumina," *Mat. Res. Soc. Symp. Proc.* **41**, 381–386 (1985).
62. A. Starke, H. Schink, J. Kolbe, and J. Ebert, "Laser induced damage thresholds and optical constants of ion plated and ion beam sputtered Al_2O_3- and HfO_2-coatings for the ultraviolet," *SPIE* **1270**, 299–302 (1990).
63. E. Masetti, A. Piegari, and A. Tirabassi, "Optical characterization of low-absorbing thin films in the visible and infrared spectrum," *SPIE* **1270**, 125–132 (1990).
64. I. H. Malitson, "Refraction and dispersion of synthetic sapphire," *J. Opt. Soc. Am.* **52**, 1377–1379 (1962).
65. V. I. Aleksandrov, V. F. Kalabukhova, E. E. Lomonova, V. V. Osiko, and V. I. Tatarintsev, "Influence of impurities and annealing conditions on the optical properties of single crystals of ZrO_2 and HfO_2," *Inorg. Mater.* **13**, 1747–1750 (1977).
66. Y. Nigara, "Measurement of the optical constants of yttrium oxide," *Japan. J. Appl. Phys.* **7**, 404–408 (1968).
67. S. Scaglione, D. Flori, I. Soymi, and A. Piegari, "Laser optical coatings produced by ion assisted deposition," *Thin Solid Films* **214**, 188–193 (1992).

Characterization of Ferroelectric Films by Spectroscopic Ellipsometry

S. TROLIER-MCKINSTRY, P. CHINDAUDOM, K. VEDAM, AND R. E. NEWNHAM

*Materials Research Laboratory,
Pennsylvania State University, University Park, Pennsylvania*

I. Introduction 249
II. Description of the SE System 254
III. Experimental Procedure 256
IV. Results and Discussion 258
    A. Epitaxial and Oriented Films on Single-Crystal Substrates 258
    B. *In-situ* Annealing Studies on Ferroelectric Films 264
        1. MIBERS Films on Sapphire 264
        2. Sol–Gel PZT Films 268
V. Relation between Film Microstructure and Electrical Properties 272
VI. Conclusions 275
    Acknowledgments 276
    References 276

I. Introduction

Ferroelectric thin films have attracted considerable interest over the past several years as potential candidates for pyroelectric sensors, ferroelectric memory elements, electrooptic switches, and miniature electromechanical transducers. While numerous techniques have been utilized to prepare multi-cation film compositions, it is clear from a review of the literature that thin-film ferroelectrics rarely display the same properties shown by bulk materials of the same composition (*1*). Four major differences can be noted in the electrical properties alone: Thin films usually show larger coercive fields, smaller remanent polarizations, lower dielectric constants, and a smeared paraelectric/ferroelectric transition (determined from the dielectric constant as a function of temperature) than do bulk ferroelectrics (See, for example, Table 1). Similar observations are apparent in most of the reported data for ferroelectric thin films. It is also clear from the literature that the electrical properties reported for a given type of film depend on the processing conditions utilized. Consequently, films produced by one technique may display markedly different properties from others of the same composition and thickness.

Several mechanisms can contribute to the differences between thin film and bulk properties. Included among these are microstructural heterogeneities, imperfect crystalline quality, mechanical stresses imposed on the film by the substrate, space charge effects, and finally, intrinsic size effects. While each of these factors can lead to *apparent* size effects, they may not reflect an *intrinsic* change in the properties with size. Thus, although there should be a fundamental limitation on the size to which ferroelectricity is stable, many other factors can mask this intrinsic limit. In particular, the first two mechanisms for size effects listed are important primarily because many film

TABLE I

REPRESENTATIVE DATA FOR THE REMANENT POLARIZATION AND COERCIVE FIELD OF $BaTiO_3$, $PbTiO_3$, AND PZT THIN FILMS

Material	$P_r(\mu C/cm^2)$	$E_c(kV/cm)$	Reference
Single-crystal $PbTiO_3$	~55–75	6.75	(2)
Laser Ablated $PbTiO_3$	60–80	50–280	(3)
Sputtered (001) $PbTiO_3$	55	75	(4)
Sputtered (001) $PbTiO_3$	35	160	(5)
Sputtered or CVD $PbTiO_3$	12	250	(6)
CVD (001) $PbTiO_3$	14.1	20.16	(7)
Bulk PZT 58/42	45	17	(8)
Sol–gel PZT	36		(9)
Sol–gel PZT 52/48	35		(11)
Laser Ablated PZT	35–45	55–60	(10)
Sol–gel PZT	18–20	50–60	(12)
Sol–gel PZT 40/60	6.6	26.7	(13)
MOD PZT 53/47	4–10	62–125	(14)
Sputtered PZT 58/42	30.0	25	(8)
Sputtered PZT 65/35, weak (100)	12.5	90	(15)
Sputtered PZT 65/35	3.6	33	(15)
Reactively Sputtered PZT 50/50	25	60	(16)
Single-crystal $BaTiO_3$	26	1	(17)
Polycrystalline $BaTiO_3$	8	3	(17)
Sol-gel $BaTiO_3$	8	25	(18)
Screen-printed $BaTi_{0.95}Sn_{0.05}O_3$	1.7–2.8	25	(19)
Sputtered (001) $BaTiO_3$	7	60	(20)
Sputtered $BaTiO_3$, weak (101)	16	20	(20)

preparation techniques produce imperfect specimens. Similar observations have been made on oxide materials for optical coatings (21–24).

Inhomogeneity in the film microstructure can take the form of incorporated porosity, surface and interface roughness, or variations in the grain size. One source of these heterogeneities is the columnar microstructure common in both vapor-deposited and chemically prepared films. Such inhomogeneities in the microstructure can be formed during the deposition process itself, though they may be altered during post-deposition annealing. In the work of Fox et al. (25), for example, it was found that ion-beam-sputtered (Pb, La)TiO$_3$ films on Pt-coated Si substrates contain microstructural features on three distinct size scales. At the finest scale the film was composed of primary particles, which grouped together to form fibers visible in the SEM. These fibers were then packed together to form a high-density amorphous film. At the largest scale, the fiber bundles were believed to be clustered into larger groupings. Upon annealing, this initial microstructure was modified both by the vaporization of some of the excess PbO in the film (often resulting in pore formation) and by crystallization of the perovskite phase. During this process, the cluster boundaries were significantly enhanced, particularly for films with high excess PbO contents. The crystallization process also altered the microstructure when crystallites grew to encompass several of the original fiber diameters (25). The resulting inhomogeneities in the films were found to play a significant role in the magnitude of the measured film dielectric constant and loss, the electrical resistivity, and the remanent polarization (26). In addition, as is shown in one of the following sections, a columnar microstructure can appreciably lower the dielectric constant and increase the coercive field of ferroelectric films. Moreover, in electro-optic devices for optical signal processing applications, entrapped porosity or surface roughness cause additional scattering and attenuation of the light signal. Within a single film, Dudkevich (27) has also shown that the grain size can vary through the thickness. Thus, for sputtered Ba$_{0.85}$Sr$_{0.15}$TiO$_3$ films, the grain size was small near the substrate (15 nm for a 4 nm thick film) and larger at the surface of thicker films (200–300 nm for a film 2,000 nm thick). Given this type of microstructural heterogeneity, it is not surprising that many properties appear to be dependent on the preparation conditions and the film thickness.

Good crystallinity is critical in ferroelectrics films if they are to display characteristic bulk properties. Many techniques for producing films can, however, result in either defective crystallites or incomplete crystallization of the ferroelectric. During sputter deposition, for example, the growing film is bombarded with high energy species. This can be problematic if the damage produced by these collisions is not removed. On the other hand, even in chemically prepared films where no bombardment occurs, poor crystallinity can result if low annealing temperatures are used.

The effect of high energy bombardment on the properties of ferroelectrics has been investigated using fast neutron irradiation (neutron energy > 50 keV). In $BaTiO_3$ single crystals, irradiation produces a high concentration of defects, which in turn lead to conversion to a metastable cubic state with a larger lattice parameter. Once this occurs, the material does not undergo a paraelectric to ferroelectric transition to temperatures as low as 78 K, even though the majority of the sample is still crystalline. In order to remove the damage, the material must be annealed to 1,000°C. Similar results have been reported for mechanically damaged ferroelectric materials (29). Such high anealing temperatures cannot be used for lead containing thin films, though, as that would lead to volatilization of the Pb. As a result, it is possible that the relatively low growth and annealing temperatures used in making some sputtered ferroelectric films is insufficient to remove all of the damage resulting from the bombardment. It is interesting that many authors working with either vacuum deposited or chemically prepared thin films report modifications of the perovskite structure with a slightly expanded cubic unit cell (30).

In one of the most complete characterizations of this effect, Surowiak et al. measured lattice strains in sputter-deposited $BaTiO_3$ films. They found that films that had experienced heavy bombardment during growth tended to have more heavily deformed crystallites and small coherent scattering sizes. Larger, less defective crystallites could be formed when the growth conditions were not as rigorous (20). Differences between the two types of films were readily apparent in the electrical properties: lower deformations values were associated with higher remanent polarizations, piezoelectric constants close to single-crystal values, and relatively narrow phase transitions. This last point was examined by Biryukov et al., who demonstrated that films with large lattice strains should be expected to show diffuse phase transitions (31). It is important to note that poorly crystallized films from any preparation method will probably display low remanent polarization, lowered dielectric and electromechanical coupling constants, and diffuse phase transitions.

It is not surprising that mechanical stresses should affect the properties of many ferroelectric thin films. The appearance of the order parameter at the transition temperature is accompanied in perovskite ferroelectrics by a spontaneous strain. The domain structure, then, should be infuenced by the strains present in thin films. This could significantly affect the reversal of domains in thin films, and hence the shape of the hysteresis loop. Forsbergh has also shown that two-dimensional stresses can stabilize the ferroelectric phase in tetragonal perovskites to higher temperatures in bulk materials (32), and this mechanism could well operate in thin films when there is good cohesion between the substrate and the film (33).

In ferroelectric materials it is also necessary to compensate the polarization at the surface of the material. This can lead to intrinsic size effects in thin films (34). Batra and Wurfel have shown that when the conductivity of the electrodes on a thin wafer is insufficient to completely compensate the polarization, the transition temperature of the ferroelectric is depressed, the spontaneous polarization levels are decreased, and for thin films, ferroelectricity becomes unstable (35,36). The importance of space charge and complete compensation of the spontaneous polarization on the properties of ferroelectric films was verified experimentally with a triglycine sulfate film electroded asymmetrically with doped silicon on one side and gold on the other (37). Since the carrier concentration of positive and negative species in the semiconductor were not equivalent, it was possible to completely destabilize the polarization direction for which compensation could not be achieved. Consequently, for 100 Hz cycling, the film behaved as a polar, not a ferroelectric material. When, however, the semiconducting electrode was illuminated with high-intensity light, more minority carriers were available to compensate the spontaneous polarization, and a full hysteresis loop could be achieved (37).

Despite this variety of mechanisms for size effects, for some compositions (particularly lead zirconate titanate) it is possible to prepare thin films with dielectric constants and remanent polarizations comparable to bulk values (11). This has been done by sol-gel processing, laser ablation, and sputtering, among other techniques. For such films, the coercive field is at least twice that of a bulk ferroelectric, perhaps because of the smaller grain sizes in films.

As progress is made in preparing high quality ferroelectric thin films, future work on size effects will probably evolve towards the investigation of intrinsic size effects. In particular, it will be interesting to learn whether the film thickness, the primary particle size in the film, or the coherent scattering region (27) is the dimension that dominates the change in properties with size. In addition, the importance of film stress and the electrical compensation of the polarization must be studied further.

Complete characterization of the structural and microstructural features of ferroelectric films would be useful in determining which mechanisms are responsible for any observed deviations from bulk properties. This in turn may affect whether thin films can be used in some of the proposed applications. Unfortunately, as is the case for transparent films used in optical coatings, non-destructive characterization of ferroelectric film as a function of thickness has, in the past, been difficult to perform. Recently we have developed spectroscopic ellipsometry (SE) as a non-destructive, non-invasive tool for depth-profiling ferroelectric thin films on both transparent and absorbing substrates. The difficulties previously inherent in measurements on all-transparent samples have been overcome via the development of a series of systematic calibrations for a rotating analyzer spectroscopic ellipsometer (*38–41*). The information derived from this technique is particularly useful in correlating the role of the processing procedure in controlling the physical structure with the net electrical properties of ferroelectric thin films. Consequently, it is possible to use SE as a characterization tool for the study of microstructure/structure/property relationships in ferroelectric thin films.

In order to investigate the role of preparation conditions in controlling the microstructure of ferroelectric films, lead-based perovskite films deposited by rf magnetron sputtering, multi-ion-beam reactive sputtering (MIBERS), and sol–gel spin-on techniques were obtained from several sources. These were then characterized optically by spectroscopic ellipsometry, and where possible the results were correlated with observations from electrical and structural investigations.

II. Description of the SE System

A schematic of the rotating analyzer spectroscopic ellipsometer is shown in Fig. 1; a complete description of the instrument has been given elsewhere (*38, 40*). A removable three-reflection achromatic compensator of the type described by King and Downs was used in measurements of transparent films on transparent substrates (*42*). Corrections for the first-order errors in the optics were made following the calibration procedure suggested by Aspnes (*43*) with the compensator removed from the optical rails. Additional sample-independent systematic corrections were employed to minimize errors due to variations in the ac/dc gain with the voltage applied to the photomultiplier tube and the dark current. Finally, an effective source correction was utilized to eliminate wavelength-dependent errors associated with the compensator. A complete description of the calibration procedures used is given elsewhere (*39, 41*). With these corrections in place, Ψ and Δ can be measured on either transparent or opaque samples with an estimated accuracy of 0.01° and 0.03°, respectively, between 300 and 800 nm.

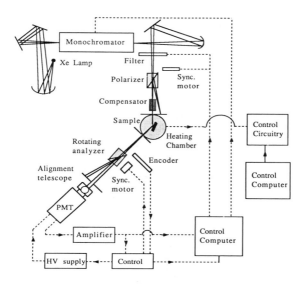

FIG. 1. Schematic of the rotating analyzer spectroscopic ellipsometer.

To extend the temperature range of the instrument, a computer-controlled windowless electrical resistance furnace was built to house the sample. As shown in Fig. 2, a kanthal-wire wrapped alumina tube was used as the heat source. Temperatures between 25 and 650°C could be achieved easily with this unit, where the upper limit was controlled by excessive deformation of the aluminum sample mount. However, a practical upper limit for *in situ* ellipsometric measurements was ~400–450°C, where glow from the furnace significantly decreased the signal/noise ratio. Typically, temperatures could be maintained within ± 2°C of the setpoint and ramps between one and 10°C/min could be achieved.

Two monolithic outer cylinders for the furnace were machined with fixed incidence angles of 70 and 80° for use with films on metallic and transparent substrates, respectively. Both these brass cylinders and the baseplate were subsequently electroplated with nickel to minimize oxidation of the copper and vaporization of the zinc at elevated temperatures. Experiments on a sapphire substrate demonstrated that there are no systematic changes in the ellipsometric parameters Δ and Ψ associated with the furnace for temperatures between 25 and 600°C. The maximum deviations for Δ and Ψ values over that temperature range were ± 0.10° and ± 0.03°. The standard deviation of the data from the average values for $\Delta(\lambda)$ and $\Psi(\lambda)$ were considerably lower, being 0.025° and 0.013°, respectively.

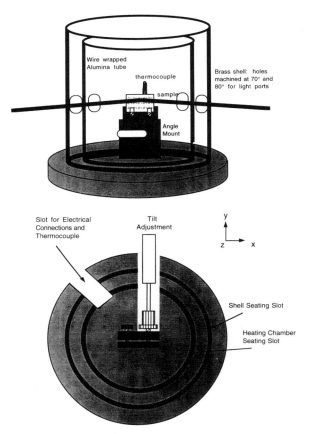

FIG. 2. Electrical resistance furnace and sample mount. (a) Side view. (b) Top view of the baseplate and sample holder. The entire baseplate can be translated along y and rotated about z to permit alignment of the sample at any temperature. Rotation of the sample about the x axis was performed with the worm gear and screw shown in the figure. This could also be adjusted at any temperature.

III. Experimental Procedure

Experimental data were collected at each wavelength successively by measuring the light intensity as a function of the analyzer position for 50 mechanical rotations of the analyzer. Incidence angles of 70° and 80°, respectively, were used for ferroelectric films on metallic and single-crystal oxide substrates. Prior to any series of measurements of films on sapphire substrates, the compensator was aligned and a straight-through spectrum was collected for the effective source correction. Then, the sample and furnace were mounted on the ellipsometer and aligned at the desired angle of incidence.

SE data were collected both on films that had been crystallized conventionally and on as-deposited films which were heat-treated *in-situ* in the ellipsometer. The *in-situ* measurements were performed to determine at what point during the processing that microstructural inhomogeneities were produced in the films. This was done by aligning the film on the sample holder and assembling the furnace around it. Data were then taken at room temperature. The chamber temperature was subsequently raised in 50° increments at 2°C/min, and additional data sets were collected following stabilization at the new temperature. Below 350–400°C, spectra were measured at the annealing temperature; above this point the sample was heated to the desired temperature, maintained for 30 minutes, and cooled below 350°C for measurement. This eliminated interference of the furnace glow with the reflected light signal. Cooling was done at 5°C/min until the furnace could no longer follow. Some of these films were subsequently annealed in a conventional furnace at 650°C for 2 h in order to match the maximum annealing temperature experienced by the conventionally annealed samples.

Following data collection, values for Δ and Ψ were modeled in order to analyze the thickness, the optical properties and the degree of inhomogeneity present within each film. Modeling of the experimental data was done under the assumption of planar interfaces between layers with all layers parallel to the substrate. Variations in the film density as a function of depth were modeled by subdividing the film into a series of layers with different volume fractions of air present in each. Bruggeman effective medium theory was then used to calculate the effective dielectric functions of the two phase mixtures. Variables in the fitting procedure included all of the layer thicknesses, the volume fraction of air present at any depth in the film, and the dispersion relation describing the optical properties of the film itself. A damped Sellmeier oscillator was used to describe the dielectric function of the films.

Reference optical property data were used to describe the substrate dielectric functions. For sapphire, reference data were taken from Malitson (*44*) and Jeppesen (*45*) for the ordinary and extraordinary indices, respectively. For the Pt-coated silicon substrates, ellipsometric spectra were collected for the bare substrate and were directly inverted to provide an effective dielectric function for the exposed metal. Modeling of the data for these substrates showed that the surfaces consisted of roughened platinum.

Typically the top 10–20 nm was composed of ~85% Pt and 15% air. Unfortunately, the degree of roughness was found to change as the substrates were heated, so there is considerable residual uncertainty in the reference effective dielectric functions for annealed Pt-coated silicon substrates.

"Best-fit" models were chosen on the basis of comparisons of the 90% confidence limits for each variable, a correlation coefficient matrix describing the interrelatedness between variables, the unbiased estimator, σ, of the goodness of fit, and calculated Δ and Ψ spectra for each model. Films on (0001) sapphire substrates were modeled using an algorithm that properly accounted for the substrate anisotropy. The routine that handled the propagation of light through anisotropic materials was written by Parikh and Allara (46) following the 4×4 matrix formalism of Yeh (47, 48). An isotropic approximation was used for ($1\bar{1}02$) sapphire substrates, however, as the angle between the inclined optic axis of the alumina and the plane of incidence of the light could not be determined accurately.

For all of the films, several geometries were tried to determine whether inhomogeneities were important in modeling the optical data. Included among the possibilities were surface roughness and/or a low density layer located either near the film/substrate interface or at some point in the middle of the film. Once an initial geometry was chosen, additional layers were incorporated into the model to mimic either graded changes or further inhomogeneities. The process was stopped when no further reduction in the unbiased estimator for the error, σ, could be achieved without increasing the 90% confidence intervals unrealistically.

IV. Results and Discussion

Several types of fully crystallized ferroelectric films were investigated. It was found that the degree of inhomogeneity present in the depth profile of the films varied considerably and was dependent both on the details of the deposition procedure and the temperature profile used during the annealing (if any). As representative examples of the extremes in microstructural quality that occur, data on two films, one nearly perfect and the other highly inhomogeneous, are presented.

A. Epitaxial and Oriented Films on Single-Crystal Substrates

Lead lanthanum titanate films prepared on (0001) sapphire substrates by rf magnetron sputtering were kindly provided by H. Adachi of Matsushita

Electric Co., Ltd. The film composition was $\sim 20/0/100$ for a $28/0/100$ powder target, with a small degree of variability between depositions. In describing the composition, the notation $x/y/1 - y$ is used to indicate the addition of x mole percent La_2O_3 to $Pb(Zr_yTi_{1-y})O_3$. These films were shown to be epitaxial with lattice matching between the $\{111\}$ planes of the perovskite lattice and the basal plane (0001) of the sapphire substrate. Previous measurements on these films demonstrated exceptional electrical, piezoelectric and electro-optic properties (49).

SE data were taken on these films as a function of temperature at a fixed angle of incidence of $80°$. It was assumed in the modeling that the film index could be treated as isotropic. This was done, despite the high degree of orientation, as $PbTiO_3$ has a fairly low initial birefringence (~ 0.03–0.01 between 400 and 700 nm) (50), and the addition of lanthanum has been shown to decrease the birefringence still further in La- modified lead zirconate titanate (PLZT) materials (51). Given the large La content in the film, it was judged that the standing birefringence would be quite low and the isotropic approximation should be a good one.

Figure 3 shows the best fit to the experimental data of the $(Pb,La)TiO_3$ film between 400 and 750 nm, in addition to the variation of all of the parameters with temperature. Structurally, the film is quite homogeneous, with only a thin layer of roughness superimposed over the bulk of the film. No reaction layer between the film and the substrate could be detected.

A comparison of the calculated refractive index (n) of the lower layer with literature data (Fig. 4) shows that the general level of n is comparable to that of single-crystal $PbTiO_3$, suggesting that the bulk of the film is dense. As is typical of perovskite ferroelectrics, the refractive index increases slightly with temperature. No anomaly in the refractive index is observed for the lead lanthanum titanate film on passing through the transition temperature range (~ 110–$140°C$) (Fig. 4b), which is reasonable in light of the comparatively small spontaneous polarization and the breadth of the transition range.

None of the physical parameters describing the thicknesses or the void fraction of air in the surface layer show significant variation with temperature as the ferroelectric–paraelectric transition is approached (the film shows a broad dielectric maximum between ~ 110 and $140°C$). Consequently, the two-layer model including surface roughness is probably physically correct and not an artifact of any losses in the material associated with its ferroelectricity (such as scattering at domain walls). On the other hand, the temperature independence of all of the physical parameters suggests that intrinsic size effects, which should be tied to the polarization and hence would vary with temperature, are not essential in fitting the optical data.

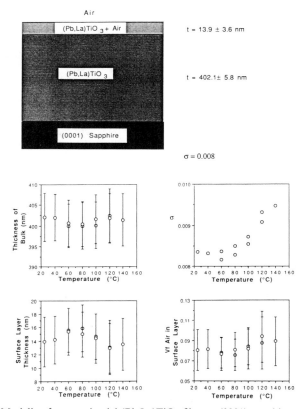

FIG. 3. Modeling for an epitaxial (Pb,La)TiO$_3$ film on (0001) sapphire.

This, in turn, implies that even with the undamaged surface of this thin film, intrinsic size effects probably do not dominate the properties of a film ~400 nm thick.

While this epitaxial lead lanthanum titanate film is dense throughout its thickness (excepting the small degree of surface roughness), not all well-oriented ferroelectric films displayed such a high degree of homogeneity. An example of the latter possibility is the sol–gel PbTiO$_3$ film on (001) SrTiO$_3$ shown in Fig. 5. This film was prepared from alkoxide precursors following a standard sol–gel procedure (52). A 0.19 M sol to which four volume percent formamide had been added was spun onto the substrate, and the organics were removed during a 10 min, 400°C pyrolysis step. Final firing was performed at 700°C for 4 h in air. While the x-ray pattern of the resultant film demonstrated excellent c-axis orientation of the PbTiO$_3$, an

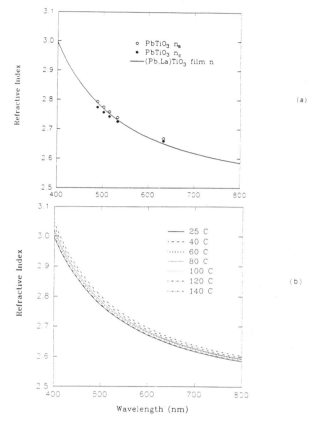

FIG. 4. Refractive index of an epitaxial (Pb,La)TiO$_3$ film on (0001) sapphire. (a) Comparison between the ellipsometrically derived refractive index for the film and literature data for PbTiO$_3$. (b) Temperature dependence of the refractive index.

SEM micrograph shows that the film surface consists of discrete grains and is, as a result, both rough and highly porous (Fig. 6). The value for the volume fraction of porosity at the surface of the film was estimated to be near 35% from the intercept method on the SEM photograph.

As can be seen in Fig. 5, this imperfect microstructure was confirmed by SE studies on the same film. For the modeling, room-temperature reference data were utilized for the SrTiO$_3$ substrate, and a polycrystalline average of reference data on PbTiO$_3$ for the film. Again, there was no indication from the ellipsometric data of any reaction layer between the film and the substrate.

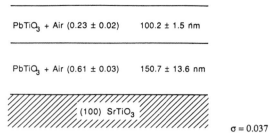

FIG. 5. Best-fit model for an inhomogeneous well-oriented PbTiO$_3$ film on SrTiO$_3$.

In all cases, fits to this film required a large volume fraction of air to be mixed with the optical properties of the PbTiO$_3$. This is consistent with the high degree of porosity evident in the SEM micrograph of the sample. It should also be noted that the ellipsometric study was performed "blind," so that the modeling was completed before access to the SEM photograph. While the agreement between the two characterization techniques is itself encouraging, it is especially noteworthy that reasonable results were obtained for a film with a very poor microstructure. This success may be due, in part, to the fine primary particle size ($\sim 50-150$ nm) in the film; that is close to the diameter/light wavelength ratio of ~ 0.25 suggested as the limit below which the Bruggeman effective medium approximation is valid in reflection measurements (53).

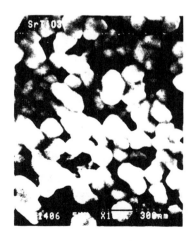

FIG. 6. SEM micrograph of the top surface of a well-oriented (001) PbTiO$_3$ on (100) SrTiO$_3$.

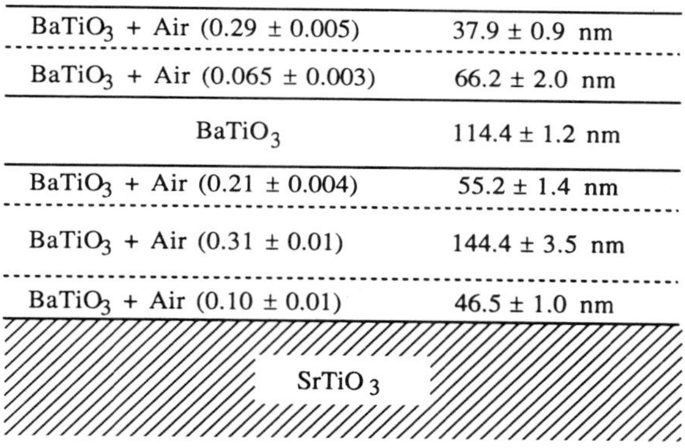

BaTiO$_3$ + Air (0.29 ± 0.005)	37.9 ± 0.9 nm
BaTiO$_3$ + Air (0.065 ± 0.003)	66.2 ± 2.0 nm
BaTiO$_3$	114.4 ± 1.2 nm
BaTiO$_3$ + Air (0.21 ± 0.004)	55.2 ± 1.4 nm
BaTiO$_3$ + Air (0.31 ± 0.01)	144.4 ± 3.5 nm
BaTiO$_3$ + Air (0.10 ± 0.01)	46.5 ± 1.0 nm
SrTiO$_3$	

FIG. 7. Best-fit model for an inhomogeneous BaTiO$_3$ film on SrTiO$_3$.

The density of this film is much lower than those of other sol–gel films investigated. It is possible that the organic phase in this film was not completely removed during the pyrolysis step. If this occurred, the film would not have collapsed completely after the pyrolysis step. When crystallization of a sol–gel film occurs before densification, the resulting film incorporates much higher porosity levels than films that are densified prior to crystallization (54). The porosity can appear either between or within individual crystallites. Consequently, it seems probable that the poor microstructure of this sol–gel PbTiO$_3$ on SrTiO$_3$ is associated with the processing, rather than an inherent limitation on the density of sol–gel ferroelectric films.

Most ferroelectric thin films possessed an intermediate degree of microstructural inhomogeneity. While the modeled microstructures of vapor-deposited films were often consistent with those expected for a columnar film, other possibilities were also found. For example, the polycrystalline BaTiO$_3$ film on SrTiO$_3$ depicted in Fig. 7 was deposited at elevated temperatures by MOCVD under conditions where the gas flow was permitted to vary during the course of the film growth. The result is a highly inhomogeneous film (38).

It should also be noted that a determination of the film inhomogeneity would be much more difficult if accurate data for Δ were not available over the full wavelength range. Consequently, use of the effective source correction and an achromatic compensator greatly extends the characterization capabilities of spectroscopic ellipsometry for transparent materials on transparent substrates. The importance of accurate measurement of Δ in depth-profiling transparent systems was previously reported by Chindaudom (38, 39).

B. IN-SITU ANNEALING STUDIES ON FERROELECTRIC FILMS

The films just discussed represent the ranges of microstructural inhomogeneities seen in ferroelectric films. In order to determine at what point during the processing that microstructural inhomogeneities were produced, several films were annealed *in situ* on the spectroscopic ellipsometer. Three examples of *in-situ* annealing will be discussed in this paper: a multi-ion-beam reactively sputtered (MIBERS) lead zirconate titanate (PZT) film on ($1\bar{1}02$) sapphire, and sol–gel PZT films on sapphire and Pt-coated silicon substrates.

1. MIBERS Films on Sapphire

Data from the *in situ* anneal of a MIBERS PZT 50:50 film on sapphire is shown in Fig. 8. Each isothermal data set illustrates the interference oscillations expected for a thin film on a substrate. As seen in Fig. 9, however, at temperatures below $\sim 400°C$, the amplitude of the oscillations is considerably damped at short wavelengths. With increasing temperature, this damping disappears, so that the envelope of the oscillations changes more smoothly with wavelength. At still higher temperatures, the magnitude of the ellipsometric angle Ψ increases significantly.

In modeling the low temperature data (below $\sim 500°C$), it was necessary to add to the oscillator describing the film a material with an abrupt increase in its light absorption coefficient in the near ultraviolet region. Without this, the damping of the interference oscillations at higher energies could not be duplicated in the modeling. Because many of the Pb-based ferroelectric thin films are non-stoichiometric following vapor deposition, this optical loss was attributed to lead oxide in the film. It was found that reference data on sputtered lead oxide (55) provided a better fit to the SE data than did evaporated lead oxide (56), as the absorption edge of the former was shifted closer to the visible frequency range. It cannot be determined from the SE data alone whether this additional loss is due to excess lead oxide in the film, or to the presene of a mixed oxidation state in the lead species. It was noted, however, that coincident with anneals at $\sim 500°C$ where the extra damping disappeared, the film changed from orange to a pale yellow color.

Figure 10 shows the SE-derived models of the film during the intermediate temperature anneals. As described above, the models below 500°C required at least some of the lossy led oxide phase to provide an acceptable fit. (All of the data at lower temperatures fit to models like that shown for 400°C.) It can be seen in Fig. 10 that between 450 and 500°C the volume

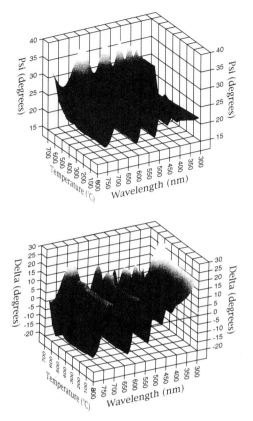

FIG. 8. SE data for the *in-situ* annealing of a MIBERS PZT film on sapphire.

fraction of lead oxide in the models decreased, starting at the surface of the film. As this occurred, the effective refractive index of the film dropped across the whole spectral range.

The progressive removal of the optically lossy lead oxide phase could be due either to evaporation of excess lead from the film or to conversion of the absorbing species into a transparent one. In work on lead lanthanum titanate films, Fox et al. (25) demonstrated that lead loss in films prepared by the MIBERS technique was initiated at temperatures on the order of 490 ± 50°C. This correlates well with the temperatures determined from the SE data. Comparable results on low temperature lead loss in thin lead zirconate titanate films have also been reported (57). The second possibility for the decrease in the concentration of the optically lossy material necesary

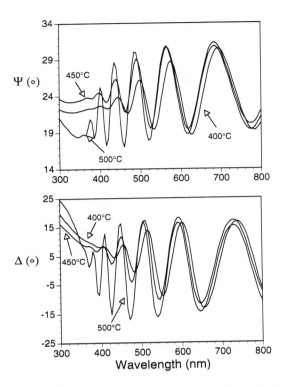

FIG. 9. Annealing of a MIBERS PZT on sapphire film showing a decrease in the high-energy damping at 500°C. All data at lower temperatures was very close to that shown for 450°C.

to fit the SE data is that the lead either a) transformed from a mixed oxidation state to a single valence or b) was assimilated into a transparent phase (like that of the evaporated PbO, the perovskite, or a pyrochlore phase). Either or both mechanisms could be operating. Similarly, although a pyrochlore phase may have formed during these lower temperature anneals, this could not be identified on the basis of the SE modeling.

Above 500°C, the film depth profile developed progressively more marked inhomogeneities. As shown in Fig. 11, the increased roughening of the film surface occurred at the same temperature that the film refractive index rose. These changes are clearly associated with crystallization of the perovskite phase. X-ray diffraction experiments confirmed that the perovskite

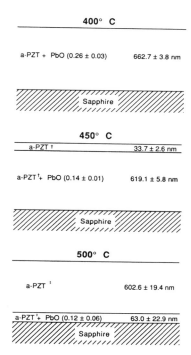

FIG. 10. Best fit models for intermediate temperature anneals of an as-deposited film on sapphire. Note that the refractive index of the "a-PZT" changed with annealing temperature as shown in Fig. 11.

phase is present in films annealed at 600 and 650°C. In addition, Fox et al. (25) reported that in the same temperature region the primary grains cluster into larger units in MIBERS lead lanthanum titanate films. Such a phenomenon may be partially responsible for the observed evolution of the inhomogeneities.

It is clear that for this film, the major degree of microstructural inhomogeneity is generated during the annealing. In fact, it has also been shown that MIBERS films deposited under the same conditions but annealed with different profiles can have very different depth profiles (40). Thus, an understanding of the role of the annealing process in controlling the film microstructure will be important in the manufacture of high-quality film-based devices. Spectroscopic ellipsometry is demonstrably a useful tool in tracking the evolution of the structure and microstructure of ferroelectric

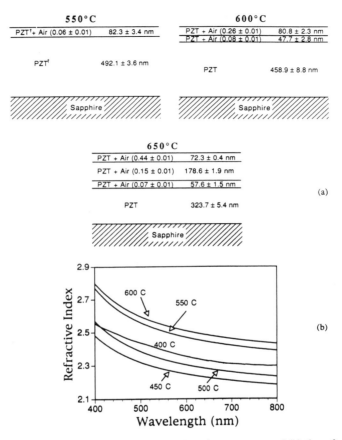

FIG. 11. Evolution with temperature of (a) the microstructure and (b) the refractive index of an as-deposited MIBERS film on sapphire during high-temperature *in-situ* annealing.

films during annealing, and so should foster the development of improved processing conditions for ferroelectric thin films.

2. *Sol–Gel PZT Films*

Similar studies were made on sol–gel $Pb(Zr_{0.5}Ti_{0.5})O_3$ films on both sapphire and Pt-coated silicon substrates. For low temperatures, the experimental curves could be fitted by allowing the thickness and optical properties of a homogeneous film to vary. In Fig. 12, a comparison is shown between the thickness as a function of annealing temperature for films that

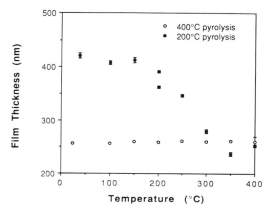

FIG. 12. Evolution in the thickness as a function of temperature for sol–gel PZT films pyrolyzed under different conditions.

had been heat-treated between spin-coatings at $\sim 200°C$ and $400°C$. For the film pre-treated at lower temperatures, a large drop in the film thickness occurred between 150 and 350°C; this temperature range was also marked by a substantial increase in the oscillator strength and position describing the effective film refractive index. The film pyrolyzed at 400°C, however, was fully collapsed and did not show any change in the film depth profile or optical properties below the burnout temperature.

For the films on both sapphire and platinum-coated silicon, σ began to rise above its average value at low temperatures for anneals between 350° and 400°C. As discussed by An et al. (58) in in-situ spectroscopic ellipsometric measurements, transitions between different models are indicated by places where σ begins to diverge from a constant value. Additional analysis of the experimental data suggests that for the PZT films, these changes did not correspond to the generation of surface roughness, interface porosity, or increased absorption in the film. At 450°C and above, however, the ongoing increase in σ could be eliminated by allowing a progressive roughening of the surface of the film on sapphire. This continued until 550°C, above which the thickness and volume fraction of air in the roughness layer were temperature-independent. The change in the microstructure was coupled with a marked increase in the film refractive index between 500 and 600°C (see Fig. 13). Independent x-ray studies on films prepared in the same manner demonstrated that the perovskite phase crystallizes in this temperature range (59). The comparatively low refractive index of the film at 600°C indicates that there is some residual porosity of the film.

The changes above $\sim 500°C$ were more difficult to follow for the film on platinum-coated silicon, largely because the optical properties of the film

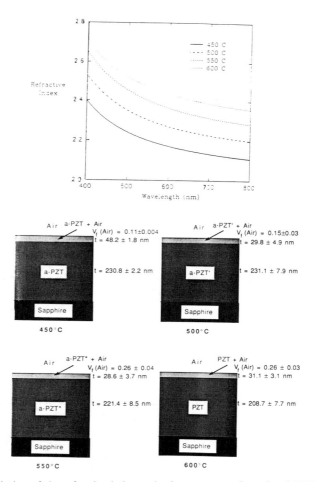

FIG. 13. Evolution of the refractive index and microstructure of a sol–gel PZT film on sapphire during high-temperature annealing.

and the substrate appeared to be evolving simultaneously. Nevertheless, it is believed that the perovskite phase crystallized somewhere in the range of 550°C. Following completion of the heating run, the film did consist primarily of the perovskite phase, with a small amount of poorly crystallized pyrochlore (as determined by x-ray diffraction).

To determine if, in fact, a reaction within the $Pt/Ti/SiO_2/Si$ substrate was possible during the extended annealing of the PZT film, several uncoated substrates were annealed *in-situ* using the same temperature profile. Figure 14 shows the experimental SE data for the substrate as a function of temperature. Data taken prior to heating could be modeled well as a layer of roughened platinum on platinum. As the penetration depth of visible light in platinum is small, it was not possible to characterize any of the layers buried deeper in the substrate. At comparatively modest temperatures ($\sim 250°C$), the microstructure of the platinum gradually altered, apparently due to coarsening of the metal grains. Nevertheless, for temperatures below 500°C, the rough surface of the platinum could be treated as a mixture of platinum and air.

Above 500°C, however, substantial and irreversible changes in the data are evident. By 550°C, both Δ and Ψ appear to display an interference

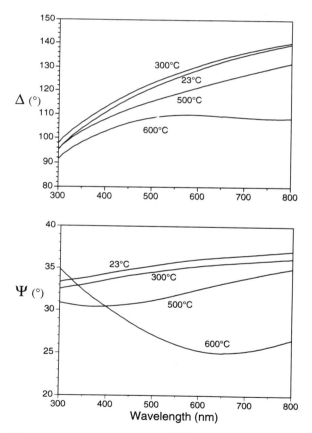

FIG. 14. Ellipsometric parameters of a Pt/Ti/SiO$_2$/Si substrate during high-temperature annealing.

fringe, suggesting the formation of a transparent/translucent overlayer on the substrate surface. This appearance of these oscillations was confirmed on a sample heated *ex-situ* using the same temperature profile. While it was not possible to conclusively identify the overlayer composition, any low dielectric constant layer such as this intermediate between a ferroelectric film and the bottom electrode would seriously degrade the apparent electrical properties of the ferroelectric. The extent of the reaction was much lower when either a shorter annealing cycle was utilized or a ferroelectric film was deposited prior to heat-treating the substrate. Consequently, this type of reaction may be negligible for more common annealing profiles, and especially for rapid thermally annealed specimens. However, even small pockets of a reaction layer produced under such circumstances would still be expected to affect the net film properties.

A summary of the SE characterization of the *in-situ* annealed films is given in Fig. 15. For all of the specimens studied, crystallization of the perovskite phase was largely complete following the 600°C anneal. Crystallization of the ferroelectric phase was also linked to the generation of surface roughness in initially smooth films. In applications where an extremely smooth surface is required, then, control of the nucleation and growth of the perovskite phase is likely to be critical.

V. Relation between Film Microstructure and Electrical Properties

Because of the presence of microstructural inhomogeneities, some ferroelectric thin films must be regarded as composite materials in which porosity (and possibly additional phases) acts as a second phase. Thus, a complete understanding of the film behavior requires a recognition of the role that the second phase distribution plays in contributing to the measured (effective) properties. Reviews of the relations between composite symmetry, connectivity, and scale on the net properties of composites have been given elsewhere (60,61). As is described in those references, addition of voids (air) to a PZT ceramic dilutes the dielectric constant in a way that depends on the geometrical arrangement of the constituent phases. Similarly, in other studies it has been shown that composite connectivity transitions dominate the electrical conductivity, dielectric constant and dielectric loss in inhomogeneous lead lanthanum titanate films containing air and PbO second phases (26).

While gross defects in the microstructure would clearly be expected to degrade the electrical properties of a thin film, even comparatively minor

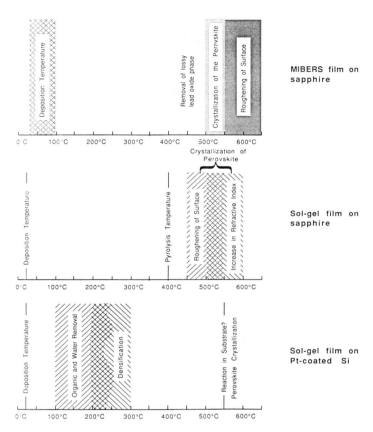

FIG. 15. Reaction schemes for annealing of several ferroelectric thin films.

inhomogeneities, such as a thin layer of low-density material near the film/substrate interface, can also have a significant effect. As described by Trolier-McKinstry et al. (40, 62), this type of low-density layer, which might be caused by the formation of a columnar or clustered microstructure, effectively introduces a low-permittivity layer in series with an otherwise perfect, high dielectric constant thin film. The result is a low net dielectric constant and a high net coercive field for the film. In addition, if the porosity in the bottom layer is distributed locally around some average value, then different regions in the film will switch at different applied fields, and the hysteresis loop becomes tilted. Finally, if in some regions the field drop across the low dielectric constant layer is sufficiently severe, it may be impossible to apply a field that will reverse the polarization of that element, and as a result, the measured remanent polarization of the film will be lowered.

It is also important to note, as shown in Fig. 16, that the presence of a low-density layer near the film–substrate interface can lead to changes in the effective properties of the film as a function of thickness that are unrelated to any intrinsic size effects in the ferroelectric. For all of the films in that figure, it was assumed that a defective layer 60 nm thick, with a dielectric constant distributed locally between 80 and 1,300, was formed immediately adjacent to the substrate. The remainder of the film thickness was assumed to be perfect, with a limiting dielectric constant of 1,300 and

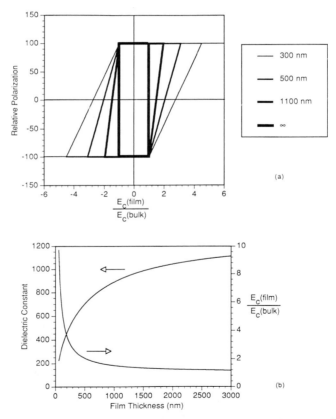

FIG. 16. The effects of a 60 nm thick low-density layer near the substrate on the electrical properties of PZT thin films. (a) Variation in the hysteresis loop with the total film thickness. (b) Variation of the dielectric constant and the coercive field with film thickness.

a well-defined coercive field of 40 kV/cm. The figure shows the effect of this inhomogeneity on the low- and high-field electrical properties of the film. In both cases, the thin defective layer plays a negligible role in determining the properties of an infinitely thick specimen. As the film thickness is decreased, however, the low-density layer exerts a progressively stronger influence on the measured properties, lowering the dielectric constant, tilting the hysteresis loop, and increasing the coercive field. Thus, it is clear that even comparatively minor inhomogeneities can seriously degrade the properties of ferroelectric thin films.

VI. Conclusions

From these studies it is clear that spectroscopic ellipsometry can be utilized to characterize inhomogeneities in ferroelectric thin films. Some degree of heterogeneity was found in most of the films examined; in all cases these were more important in modeling the ellipsometric spectra than were intrinsic size effects. The inhomogeneities present are a strong function of the film preparation conditions and are varied both by the deposition parameters and by the post-deposition annealing. However, the presence of microstructural imperfections is not necessarily linked to the existence of defects in the crystal structure of ferroelectric films. As a result, well-crystallized, and even well-oriented, films can display poor microstructures. Thus, in developing ferroelectric thin films for device applications, it is imperative that the role of processing in controlling the film microstructure (and hence the film properties) be investigated.

No evidence for a reaction between Al_2O_3 or $SrTiO_3$ substrates and the lead-based ferroelectric films was detected at typical processing temperatures. It was found, however, that long-term annealing of $Pt/Ti/SiO_2/Si$ substrates resulted in marked changes in the ellipsometric spectra that appeared to be indicative of formation of a thin insulating overlayer. While the reaction appeared to slow either if shorter annealing cycles are used or if the substrate is coated with a film prior to high-temperature exposure, formation of small reaction zones could degrade the apparent electrical properties of the film. Deterioration in the effective film properties would be more severe for specimens prepared on substrates exposed to long-term high-temperature heat treatment during the deposition process.

It has also been demonstrated that spectroscopic ellipsometry can track the evolution of crystallinity and microstructural inhomogeneities during annealing of as-deposited films. One of the advantages of performing these studies on transparent materials is that the entire depth of the film can be sampled (and characterized) at once. In these studies, crystallization of the perovskite phase was shown to be largely complete after half an hour at 600°C for both MIBERS and sol–gel PZT 50/50 films (62). For the prolonged heating cycles utilized during *in-situ* annealing of the ferroelectric films, roughening of the film surface was coincident with this crystallization and can probably be attributed to the growth of crystal nuclei. In addition, lower-temperature phenomena such as burnout of organic phases, film densification, and changes in the lead species present could be identified.

The inhomogeneities present in many of the ferroelectric films alter the net electrical and optical properties shown by the films. In particular, low-density regions near the film–substrate interface significantly increase the coercive field and lower the measured dielectric constant values of well-crystallized films. In addition, if the dielectric constant of the low-density layer is too low, the excessive field drop may prevent switching of that region of the film. This would reduce the measured remanent polarization value. Optically, microstructural heterogeneities (including surface roughness) would lead to additional scattering of the signal and could limit low-loss propagation in electro-optic devices. Given the close correlation between the microstructure and properties of ferroelectric thin films, preparation of high-quality film-based devices will require study of microstructure/processing relationships.

Finally, it has also been demonstrated that many of the apparent size effects reported for ferroelectric thin films are probably associated with either poor crystallinity or a defective microstructure, rather than intrinsic changes in the ferroelectric properties with film thickness.

Acknowledgments

The authors would like to thank Dr. H. Adachi of the Matsushita Electric Corp., Dr. K. Kushida of the Hitachi Central Research Laboratory, Dr. H. Hu, and Dr. J. Chen of the Penn State Materials Research Laboratory for providing thin-film samples for study. Extensive discussions with Dr. G. R. Fox and Dr. K. R. Udayakumar are also thankfully acknowledged.

References

1. R. E. Newnham, K. R. Udayakumar, and S. Trolier-McKinstry, *in* "Chemical Processing of Advanced Materials" (L. L. Hench and J. K. West, eds.), pp. 379–393. John Wiley and Sons, New York, 1992.

2. V. G. Gavrilyachenko, R. I. Spinko, M. A. Martynenko, and E. G. Desenki, *Sov. Phys.–Sol. State* **12**, 1203–1204 (1970).
3. H. Tabata, O. Murata, T. Kawai, S. Kawai, and M. Okuyama, private communication.
4. E. Yamaka, H. Watanabe, H. Kimura, H. Kanaya, and H. Okhuma, *J. Vac. Sci. Technol.* **A6**(5), 2921–2928 (1988).
5. K. Iijima, Y. Tomita, R. Takayama, and I. Ueda, *J. Appl. Phys.* **60**, 361–367 (1986).
6. M. Okuyama and Y. Hamakawa, *Ferroelectrics* **63**, 243–252 (1985).
7. S.-G. Yoon, H. Y. Lee, and H. G. Kim, *Thin Solid Films* **171**, 251–262 (1989).
8. K. Sreenivas and M. Sayer, *J. Appl. Phys.* **64**, 1484–1493 (1988).
9. D. A. Payne, *ISAF 1991 Abstract* 12.1 (1990).
10. J. Lee, L. Johnson, A. Safari, R. Ramesh, T. Sands, H. Gilchrist and V. G. Keramidas, *Appl. Phys. Lett.* **63**, 27–29 (1993).
11. K. R. Udayakumar, J. Chen, S. B. Krupanidhi, and L. E. Cross, paper presented at ISAF 1990 (1990).
12. S. K. Dey and R. Zuleeg, *ISAF 1990 Abstracts*, 12.2 (1990).
13. G. Yi, Z. Wu and M. Sayer, *J. Appl. Phys.* **64**, 2717–2724 (1988).
14. S. A. Mansour, D. A. Binford, and R. W. Vest, *Integrated Ferroelectrics* **1**, 43–56 (1993).
15. A. Croteau, S. Matsubara, Y. Miyasaka, and N. Shohata, *Jpn. J. Appl. Phys.* **26 suppl. 26-2**, 18–21 (1987).
16. H. Hu and S. B. Krupanidhi, *J. Appl. Phys.* **74**, 3373–3382 (1993).
17. B. Jaffe, W. R. Cook, Jr., and H. Jaffe, "Piezoelectric Ceramics," Academic Press Ltd., India, 1971.
18. T. Hayashi, N. Ohji, K. Hirohara, T. Fukunaga, and H. Maiwa, *Jpn. J. Appl. Phys.* **32**, 4092–4094 (1993).
19. M. Loposzko, M. Pawelczyk, M. Urbanska, and Z. Surowiak, *Thin Solid Films* **69**, 339–345 (1980).
20. Z. Surowiak, A. M. Margolin, I. N. Zakharchenko, and S. V. Biryukov, *Thin Solid Films* **176**, 227–246 (1989).
21. K. H. Guenther, *Appl. Opt.* **23**, 3612 (1984).
22. K. H. Guenther, *Appl. Opt.* **23**, 3806 (1984).
23. H. A. Macleod, *SPIE* **325**, 21 (1982).
24. H. A. Macleod, *Appl. Opt. and Opt. Eng.* **10**, Academic Press, New York, 1987.
25. G. R. Fox, S. B. Krupanidhi, K. L. More, and L. F. Allard, *J. Mat. Res.* **7**, (1992).
26. G. R. Fox and S. B. Krupanidhi, *J. Mat. Res.* **8**, 2203–2215 (1993).
27. V. P. Dudkevich, V. A. Bukreev, VI. M. Mukhortov, Yu. I. Golovko, Yu. G. Sindeev, V. M. Mukhortov, and E. G. Fesenko, *Phys. Stat. Sol.* (a) **65**, 463–467 (1981).
28. M. C. Wittels and F. A. Sherrill, *J. Appl. Phys.* **28**, 606–609 (1957).
29. M. Schoijet, *Brit. J. Appl. Phys.* **15**, 719–723 (1964).
30. S. Naka, F. Nakakita, Y. Suwa, and M. Inagaki, *Bull. Chem. Soc. Japan* **47**, 1168–1171 (1974).
31. S. V. Biryukov, V. M. Mukhortov, A. M. Margolin, Yu. I. Golovko, I. N. Zakharchenko, V. P. Dudkevich, and E. G. Fesenko, *Ferroelectrics* **56**, 115–118 (1984).
32. P. W. Forsbergh, Jr., *Phys. Rev.* **93**, 686–692 (1954).
33. G. A. Rosetti, Jr., L. E. Cross, and K. Jushida, *Appl. Phys. Lett.* **59**, 2524–2546 (1991).
34. K. Binder, *Ferroelectrics* **35**, 99–104 (1981).
35. I. P. Batra, P. Wurfel, and B. D. Silverman, *Phys. Rev.* **B8**, 3257–3265 (1973).
36. P. Wurfel, I. P. Batra and J. T. Jacobs, *Phys. Rev. Lett.* **30**(24), 1218–1221 (1973).
37. P. Wurfel and I. P. Batra, *Phys. Rev.* **B8**, 5126–5133 (1973).
38. P. Chindaudom, S. Trolier-McKinstry, and K. Vedam, in press (1994).

39. P. Chindaudom and K. Vedam, *Appl. Opt.*, **32**, 6391–6398 (1993).
40. S. Trolier-McKinstry, H. Hu, S. B. Krupanidhi, P. Chindaudom, K. Vedam, and R. E. Newnham, *Thin Solid Films*, **230**, 15–27 (1993).
41. P. Chindaudom and K. Vedam, this volume.
42. R. J. King and M. J. Downs, *Surf. Sci.* **16**, 288–302 (1969).
43. D. E. Aspnes, *J. Opt. Soc. Am.* **64**, 812–819 (1974).
44. I. H. Malitson, *J. Opt. Soc. Am.* **52**, 1377–1379 (1962).
45. M. A. Jeppesen, *J. Opt. Soc. Am.* **48**, 629–632 (1958).
46. A. N. Parikh and D. L. Allara, *J. Chem. Phys.* **96**, 927–945 (1992).
47. P. Yeh, *J. Opt. Soc. Am.* **69**, 742–756 (1979).
48. P. Yeh, *Surf. Sci.* **96**, 41–53 (1980).
49. H. Adachi, T. Mitsuyu, O. Yamazaki, and K. Wasa, *J. Appl. Phys.* **60**, 736–741 (1986).
50. W. Kleeman, F. J. Shafer, and D. Rytz, *Phys. Rev.* **B34**, 7873–7879 (1986).
51. C. E. Land, P. D. Thacher, and G. H. Haertling, in "Applied Solid State Science: Advances in Materials and Device Research" (R. Wolfe, ed.), Vol. 4, pp. 137–233. Academic Press, New York, 1974.
52. K. Kushida, K. R. Udayakumar, S. B. Krupanidhi, and L. E. Cross, *J. Am. Ceram. Soc.* **76**, 1345–1348 (1993).
53. W. G. Egan and D. E. Aspnes, *Phys. Rev.* **B26**, 5313–5320 (1982).
54. J. L. Keddie and E. P. Giannelis, *J. Am. Ceram. Soc.* **74**, 2669–2671 (1991).
55. E. P. Harris, P. S. Hauge, and C. J. Kircher, *Appl. Phys. Lett.* **34**, 680–682 (1979).
56. A. E. Ennos, *J. Opt. Soc. Am.* **52**, 261–264 (1962).
57. S. B. Krupanidhi, H. Hu, and V. Kumar, *J. Appl. Phys.* **71**, 376 (1992).
58. I. An, H. V. Nguyen, N. V. Nguyen, and R. W. Collins, *Phys. Rev. Lett.* **65**, 2274–2277 (1990).
59. J. Chen, K. R. Udayakumar, K. G. Brooks, and L. E. Cross *J. Appl. Phys.* **71**, 4465 (1992).
60. T. R. Gururaja, A. Safari, R. E. Newnham, and L. E. Cross, in "Electronic Ceramics: Properties, Devices, and Applications" (L. M. Levinson, ed.), pp. 92–128. Marcel Dekker, New York, 1988.
61. R. E. Newnham and S. Trolier-McKinstry, *J. Appl. Cryst.* **23**, 447–457 (1990).
62. S. E. McKinstry, "Characterization of Ferroelectric Surfaces and Thin Films by Spectroscopic Ellipsometry," Ph.D. thesis, The Pennsylvania State University (1992).

Effects of Optical Anisotropy on Spectro-ellipsometric Data for Thin Films and Surfaces

ATUL N. PARIKH

*Department of Materials Science and Engineering
The Pennsylvania State University
University Park, Pennsylvania*

AND

DAVID L. ALLARA

*Department of Materials Science and Engineering and
Department of Chemistry
The Pennsylvania State University
University Park, Pennsylvania*

I. Introduction 279
II. The Generalized Approach 283
   A. Definition of the Fundamental Electromagnetic Problem 283
   B. Implementation of the Yeh 4 × 4 Transfer Matrix Method 290
III. Comparison with Available Data and Previous Methods 291
IV. Simulations of Anisotropic Effects 295
   A. Construction of a Model Anistropic Medium from
      an Ensemble of Oriented Oscillators 295
      1. Selection of an Oscillator Model 295
      2. Generation of Optical Function Tensors 298
   B. Single-Interface Ambient–Substrate Systems 300
      1. Single-Wavelength, Variable-Angle Measurements 301
      2. Spectroscopic Measurements 305
      3. Variation of the Ambient Medium 308
   C. Multiple-Interface, Ambient–Film–Substrate Systems 309
   Acknowledgment 312
   References 313

I. Introduction

The technique of ellipsometry has been proven to be an effective method for quantitative characterization of bulk solid surfaces, thin films, and interfaces for a variety of materials including metals, semiconductors, and organics (*1–4*). The advantages of ellipsometry (*1–4*) are the high accuracy and precision of the measurements with a high sensitivity to small perturbations in surface structure, the ability to carry out experiments *in situ* in gas or liquid media, the non-destructive character of the measurement and the amenability to detailed, quantitative characterization using classical optical theory. Significant advances have been made in recent years in the instrumentation, and it is now possible to obtain high-precision data with very fast collection times (*5*) over a range of spectral frequencies (spectroscopic ellipsometry, SE) of the incident light (*6–8*) and at a range of incident angles (*9–11*) (multiple angle ellipsometry, MAI, or variable angle spectroscopic ellipsometry, VASE) on a routine basis. These advances have affected diverse applications, including surface modifications (*12*), e.g., semiconductor passivation, monolayer adsorption, corrosion and catalysis; optical properties of ultrathin films, both organic and inorganic (*13,14*); and real-time studies of film growth (*15–19*). In order to take full advantage of these experimental capabilities, it is necessary that theoretical methods be available for accurate analyses of the data in terms of material properties. In particular, it is necessary to make quantitative connections between the experimental parameters, e.g., wavelength of the incident light and the angle of incidence; the relevant material parameters, e.g., dielectric functions and film thickness; and the observed polarization measurements. The fundamental equations of electromagnetic interactions that describe the necessary relationships are well known, but rarely is it possible to obtain convenient analytical expressions, so the data generally are analyzed by exhaustive numerical simulation procedures. The typical analysis approach applies classical electromagnetic theory, in the form of boundary value equations, to an idealized model of the experiment. The simplest models assume a collimated monochromatic beam of light impinging at a fixed angle of incidence upon a sample consisting of stacks of variable thickness, parallel layers of homogeneous, isotropic media with infinitely sharp interfaces (*1*). However, typical samples inevitably never fit such an ideal description. In practice, surfaces and interfaces are rough, the materials exhibit non-uniformities in density and compositional distributions, and often there is a distinct directional or anisotropic character to the optical and electrical response of the material. Recent advances have addressed these problems by altering the simulation models to accommodate surface roughness (*20*), heterogeneously structured materials (*21–23*) and refractive index gradients (*24,25*). However, one important aspect that has yet to receive satisfactory attention is optical anisoptopy. In many applications, the materials are sufficiently anisotropic that neglect can cause serious error in the analysis of the ellipsometric data. There are many examples of anisotropy that arise in materials applications, and several illustrations follow. Crystalline materials such as GaSe, Bi_xTe_y, MoS_2, SbSI, mica, topaz, and graphite exhibit strong optical anisotropies at their surfaces (26). In thin films of otherwise isotropic materials, optical anisotropy can arise when the film thickness is decreased to the point that the electronic states of the material exhibit quantum confinement in the out-of-plane direction while remaining isotropic in the in-plane direction (27). In vacuum deposition of dielectric thin films, the growth can occur via columns that have their long axes oriented in the direction of the evaporation source. If the source flux is arranged perpendicular to the substrate, normal columnar micro-structures can result, and as the flux angle becomes more oblique, the columnar micro-structures become more tilted. The optical properties of such thin films almost invariably demonstrate high degrees of optical anisotropies (28). Significant anisotropies are also known to arise at electrode surfaces as a result of functionalization and electro-deposition processes (29). In gas–solid chemisorption processes, e.g., oxygen on copper, anisotropic surface structures have noted (12). Finally, in recent applications of organic films—including Langmuir monolayers of fatty acids at air–water interfaces, Langmuir–Blodgett films on both amphiphiles and biological molecules, and liquid-crystal assemblies—the uniform, high orientational order of the molecules (30) can give rise to anisotropic optical responses. In all these examples and other related ones, it is clear that accurate ellipsometric characterization of the material/thin film/surface structures will depend critically upon rigorous inclusion of anisotropy in simulation models.

Efforts to accommodate optical anisotropy in analytical ellipsometric methods have been reported over the last two decades (31–43). In general, these approaches have involved limited types of anisotropies and experimental measurements. Early efforts to model ellipsometric data of anisotropic systems required the presence of uniaxial symmetry, with the principal axis parallel or perpendicular to the incident plane (31–36). Many more recent approaches are based on simplified, closed-form expressions obtainable under constraints that allow the use of Drude's equations (44). The practicality of these specific approaches is obvious: The basic electromagnetic equations are simplified considerably and easily formulated, sometimes even analytically, and the number of unknowns, particularly the number of independent terms in the dielectric tensor, is reduced with a corresponding reduction in the number of required independent experiments. Aspnes (42) has recently presented an approximate method for analysis of complex reflectance ratios measured in ellipsometric measurements of biaxial crystals. In this method, anisotropies are expressed as small-scale perturbations to an otherwise isotropic, diagonalized dielectric tensor. All of these treatments are applicable only for cases in which the character of the anisotropy can be shown to fit the simplifying constraints of the specific treatment. However, more often than not, an independent knowledge of the character of anisotropy (uniaxial or biaxial, direction of principal axis, etc.) is unavailable, and the application of more generalized models is imperative. In this regard, a serious limitation of these approaches is that there appears to be no recognizable way in which they can easily be generalized to apply to more complicated situations. Thus, one is left with the task of constructing an approach based on fundamental relationships stripped of simplifying special cases.

Resolution of the preceding problem is certainly possible, in principle, since the generalized boundary-value equations of the electromagnetic interactions inherently contain anisotropic response (45). Rigorous formalisms have been developed to cover the cases of multilayer film stacks, and the most amenable of these for the problem at hand are based on 4 × 4 transfer matrix methods originally developed by Berreman (46) and by Yeh (47). Further adaptations have been made to specific applications (48), and some have been developed for applications to ellipsometric analysis. De Smet (34) has adopted Berreman's differential matrix method to determine null polarizer-analyzer settings for optically active and biaxial surfaces. Debe and Field (43) recently have reported the development of an algorithm based on Berrman's treatment and its application to the ellipsometric characterization of phthalocyanine thin films. They used the method to derive the orientation of the principal axes of crystalline films deposited under different conditions. However, the details of their specific algorithm were not presented. Recently, we (49) have adapted Yeh's matrix formalism and constructed algorithms that cover the most generalized cases, including full biaxial symmetry and multiple, lossy layers. We have applied this treatment to numerical simulations of reflectivity spectra in the infrared region and achieved quantitative agreement with experimental data (50). At present this appears to be the only generalized treatment available for application to ellipsometric studies of anisotropic materials. Accordingly, we focus on this treatment and in this article describe an extension to ellipsometric studies of anisotropic materials.

The article is constructed as follows. In Section II, the basic 4 × 4 matrix formalism of Yeh is reviewed. Next, an adaptation to computationally tractable algorithms is presented. In Section III, selected experimental observations are modeled in order to demonstrate the validity of the calculations. In Section IV, numerical illustrations of the effects of optical anisotropy are given in which hypothetical models are applied to selected classes of ellipsometric experiments in order to generate useful physical insight into the variety of possible anisotropic effects. Finally, practical recommendations are made with regard to the design of ellipsometric experiments that can be useful in characterizing optical anisotropy.

II. The Generalized Approach

A. Definition of the Fundamental Electromagnetic Problem

The general reflection ellipsometry experiment of interest here is schematically shown in Fig. 1. An incident beam of collimated light is reflected from

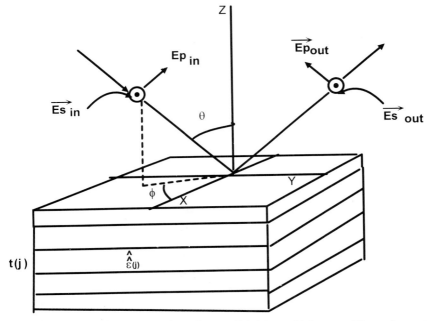

FIG. 1. A schematic representation of a parallel-slab, multiphase, multilayered sample consisting of a total of N phases with infinitely sharp interfaces. An arbitrarily selected jth slab is characterized by a uniform thickness t_j and a complex dielectric function $\hat{\varepsilon}_j$. The xy plane lies in the plane of the sample surface. The incoming light beam is traveling in the semi-infinite ambient medium 1 with the xy projection at the azimuthal angle ϕ to the x angle. The plane of incidence is defined by the incoming and reflected beams and contains the angle of incidence θ. The definitions of the directions of the electric fields for s- and p-polarizations are shown for the incident and reflected beams as perpendicular and parallel to the incidence plane, respectively.

the surface of a stack of semi-infinite, parallel, planar slabs. The angle between the incoming beam and the surface normal defines the *angle of incidence* θ, which is contained in the *plane of incidence*. The coordinate system is chosen for convenience to have the slab surface as the xy-plane and the plane of incidence as the xz-plane. The state of polarization of the electric field of the incident beam is defined in terms of the direction of the electric field vector relative to the incidence plane. It is convenient to define two special *linear polarization* states, s and p, for which the electric field vector is aligned perpendicular or parallel to the plane of incidence, respectively. The most general case of a polarization state is *elliptical polarization*, in which the electric field vapor can be decomposed into component s- and p-states with both an *amplitude* and a *phase* difference between them. When the phase difference is zero, the light is linearly polarized. The basic ellipsometry experiment consists of measuring the change in the ellipticity of the polarization state upon reflection. The data is interpreted in terms of formal relationships between the changes in these polarization states and material properties via the electromagnetic theory of the interaction of light with matter. These relationships are presented next.

Before proceeding with the presentation of equations, it is necessary to briefly discuss conventions and representations. In order to be consistent we have followed throughout the standard conventions used in ellipsometry (52). A real scalar quantity s is represented in normal typeface (s), when complex with tildes ($\tilde{s}$). *A vector quantity* v is expressed in bold typeface (**v**) and with an arrow ($\vec{v}$) when complex. A tensor t is expressed as $\hat{t}$, its complex elements t_{jk} as $\hat{t}_{jk}$, and real elements as t_{jk}.

For a linearly polarized infinite, homogeneous, plane wave of vacuum wavelength λ propagating through a non-absorbing medium with associated refractive index n_1, the time-dependent character of the electric field vector is given by

$$\vec{\mathbf{E}} = \vec{\mathbf{A}}^\circ e^{-i(\omega t - \mathbf{K} \cdot \mathbf{r} + \varphi)} = \vec{\mathbf{E}}^\circ \cdot e^{-i(\omega t - \mathbf{K} \mathbf{r})}, \tag{1}$$

where $\vec{\mathbf{A}}^\circ$ is the vector defining the maximum amplitude and direction of the electric field, ω is the angular frequency of the light, **K** is the wavevector defined as having magnitude $|\mathbf{K}| = 2\pi n_1/\lambda$ and a direction along the wave propagation axis, **r** is the location vector of a point along the wave, and φ is a phase angle to relate the phase of the oscillating electric field to some reference. It is further convenient to define the complex vector $\vec{\mathbf{E}}^\circ$, which carries the plane factor $e^{-i\varphi}$. The simplest, and most typical, reflection experiment is one in which s- or p-polarized light reflects off the surface (Fig. 1) of an isotropic material with conservation of the polarization direction. The complex electric fields of the incoming and reflected outgoing linearly polarized waves at the surface are each represented by Eq. (1), but with different amplitude and phase factors, $|\vec{E}^\circ|$ and φ. The complex reflectivity for this special case then is defined as the time-independent ratio of the incoming and outgoing electric field vectors at the specific point of reflection:

$$\frac{|\vec{E}_{out}|}{|\vec{E}_{in}|} = \frac{|\vec{E}^\circ_{out}|}{|\vec{E}^\circ_{in}|} = \frac{|\vec{E}^\circ_{out}|}{|\vec{E}^\circ_{in}|} e^{(\varphi_{out} - \varphi_{in})} = \tilde{r}. \tag{2}$$

The complex scalar $\tilde{r}$, a convenient abbreviated notation for complex electric field ratios, is usually known as the Fresnel coefficient and is designated as $\tilde{r}_s$ or $\tilde{r}_p$, depending on the polarization state. The preceding special case describes the vast majority of standard, non-ellipsometric reflection experiments.

However, in the ellipsometry experiment the general case of the reflection of elliptically polarized light must be treated. In this case the complex electric field vectors can be expressed as linear combinations of s- and p-polarization states:

$$\vec{E}_{in} = [\vec{E}^\circ_{s\,in} + \vec{E}^\circ_{p\,in}]e^{-i(\omega t^- - \mathbf{k}\cdot\mathbf{r})}, \tag{3a}$$

$$\vec{E}_{out} = [\vec{E}^\circ_{s\,out} + \vec{E}^\circ_{p\,out}]e^{-i(\omega t - \mathbf{k}\cdot\mathbf{r})}. \tag{3b}$$

In these equations the complex $\mathbf{E}^\circ$ amplitude vectors for either propagating wave can have different absolute magnitudes, $|\vec{E}^\circ_s|$ and $|\vec{E}^{lim}_p|$, and phase angles, φ_s and φ_p. The polarization states are conveniently defined in terms of the ratio of the complex electric field components along the s and p directions:

$$\tilde{\chi} = \vec{E}^\circ_p / \vec{E}^\circ_s = (|\vec{E}^\circ_p|/|\vec{E}^\circ_s|) \cdot e^{-i(\varphi_p - \varphi_s)}, \tag{4}$$

where the complex quantity $\tilde{\chi}$ specifies the ellipticity of the light. In essence, ellipsometric parameters furnish a relationship between $\tilde{\chi}_{in}$ and $\tilde{\chi}_{out}$. There are two special cases of $\tilde{\chi}$ values. When $|\vec{E}^\circ_s|$ or $|\vec{E}^\circ_p| = 0$, $\tilde{\chi} = \infty$ or 0 and the light reverts accordingly to s- or p-polarization. For $|\vec{E}^\circ_s|$ and $|\vec{E}^\circ_p| \neq 0$ when $\varphi_s = \varphi_p$, $\tilde{\chi}$ is real and the light reverts to a state of arbitrary linear polarization. When $|\vec{E}^\circ_s| = |\vec{E}^\circ_p|$ and $|\varphi_s - \varphi_p| = \pi/2$, $\tilde{\chi} = 1$ and the light is circularly polarized. All other cases are elliptic. Generally, one considers just these relative phase relationships between the s and p fields at any given instant and location and ignores the $e^{-i(\omega t - \mathbf{k}\cdot\mathbf{r})}$ propagation term until needed. The most general linear relationship between the complex electric field components of input and output light at the sample surface can be expressed conveniently in terms of a matrix operation:

$$\begin{bmatrix} \vec{E}_{s\,out} \\ \vec{E}_{p\,out} \end{bmatrix} = \hat{\tilde{r}} \begin{bmatrix} \vec{E}_{s\,in} \\ \vec{E}_{p\,in} \end{bmatrix}, \tag{5}$$

where $\hat{\tilde{r}}$ is a 2 × 2 Fresnel reflectance matrix,

$$\hat{\tilde{r}} = \begin{pmatrix} \tilde{r}_{ss} & \tilde{r}_{sp} \\ \tilde{r}_{ps} & \tilde{r}_{pp} \end{pmatrix}, \tag{6}$$

$$\tilde{r}_{ij} = |\vec{E}_{j\,out}|/|\vec{E}_{i\,in}|, \tag{7}$$

with the element $\tilde{r}_{ij}$ defined as the Fresnel reflectance coefficient for the incident and reflected beams of i- and j-polarization, respectively. These coefficients are defined exactly as given in Eq. (2), except that the incident and reflected waves can have arbitrarily different polarization states, thus allowing representation of the fully general case in which s and p states are both present and can interconvert upon reflection. The values of the matrix elements carry the details of the physical mechanism of the interaction of the light with the sample. A simpler class of reflection experiments than ellipsometry involves just the measurement of the power reflectivity of the light. The latter is defined in terms of the ratio of the absolute squares of the complex electric fields:

$$\hat{R} = |\vec{E}_{out}|^2/|\vec{E}_{in}|^2 = (\vec{E}_{out}^* \cdot \vec{E}_{out})/(\vec{E}_{in}^+ \cdot \vec{E}_{in})$$
$$= \tilde{r}^* \cdot \tilde{r} = I_{out}/I_{in}, \tag{8}$$

where $\vec{E}^*$ denotes a complex conjugate and the I's represent the physically measurable intensities of the electric field. In this case one can define a real 2 × 2 *reflectivity matrix*, $\hat{R}$, corresponding to the complex matrix r in Eq. (5), with elements $R_{ij} = \tilde{r}_{ij}^* \cdot \tilde{r}_{ij}$:

$$\hat{R} = \begin{pmatrix} R_{ss} & R_{sp} \\ R_{ps} & R_{pp} \end{pmatrix}. \tag{9}$$

A conventional ellipsometry experiment measures the change of polarization upon reflection from a sample surface (*1*). The most popular representation of ellipsometric data is in terms of the complex ellipsometric function:

$$\tan \psi e^{i\Delta} \equiv \tilde{\rho} \equiv \frac{\tilde{\chi}_{out}}{\tilde{\chi}_{in}} = \frac{\vec{E}_{p\,out}^\circ/\vec{E}_{s\,out}^\circ}{\vec{E}_{p\,in}^\circ/\vec{E}_{s\,in}^\circ} \tag{10}$$

Ellipsometry data are directly related to the ellipsometry change on reflection and the angular quantities Ψ and Δ, the amplitude ratio angle and the angular phase shift, respectively, between the incident and reflected complex electric fields at the surface. These quantities are then utilized to compute the complex reflectance ratio, $\tilde{r}_{pp}/\tilde{r}_{ss}$ (*51*), as defined in Eqs. (5) and (6). This operation requires the assumption that the $\bar{\mathbf{E}}_s$ and $\bar{\mathbf{E}}_p$ components never interconvert upon reflection, *viz.*, $\tilde{r}_{sp}$ and $\tilde{r}_{ps}$ are zero, and, in fact, such an assumption is always valid except in very limited cases of extremely anisotropic media (see below). The most common experiment is null-ellipsometry (*1*), in which the azimuthal angles of polarizing elements placed immediately before and after the sample, P and A, respectively, are adjusted to achieve the condition of minimum light intensity at the detector. For a typical configuration, $\Delta = A$ and $\Psi = 90° - P$. In general, Ψ and Δ are not obtained quite as directly for other common ellipsometer configurations, e.g., rotating analyzer–fixed polarizer, and discussions of the details of data analysis for these experiments have been summarized (*6*).

Once the ellipsometry experiments has provided values of the Fresnel coefficients $\tilde{r}_{ij}$, the next step is to relate these quantities to the material properties and the structure of the sample. This is done by constructing mathematical models that represent the interaction of the light with the sample for the specific experimental conditions, fully specified by the frequency of the light, the angle of incidence θ and the sample alignment in the beam, specified by an azimuthal angle ϕ between the plane of incidence of the light and some important axis associated with the sample structure, e.g., a crystal axis. The standard models are based on the classical electromagnetic theory of light interacting with an ideal, stratified-layer structure of isotropic, homogeneous media (*45*). In this approach, the sample is completely described, as shown schematically in Fig. 1, by specifying for each layer i, the associated thickness t_j and an optical response function, $\tilde{\varepsilon}_j$ or, alternatively, $\tilde{n}_i$. The latter two scalar quantities are the complex dielectric function and the complex refractive index, either of which equivalently describes the intrinsic optical response, and are defined following standard convention (*52*) as

$$\tilde{\varepsilon} = \varepsilon_1 - \mathbf{i}\varepsilon_2 \qquad (11)$$

and

$$\tilde{n} = n - \mathbf{i}k, \qquad (12)$$

where $\mathbf{i} = \sqrt{(-1)}$. The relationship between these two quantities is

$$\tilde{\varepsilon} = \tilde{n}. \qquad (13)$$

In the general case of anisotropic media, these quantities are second-rank tensors ($\hat{\varepsilon}$ or $\hat{n}$), and for isotropic media, they of course reduce to scalars.

The essential electromagnetic theory is based on the boundary value relationships of Maxwell's equations (44), which serve to relate th complex Fresnel coefficients, derived via Eq. (7), to the layer thicknesses and the frequency-dependent optical response functions. The material properties of each layer are accessed by additional relationships between the optical functions and the material structure and composition, a subject dealing with condensed-matter properties and outside of the scope of the present discussion. A central problem that arises in the preceding strategy is that although the relevant electromagnetic theory is explicitly defined in detail, in practice, for the most general experimental configurations and sample types, and theory quickly leads to such cumbersome mathematical expressions for extraction of the layer thicknesses and optical functions from the ellipsometric parameters that analytical solutions are often hopelessly intractable. This problem is generally circumvented by replacing analytical solutions with numerical computation algorithms based directly on the analytically intractable, core electromagnetic equations. This approach involves interative comparisons of simulated data, based on trial solutions, with the experimental data and convergence based on some best-fit criterion (53). However, even with the latter approach, the development of computational algorithms involving the most general experimental conditions and sample types has remained a challenging problem, and generalized approaches have not been available.

Based on the foregoing analysis problems, the emphasis in ellipsometry experiments has been on simplified cases of special importance. The most prevalent simplification has been the consideration of only isotropic media, a simplification that allows the optical response tensors to be replaced by scalars and leads to an enormous simplification in the computations and analytical formulations. Effectively the only experiments of anisotropic materials considered have been those using simplified sample structures, limited experimental conditions and simplifying theoretical approximations. In order to avoid these simplifications, the underlying relationships that need to be brought into workable forms are those of the propagation of an infinite electromagnetic plane wave through layered, anisotropic media for which each layer is described by a thickness and a complex dielectric (or optical function) tensor.

The most rigorous and comprehensive approaches to this problem appear to be the 4×4 transfer matrix methods in which the $\vec{E}$ and $\vec{B}$ (magnetic) field vectors at any depth location in the sample are determined relative to the values at the outer boundaries of the sample by a sequential application of matrix operators across each intervening interface and through each intervening phase. Of specific importance for ellipsometry is that these methods account rigorously for polarization states of the beam upon interaction with the sample (*44–50*). A unique effect to be accounted for in anisotropic media is polarization mixing, in which a linearly polarized incident beam in the pure p or s state exists as a mixture of p and s states. The two most general approaches are those by Yeh (*47*) and by Berreman (*46*). Both treatments begin by applying the continuity of the tangential components of the $\vec{E}$ and $\vec{B}$ fields across the interfaces. Further, the same functional form of the matrix transforms applies to the two fields. This is an advantage because the tangential $\vec{B}$-field component can be used to calculate the perpendicular $\vec{E}$-field component at any point, and thereby the need to consider perpendicular field components is eliminated. This simplification directly reduces the electromagnetic variables by half and leads to the 4×4 matrix formalisms of the boundary value relationships of the field vectors. For purposes of deriving analytical expressions for simplified cases, the matrix formulation approach is not as useful as other more phenomenological approaches (*42*). The value of the 4×4 matrix methods is that they are quite amenable to numerical computations and form the basis of accurate calculations for situations in which all the experimental and material parameters are specified numerically. However, from a variety of such calculations, one can test approximations used to derive simplified analytical expressions and thereby check the validity of the expression over a given range of experimental conditions. An important point is that the matrix methods are sufficiently general that they describe any experiment regardless of the wavelength of the incident light. Thus, they are equally applicable to infrared spectroscopy as well as x-ray reflectivity measurements. Of course, each spectral region will have its own unique sets of dielectric functions to describe the optical effects. In this paper we concentrate on the ultraviolet/visible (*uv*-vis) wavelength region, which is traditionally associated with ellipsometry.

Berreman's treatment focuses on solving the first-order Maxwell's equations with constitutive equations of the film material. Electric field polarization mixing at the interfaces is accounted for by considering the total field as the weighted sum of partial modal fields present within the film layers. Yeh's formalism, on the other hand, deals directly with the partial modal fields. This leads to greater flexibility in handling boundary conditions at interfaces, and thus affords greater convenience for computing the reflectivities. However, Yeh's formalism is applicable only to magnetically isotropic phases. For materials with magnetic anisotropy and for optically active materials, Berreman's formalism is applicable. Since the films of general concern for optical characterization are usually non-magnetic and optically inactive, we have chosen to employ the more systematic and readily adaptable formalism of Yeh in our computations.

The most useful applications of reflection (or transmission) measurements of thin films involve characterization of absorption features, e.g., electronic transitions in the *uv*-visible region and vibrational excitations in the IR region. For absorbing media, the electromagnetic propagation involves a loss in power, and consequently any theory to be used must accommodate lossy media. Yeh did not explicitly consider lossy media in his treatment, but we have shown that this extension is readily accommodated (49).

B. Implementation of the Yeh 4 × 4 Transfer Matrix Method

The detailed development of the matrix equations to handle anisotropy is essentially that of Yeh, with the exception that Yeh's treatment has been expanded to accomodate complex optical (or dielectric) functions. The basic electromagnetic principals applying to these equations can be found elsewhere (47). For brevity, the Fortran codes developed are not presented here, but can be obtained from the authors (DLA) upon request.

The solution of the electromagnetic boundary value equations leads to the following relationship between the electric fields in the final medium (N) and the first medium (1), which contains the light source:

$$\begin{pmatrix} A_s(1) \\ B_s(1) \\ A_p(1) \\ B_p(1) \end{pmatrix} = \tilde{M} \begin{pmatrix} C_s(N) \\ D_s(N) \\ C_p(N) \\ D_p(N) \end{pmatrix}. \qquad (14)$$

In this equation, A, B, C, and D are the electric field intensities characterized by either s or p-polarization directions. The A and C fields are associated with waves moving in the direction of phase 1 to phase N, and conversely the B and D fields involve waves moving in the reverse direction. The detailed nature of the 4 × 4 matrix $\tilde{M}$ is explicitly defined in terms of the experimental geometry, the sample structure, the sample optical functions, and the frequency of the light. Details of the calculations of the matrix elements in terms of these quantities are given elsewhere (49).

Solution of the boundary value equations results only in setting relative values of the fields with respect to their location along the z-axis in the sample (refer to Fig. 1). These values are referenced to the fields in the Nth phase at the last boundary $[N/(N-1)]$, where it is required that the modal field amplitudes, $D_s(N)$ and $D_p(N)$, vanish since there can be no back-reflection from this infinite phase, and where the value of the outgoing fields, $C_s(N)$ and $C_p(N)$, can be set to unity for convenience. The complex reflectivities then can be computed easily by taking the ratio of the reflected amplitudes to the incident amplitudes, respectively, for a desired input polarization of the beam. For instance, the fractional complex amplitude of an s-polarized input beam that appears as a reflected s-polarized output beam can be computed as follows:

$$\tilde{r}_{ss} = \left[\frac{C_s(1)}{A_s(1)}\right]_{(A_p(1)=0)} = \frac{m_{21}m_{33} - m_{23}m_{31}}{m_{11}m_{33} - m_{13}m_{31}}, \quad (15)$$

where m_{jk} represents the corresponding element of the matrix $\tilde{M}$. Similarly, all the coefficients of the complex reflectance matrix ($\tilde{r}_{ss}$, $\tilde{r}_{sp}$, $\tilde{r}_{ps}$, $\tilde{r}_{pp}$) coefficients can be calculated. The power reflectivities and ellipsometric parameters $\tilde{r}_{pp}/\tilde{r}_{ss} = \tan\Psi e^{i\Delta}$ then can be calculated easily:

$$\tilde{r} \cdot \tilde{r}^* = |\tilde{r}|^2 = R, \quad (16)$$

$$A_e = \text{Re}(\tilde{r}_{pp}/\tilde{r}_{ss}); \quad B_e = \text{Im}(\tilde{r}_{pp}/\tilde{r}_{ss}), \quad (17)$$

$$\Psi = \tan^{-1}(\sqrt{A_e^2 + B_e^2}), \quad \Delta = \tan^{-1}\left(\frac{A_e}{B_e}\right) \quad (18)$$

The calculations are then repeated for the entire range of frequencies of interest to provide a simulated ellipsometric spectrum. Simulations of variable angle and variable thickness measurements follow similarly.

III. Comparison with Available Data and Previous Methods

In order to establish the validity of the algorithms developed in previous sections, we have performed simulations of ellipsometric data taken from a variety of studies available in the published literature. In all cases for which direct comparisons can be made, the simulated and published results are in good agreement. In all comparisons care was taken to account for differences in conventions for specification of ellipsometric quantities.

Jones and co-workers have modeled the experimentally determined ellipsometric responses for bismuth tellurium sulfide crystals using the Fresnel reflectance coefficients for uniaxially anisotropic surfaces with the orientation of the optic axis along the surface normal (40). Using their values of

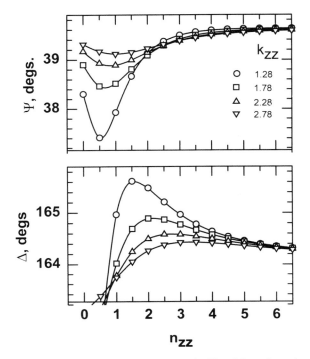

FIG. 2. The dependence of the ellipsometric angles Ψ and Δ on the real and imaginary parts of $\hat{n}_{zz}$ (n_{zz} and k_{zz}) calculated by the present approach (—) and by two-phase Fresnel's equations used in Ref. 40 (discrete symbols) for a bismuth tellurium sulfide crystal in air with 546.1 nm light at an incident angle of 45°. The optical constants of the crystals were taken directly from Ref. 40. Four values of k_{zz} were considered: ○, 1.28; □, 1.78; △, 2.28; and ▽, 2.78. Excellent agreement between the two calculation procedures is evident.

the principal components of the optical function tensor, $\hat{n}_{xx} = \hat{n}_{yy}$ and $\hat{n}_{zz}$, we have accurately reproduced their ellipsometric computations for as many as 56 different combinations of the tensor elements of the optical function tensors, as shown in Fig. 2. The continuous lines in Fig. 2 are constructed from our calculations based on at least 40 computed points at interpolated values of the variable tensor elements for each curve.

Ayoub and Bashara (54) have reported calculations of the changes in ellipsometric parameters associated with the formation of a Langmuir film of palmitic acid n a water surface. Fresnel reflectance coefficient expressions were derived based on a model structure of a single uniaxial layer sandwiched between two semi-infinite isotropic bounding phases. Using Ayoub and Bashara's values of ordinary and extraordinary refractive indexes of the LB film and the values of the dielectric functions of water, we have calculated the changes in ellipsometric angles with our algorithms. A comparison of the two calculations demonstrates quantitative agreement.

Ellipsometric characterization, including simulations using Schopper's extension of Fresnel's equations for non-oblique principal axes ($\hat{n}_{ij} = 0$, $i \neq j$), of a series of variable thickness ($t = 0–4{,}000$ Å), biaxially oriented films of anthracene molecules on an isotropic borosilicate glass substrate have been reported by Elsharkawi and Kao (*41*). All the parameters of optical properties and experimental geometry were taken from their report and applied to computations using the present generalized approach. The results of our simulations, in thickness increments of 100 Å in the given thickness range, are shown in Fig. 3 as a continuous curve, together with selected data taken from their original report. Our simulations overlap all of Elsharkawi and Kao's data, exactly duplicating the spiral-like Ψ, Δ pattern obtained in their computations. Similar calculations of hypothetical, uniaxial Langmuir–Blodgett (LB) films ($\hat{n}_{xx} = \hat{n}_{yy} = 1.55$, $\hat{n}_{zz} = 1.45$) of variable thickness on silicon substrates (without consideration of any oxide overlayer) ($\hat{n}_3 = 4.3 + i0$) for 632.8 nm light at an angle of incidence of 68° have been reported by den Engelsen (*33*). He considered the case of uniaxially anisotropic films with the principal axis along the surface normal. Our calculations for the identical thickness range, 0–6,400 Å, with 100 Å increments, generates identical 2.75 spiral loops in the Δ–Ψ plane, where each loop represents one complete (360°) circle of a spiral, as reported by den Engelsen. The actual results are not shown here in the interest of brevity.

Finally, we have attempted to simulate the spectroscopic ellipsometric experiments conducted by Debe and Field (*43*) for evaporated films of metal-free phthalocyanine films supported on planar, evaporated copper thin film substrates. It seemed important to consider this data since Debe and Field also included a rigorous computational model based on Berreman's approach (see Section II.A). However, while the optical tensor elements for phthalocyanine films for two different orientations were reported by the authors, the substrate optical constants were not explicitly given. We therefore selected literature values for evaporated copper films (*55*). Regardless of the reported values selected from the variety available, rather large discrepancies between our results and those reported by Debe and Field always resulted in the shorter-wavelength region of the spectral

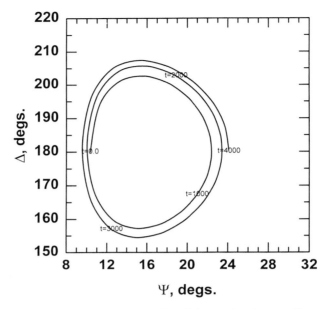

FIG. 3. The dependence of Δ and Ψ on the thickness of anthracene films supported on borosilicate glass for measurement at a wavelength of 546.1 nm. The experimental data of Elsharkawi and Kao (41) are plotted as reported in the form of a continuous curve drawn through a number of discrete experimental data points. Our calculated points were based on the optical constants of the anthracene film and glass substrates reported by Elsharkawi and Kao. A total of 400 calculated points using the present method fall directly on the reported curve with deviations too small to be visible to the eye. The pattern is in direct quantitative correspondence with that reported in Fig. 1 of Ref. 41.

range. The reasons for these discrepancies remain unknown at present, so no direct comparison can be made between our method and the Debe–Field study.

In summary, for the variety of ellipsometry studies arbitrarily selected from the literature, as illustrated earlier, and for others not presented, the quantitative agreement between our results and the published ones attest to the validity of our approach, including both the theory and computational algorithms. It must be pointed out that while the best tests would be those of samples with full biaxial anisotropy and with an arbitrary orientation of the principal axes (such that 3 × 3 dielectric tensors would exhibit significant off-diagonal elements), the reported cases inevitably have been selected by the original authors to be of lower symmetry and/or selected orientation in order to achieve experimental and interpretational convenience in the author's studies. The hope of the rigorous computational methods, of course, is that analyses of these more complex experiments can be made tractable and compromising simplifications in experimental systems would not be necessary. With this base we now proceed to a series of numerical simulations of hypothetical experiments in order to examine the effects of anisotropy on a variety of typical materials structures that might be analyzed using ellipsometry methods. With these simulations we hope to show that ellipsometric characterization of highly anisotropic samples is indeed feasible on a quantitative basis.

IV. Simulations of Anisotropic Effects

The specific examples presented will focus on anisotropic organic thin films, a reflection of both our particular interest in these materials and their accelerating popularity. However, it is stressed that the calculational methods and the analysis strategies are entirely general and can be employed for ellipsometric experiments on any type of material. We have used the simulations to flesh out physical insight about the various effects due to optical anisotropy in selected popular experimental approaches. No attempts were made to invert the ellipsometric data to derive back the original sample structures using iterative simulation, best-fit strategies; this constitutes an independent and important problem in its own right.

A. Construction of a Model Anisotropic Medium from an Ensemble of Oriented Oscillators

1. Selection of an Oscillator Model

In order to illustrate the effects of optical anisotropy systematically, hypothetical media have been generated from ensembles of a single, noninteracting hypothetical Gaussian oscillator with an associated transition electric dipole moment vector, $\mathbf{p}$ (a more detailed discussion of this quantity will be given in the next section). All oscillators in a given ensemble have identical values of $|\mathbf{p}|$, and the media differ only in the relative alignment of the individual $\mathbf{p}$ vectors. First we discuss the simplest case of the creation of isotropic medium for which the individual $\mathbf{p}$ vectors are randomly oriented, and accordingly the optical functions of the medium are specified in terms of scalar quantities. In the following discussions, this response is expressed in terms of the complex refractive index (Eq. (11)). The medium has been assigned a maximum intensity value of $\text{Im}[\hat{n}(v_{\max})] = k_{\max} = 2.5$ at a frequency of $v_{\max} = 2 \times 10^4 \text{ cm}^{-1}$ (500 nm, 2.5 eV) and a full width at half maximum of $\Gamma_{1/2} = 2 \times 10^3 \text{ cm}^{-1}$ (50 nm, 0.25 eV). The values of $k_{\max}$ correspond to an extinction coefficient of $1 \times 10^5 \text{ l mol}^{-1} \text{ cm}^{-1}$, representative of values for strong dyes, for example, polymethine dyes (56). The value of $\Gamma_{1/2}$ is typical of discrete lines observed in thin films. Wider lines were not considered, since little additional insight would be gained for the added computational burden of larger frequency ranges, keeping constant resolution. The spectrum of $\text{Re}[\hat{n}(v)] = n(v)$ was obtained from $k(v)$ by the standard Kramers–Kronig (KK) transformation (57):

$$n(v_i) = n_\infty + P \int_{v_a}^{v_b} \frac{k(v) v \, dv}{v^2 - v_i^2}. \tag{19}$$

To account for the finite range of frequencies in the numerical integration, a constant value of $\tilde{n} \equiv n_\infty = 2.0 + 0i$ was assumed outside of the range $10,000 < v < 50,000 \text{ cm}^{-1}$ (1,000–200 nm), and the integration was carried out within this frequency range. A plot of the real and imaginary parts of the resultant complex refractive index spectrum is shown in Fig. 4.

To construct an anisotropic medium, an ensemble of oscillators was generated with the transition dipole vectors all aligned parallel. The optical response of such a medium is described by an optical function tensor $\hat{n}(v)$ in which the tensor elements $\hat{n}_{jk}$ express the fact that when the electric field vector of the light is orthogonal to **p**, a situation in which no excitation of the oscillators can occur, $\text{Re}(\hat{n}_{ii}) = n_\infty = 2.0$, $j = 1, 3$, while for a parallel arrangement, $\text{Re}(n_j)$, $j = 1, 3$ is given by the standard KK transform given in Eq. (19). The explicit directions for calculation of the $\hat{n}_{jk}$ tensor elements, including, wherever applicable, off-diagonal elements, is developed in the next section.

Two prototypical cases of common experimental situations for ellipsometric analyses are considered: (1) reflection from the surface of a bulk medium, and (2) a substrate-supported thin film of material. The generalization to more complex multilayered structures, even including cases where the incident medium is anisotropic, is straightforward.

The preceding oscillator ensembles allow three limiting types of media to be considered: (1) isotropic, (2) anisotropic with all the oscillator **p** vectors aligned in a parallel, but randomly distributed orientation to the surface

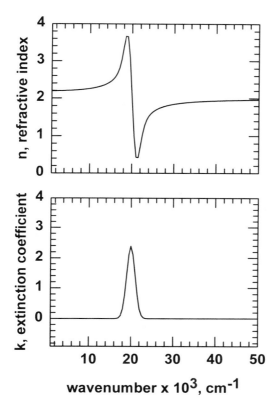

FIG. 4. Optical functions of an isotropic medium constructed from a randomly oriented ensemble of identical Gaussian oscillators. The spectrum of k, the imaginary part of the optical function, shows $k_{max} = 2.5$ at 20,000 cm^{-1} (500 nm; 2.5 eV), the resonance frequency of the oscillator. The full-width–half-maximum is 2,000 cm^{-1} (50 nm; 0.25 eV). The real part of the optical function, n, was calculated using a Kramers–Kronig transformation (see text for details).

plane (xy-plane in Fig. 1), and (3) anisotropic with all the **p** vectors aligned in a perpendicular orientation to the surface plane (parallel to the z-axis in Fig. 1). Cases 2 and 3 represent uniaxial or biaxial symmetry, depending upon the specific azimuthal orientation of **p** with respect to the plane of incidence (the angle θ, as defined in Fig. 1). For a random distribution of azimuthal angles of the individual oscillator **p** vectors, the explicit definition of a uniaxial system, the averaged ellipsometric response will correspond to the special case of a biaxial system with a specific azimuthal angle (as defined in Fig. 1) of (45° + n90°) where n is an integer; all other angles give rise to biaxial response. In case 2, the in-plane medium, all simulations were done with the specific case in which the azimuthal angle was fixed at 45°, equivalent to a distribution of **p** vectors randomly in the surface plane. A last point of definition is that all of these media exhibit principal optic axes with the major one containing values of $k \neq 0$ along the transition moment direction of the oscillator and with the other two axes, associated with values of $k = 0$, orthogonal to the major axis. Of course, when these media are turned at fully oblique angles to the plane of the incident light they will give rise to responses expected for a biaxial crystal.

With regard to the latter consideration, more complex cases certainly can be considered in which two or more types of oscillators, with different spectral responses, are present with variable orientations and in variable compositions in different layers in multilayered stacks. However, the general principles are identical to the limiting cases considered later, and inclusion of such complex examples would serve mostly to weigh the presentation down with details while adding little to the understanding of the principles involved. Such complexities are best reserved for analyses of actual complicated experimental systems of focused interest.

2. Generation of Optical Function Tensors

This section considers the relationship between the optical function tensors of a medium and the properties of the oscillators that make up the medium. In the dipole approximation (58), the power absorbed when an electromagnetic field oscillating at a given frequency interacts with an electric dipole is given by the proportionality

$$I \propto |\mathbf{E}_{loc} \cdot \mathbf{p}|^2, \tag{20}$$

where I is the power absorbed or the rate of the excitation, $\mathbf{E}_{loc}$ is the electric field vector in the local vicinity of the molecule and **p**, in quantum-mechanical terms, is the electric dipole moment transition matrix element for the given excitation at the given frequency (61). The unit vector $\mathbf{p}/|\mathbf{p}|$ defines the direction of the transition dipole and, of course, $|\mathbf{p}|$ defines the strength of a single isolated oscillator. For a condensed-phase medium of parallel aligned oscillators, the value of $k = \text{Im}(\mathbf{n})$ defines the intensity of the transition, and so it is convenient to use k rather than $|\mathbf{p}|$ as the basis for intensity, since k often can be derived experimentally from spectroscopic measurements on the pure material, whereas $|\mathbf{p}|$ is usually thought of as a property of an isolated species. Therefore, for the purposes at hand it is convenient to form a vector **k** that carries both the direction of **p** and the intensity of k. In a real physical material, of course, it would need to be shown that the direction of **p** along some molecular (internal) coordinate system is not perturbed by matrix effects relative to the direction for an isolated oscillator. Since the intensity of a transition is usually derived from measurements on an isotropic sample, the following relationship is convenient:

$$|\mathbf{k}| = 3.0 k_{\text{iso}}, \qquad (21)$$

where $\mathbf{k}$ represents the effective response of the material when all the oscillators are aligned in the direction of the exciting electric field, and k_{iso} is the value of $\mathbf{k}$ for random orientation of the oscillators in an isotropic sample. To form a tensor, the vector quantity $\mathbf{k}$ is decomposed into components along the external coordinate axes (59). The imaginary parts of the diagonal elements of the optical function tensor of the medium composed of these oscillators is generated by the equations

$$k_{xx} = |\mathbf{k}| \sin^2 \alpha \;\; \cos^2 \beta,$$
$$k_{yy} = |\mathbf{k}| \sin^2 \alpha \sin^2 \beta,$$
$$k_{zz} = |\mathbf{k}| \cos^2 \alpha, \qquad (22)$$

where α is the tilt angle of the transition dipole from the surface normal (z) and β is the azimuthal angle in the surface plane (xy), as shown in Fig. 5. If

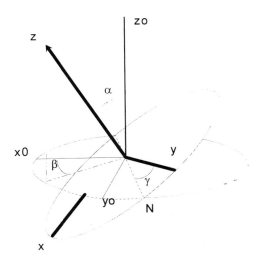

FIG. 5. A graphical description of the Eulerian angles α, β, and γ, used to define the oblique orientations of the principal axes of a crystal (x, y, z) with respect to the ellipsometer coordinate system (x_0, y_0, z_0): α is the z, z_0 angle; β is the angle between x_0 and the projection of z in the x_0, N plane, where N is the interaction line between the x_0, y_0 and x, y planes; and γ is the angle between y and N.

more than one type of independent oscillator were present in the medium, the total value of k_{ee} at a given frequency, where $e = x, y$, or z, would be given by the sum of k_{ee}^j terms over all oscillators, where j is the index for the jth type of oscillator.

Each complex diagonal element $\hat{n}_{ee}$ of the tensor $\hat{n}$ is given by $\text{Re}(\hat{n}_{ee})$, $\text{Im}(\hat{n}_{ee})$ or n_{ee}, and k_{ee}. A useful approximation is that $\text{Re}(\hat{n})$ consists of an oscillator contribution, $n_{\text{osc}}(v)$, which is direction- and frequency-dependent, and an isotropic background response, $n_\infty(v)$, which varies slowly or is constant with frequency. The latter quantity is essentially the constant specified in the Kramers–Kronig transform (Eq. (19)), and the former is the quantity given by the integral term. Once the diagonalized tensor elements are known, the non-diagonal tensor for any other orientations of the oscillator, as given by $\mathbf{k}$, can be obtained by methods of simple matrix rotation,

$$\hat{\hat{\varepsilon}}_{3 \times 3} = R^{-1} \hat{\hat{\varepsilon}} R, \tag{23}$$

where R is the rotation matrix defined as (*59*)

$$R = \begin{pmatrix} \cos\alpha \cos\beta \cos\gamma - \sin\beta \sin\gamma & \cos\alpha \sin\beta \cos\gamma + \cos\beta \sin\gamma & -\sin\alpha \cos\gamma \\ \cos\alpha \cos\beta \sin\gamma - \sin\beta \cos\gamma & -\cos\alpha \sin\beta \sin\gamma + \cos\beta \cos\gamma & \sin\alpha \sin\gamma \\ \sin\alpha \cos\beta & \sin\alpha \sin\beta & \cos\alpha \end{pmatrix},$$

(24)

for which α, β, and γ are, respectively, tilt, azimuth, and twist in terms of the Eulerian coordinates as defined in Fig. 5.

B. SINGLE-INTERFACE AMBIENT–SUBSTRATE SYSTEMS

Single interface, ambient/substrate structures form the simplest, and probably one of the most extensive, classes of materials structures that can be investigated by ellipsometry. Ellipsometric characterization of the structures and electronic states of adsorbate-free surfaces of metals and semiconductors under ultrahigh-vacuum conditions is of fundamental importance in surface physics and is relevant to applications such as electronic devices. Ellipsometric characterization of liquid/solid interfaces similarly is an important goal in surface chemistry and has applications such as electrochemistry. Analyses of these single-interface structures have involved both single-wavelength and spectroscopic measurements, fixed and multiple incidence angles and variables ambient phases, e.g., different liquids or gases. The measurements typically are utilized to determine the optical functions of the materials and/or to characterize surface (interface) structure in terms of deviations from a perfect, sharp interface. The latter implies, of course, that a discrete interface layer exists, and thus the final model will actually be three-phase, but the basic analysis begins with the two-phase system. In the interpretation of the data it must be realized that anisotropy in the material can exert strong influences on the measured ellipsometric responses and thus, if not taken into proper account, will lead to inaccurate descriptions of the sample characteristics. In order to gauge the magnitude of these effects and to provide some useful physical insight as to their basis, simulations based on the single oscillator model (Section IV.A) for selected limiting structures are presented next. While the simulation examples have been selected to be simple for illustration purposes, the results are of general applicability to more complex anisotropic structures, and the computational algorithms can handle significantly more complex systems as needed, with no formal limitations on the number of independent phases and their optical tensor values to be considered.

1. Single-Wavelength, Variable-Angle Measurements

Simulations of General Angle Dependence: The first task chosen is the simulation of the general ellipsometric response of a two-phase system as a function of incidence angle. This provides the simplest case of a multiplexed ellipsometry experiment, since the optical or dielectric functions are invariant with the angle of the incident light, in contrast to the spectroscopic experiment in which the functions themselves vary with the energy of the light. Figure 6 shows simulated Ψ, values at $\lambda = 500$ nm (20,000 cm^{-1}, 2.5 eV) for a series of incidence angles between 5 and 89° for each of the three limiting oscillator orientations, isotropic, in-plane and out-of-plane (see Section IV.A). The variation of the values of the s and p reflectivities, R_s and R_p, with the incidence angle, θ, is displayed in Fig. 7. Whereas R_s increases monotonically with θ, R_p displays a characteristic curve with a single minimum, defined as the pseudo-Brewster's angle θ_{pB}. The true Brewster's angle, θ_B, is defined as the angle for which $R_p = 0$ upon reflection from a non-absorbing ($k = 0$) medium. For the limiting in-plane ($\alpha = 90°$ and $\beta = 45°$ in Eq. (22)) and out-of-plane ($\alpha = 0°$ and $\beta = 45°$ in Eq. (22)) orientation anisotropic cases, sizable differences in the angular dependence of R, Δ and Ψ are observed. In particular, the location of θ_{pB} and the associated R_p values are quite sensitive to oscillator orientation and suggest that the former represents an optimal angle of incidence to probe anisotropic effects. The variability of the value of θ_{pB} is also reflected in the angle-dependent Ψ plots (Fig. 6). An important summary point is that the anisotropy-induced changes in Δ and Ψ are pronounced enough that

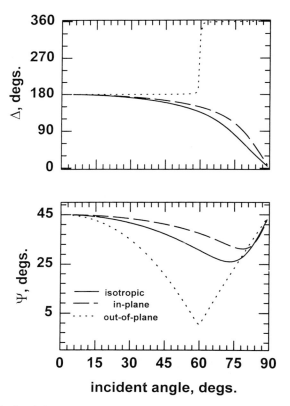

FIG. 6. Calculated dependence of the ellipsometric angles Δ and Ψ on the incident angle for a two-phase system consisting of an oscillator medium and air. The wavelength of incident light was chosen as 500 nm to coincide with the maximum extinction coefficient value of the medium. Three extremes in oscillator anisotropy were used: isotropic, principal axis along z (in-plane) and principal axis in the xy plane (out-of-plane) (for details see text). The angle of incidence θ was varied in steps of $2°$ between $5°$ and $89°$.

experiments involving multiple-angle measurements should be an effective means of characterizing structure-induced film anisotropy.

A Test of a Common Approximation: In order to interpret experiments of the preceding type, simplifying approximations have been made that allow analytical treatment of the data. With the ability of make rigorous numerical calculations, these approximations can be tested. One common approximation, justified both on theoretical and experimental grounds, is that for

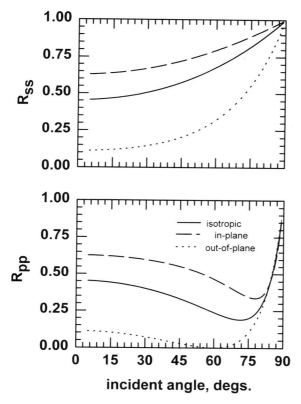

FIG. 7. Dependence of the power reflectivities, R_{ss} and R_{pp}, on the angle of incidence for the cases in Fig. 6.

uniaxial crystals with the principal optic axis oriented in the plane of incidence, the value of the principal dielectric tensor component along the z axis ($(\hat{\varepsilon}_{zz})$ plays a relatively unimportant role in controlling the values of the ellipsometric parameters, Δ and Ψ (42). In order to test this assertion through simulation, a dielectric tensor was constructed for k_{max} at 500 nm for an out-of-plane oriented crystal. Figure 8 shows the changes in Ψ and Δ, at a fixed 45° angle of incidence (well off θ_{pB} for all cases), as functions of variable $n_{xx}(=n_{yy})$, n_{zz} and k_{xx} and k_{zz}. The results show that the calculated variations of Ψ and Δ are minor for variations in the tensor elements zz and are significantly more pronounced when the elements xx are varied. This result then verifies the approximations explicitly for the conditions chosen and further implies its general validity over a range of similar conditions.

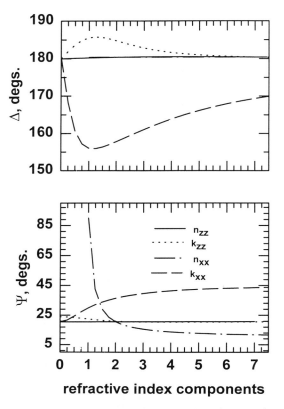

FIG. 8. Dependence of the ellipsometric angles Δ and Ψ on the real and imaginary components of the complex optical function tensor elements $\hat{n}_{zz} n_{zz}$ and k_{zz}) and $\hat{n}_{xx}$ (n_{xx} and k_{xx}) for the out-of-plane case in Fig. 6. Each trace represents the changes that occur when the indicated tensor element is varied from 0 to 7.5 in steps of 0.25 while maintaining all other elements constant (see text for details). The incident angle was chosen as 45° (see text for details).

Brewster's Angle for Anisotropic Substrates: The foregoing sets of calculations contain an important point with respect to the observation of near-zero values of R_p at specific angles. While non-absorbing media lead to a value of $R_p = 0$ at θ_B, it is generally believed that for absorbing media R_p always exhibits nonzero minimal values, observed at θ_{pB}. However, the preceding simulations show that for the model anisotropic medium with an out-of-plane orientation of the oscillators, for which the absorbance or loss is entirely contained in the zz tensor element alone, values of $R_p = 0$ arise at θ_{pB}. Similarly behavior of the values of R_p for uniaxially oriented crystals have been recently reported (*60,61*). These results demonstrate the unexpected effects that may arise when anisotropy is present in a substrate and suggest that simulations will be extremely helpful in studies of the variations of reflectivities and ellipsometric parameters in complicated cases of phases exhibiting biaxial anisotropy and other more complicated cases, such as stacks of anisotropic layers, heretofore not explorable.

Effects of Oblique Orientations of Crystal Principal Axes with the Plane of Incidence: In cases of the investigations of a material with known directions of the optical axes, it is easy to align the sample to minimize the complexities involved in interpreting the ellipsometric data. That is, the sample can be aligned with respect to the plane of incidence of the light in order to allow consideration of only the diagonal elements of the optical (or dielectric) tensor in the data analysis. However, in the more general case, such alignment may not be physically possible for some reason, or the exact directions of the principle axes may not be known. In these cases the off-diagonal elements become important. In order to understand the relative role of the latter, several cases of the oscillator model are considered in which the principal optic axis is set at oblique angles with respect to the plane of incidence of the light. For the calculations, the oscillator axis is set at a variable angle of α with respect to the surface normal (z axis, see Fig. 5). This situation corresponds to the case of a uniaxial, optically anisotropic crystal cut at an oblique angle α (*62*) to the principal axis and then mounted in the ellipsometer for analysis. In such cases, referring to Fig. 5 and Eq. (24), the dielectric (or optical function) tensor components along xz and zx ($\hat{\varepsilon}_{xz}$ and $\hat{\varepsilon}_{zx}$) become finite. The dependence of the ellipsometric parameters as a function of α is shown in Fig. 9. Significant, measurable changes in Ψ and Δ values are observed at θ_{pB} and demonstrate the sensitivity of θ_{pB} to the tilt of the principal axis. These results demonstrate the advantages that can be gained by using simulations in measurements involving obliquely aligned, anisotropic crystals.

2. Spectroscopic Measurements

While the foregoing calculations explore the effects of angle variation on ellipsometric measurements of anisotropic materials, the more common form of multiplexed ellipsometry is that of variable wavelength or energy. The most typical range of wavelengths is over the visible region from ~ 350 to 850 nm, corresponding to an energy range of 3.5 to 1.5 eV or 2.9×10^4 to 1.2×10^4 cm^{-1}. To cover the broadest spectral range, several sets of simulations were carried out over the energy range of $1.0-5.0 \times 10^4$ cm^{-1}

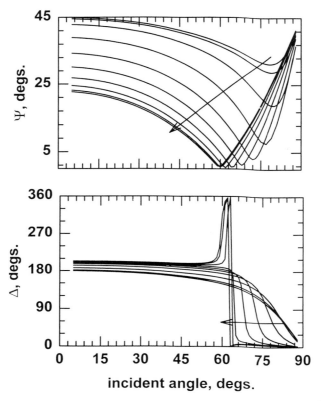

FIG. 9. The dependence of calculated ellipsometric angles Δ and Ψ as a function of the incidence angle for varied oblique angles α of the oscillator axis with respect to the plane of incidence of the light. The geometry of the experiment is same as the one described in Fig. 6. The direction of the arrow indicates increasing values of α from 0 to 90 in steps of 10° (see text for details); α = 0 and 90° corresponds to the cases of orientation of the oscillator axis in-plane and out-of-plane, respectively.

(1.25 to 6.3 eV). A 3-D plot of calculations of Ψ for a variable-angle spectroellipsometry (VASE) experiment is shown in Fig. 10 for the three limiting orientations (Section IV.A) of the ensemble of independent oscillators. Comparison for three different crystal orientations gives very different Ψ-energy–angle surfaces. Similarly, Δ-energy–angle surfaces are also vastly different, but are not shown here in the interest of brevity. The richness of these data, together with the rigorous capabilities of the simulation methods, should allow a thoroughly exhaustive analysis of an anisotropic substrate

EFFECTS OF OPTICAL ANISOTROPY ON SPECTRO-ELLIPSOMETRIC DATA 307

(a)

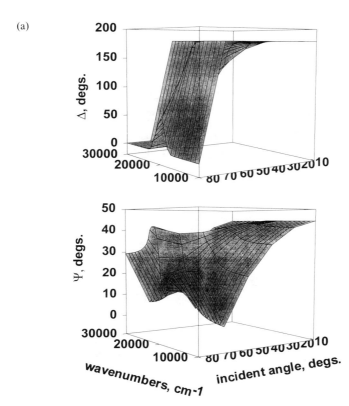

(b)

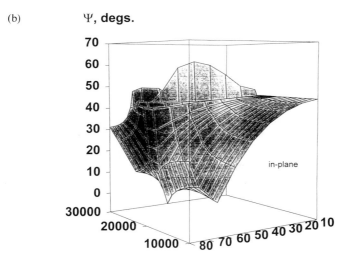

(c) Ψ, degs.

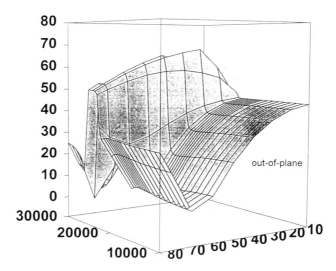

FIG. 10. Three-dimensional plots of calculated ellipsometric angles Ψ as functions of the frequency of the light and the angle of incidence for the two-phase systems described in Fig. 6 for the three diferent orientations of the oscillator axis: (a) isotropic; (b) in-plane; and (c) out-of-plane. The calculations of Δ values are also shown in (a) for the isotropic medium. Since the dependence of Δ on frequency and angle of incidence is rather smooth and uninformative, relative to the behavior of Ψ, as seen in (a) the Δ curves for (b) and (c) are not shown.

by spectroscopic ellipsometry, including even the addition of the variable-angle mode.

3. Variation of the Ambient Medium

Ambient media were selected to be isotropic and non-absorbing with refractive index values varying from 1.0 to 3.5, a range that contains a shift from a rarer to a denser ambient medium with respect to the substrate. The sample consisted of the oscillator medium with out-of-plane orientation (see the foregoing). Figure 11 shows the dependence of Ψ on incidence angle for selected values of ambient optical constants. The shifts in Brewster's angles and the exact trajectory of Ψ as functions of incident angle are both highly sensitive to the refractive index of the ambient medium for a given orientation of the oscillator. Similar calculations were done for the other

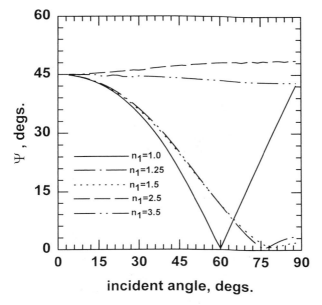

FIG. 11. Dependence of the ellipsometric angle Ψ on the incident angle for the two-phase system described in Fig. 6 for different selected non-absorbing ambient phases. The calculations were performed for five different values of the ambient refractive index ($\tilde{n}_1$) shown in the figure.

two extremes in oscillator orientations. These dependencies on angle of incidence and ambient refractive indices ($\tilde{n}_1$) were further found to be significantly different for different orientations of the oscillators in the anisotropic media. These calculations are not presented here, but it is pointed out that they reveal clearly an advantage to multiple ambient medium experiments in resolving film anisotropies.

C. Multiple-Interface, Ambient–Film–Substrate Systems

The most general sample type in ellipsometric measurements consists of multiple-phase systems with an ambient overlayer surrounding a multiple film stack. This description encompasses a large variety of sample systems, including substrate-bound ultrathin films of organic, inorganic and metallic characters; parallel-layered stacks of different refractive index transparent media, often approximations to surfaces with gradients of optical constants;

and freely suspended thin films. The use of ellipsometry in these types of samples supplies important characterization information for a diverse variety of currently popular studies, including the growth and formation kinetics of vapor-deposited films on solid supports, self-assembled monolayers and Langmuir–Blodgett films, electrochemistry, corrosion and optical devices, to name a few. All the ellipsometric approaches discussed earlier, including multiple-angle, spectroscopic and multiple ambient media, are useful in these systems. An additional approach applicable in these samples is the one involving *multiple or variable thickness* measurements (*18*). In the simplest case, the optical functions and physical structure of the medium are invariant with film thickness, *viz.*, the medium is of constant atomic or molecular structure regardless of the thickness. In more complex cases, the properties themselves may vary with thickness, either in the form of a smooth gradient or with abrupt discontinuities. In either case, the approach provides a unique set of ellipsometric data useful for detailed characterization. In order to illustrate the general value of this approach, we have employed a simple example of a two-interface, substrate/film/ambient structure. The film consists of the single oscillator medium developed earlier (Section IV.A) with all three limiting orientation cases considered. The substrates selected are a metal, Cu, a semiconductor, Si with no oxide, and a transparent, biaxial crystal, SbSI. All calculations were performed at a 70° angle of incidence at the film oscillator resonance frequency of $20,000\,\text{cm}^{-1}$ (500 nm). The Cu and Si are assigned a single scalar value of the optical function of $1.141 + 2.473i$ and $4.398 + 0.073i$ at 5,000 nm (*55*), and the optical function tensor for SbSI was obtained from Yariv and Yeh (*26*). The SbSI crystal was assumed to be aligned in the ellipsometer to produce nonzero values only for the diagonal elements such that $\hat{n}_{xx} = 2.7 + 0i$, $\hat{n}_{yy} = 3.2 + 0i$, and $\hat{n}_{zz} = 3.8 + 0i$. The calculations were performed for the three different orientations of the oscillator medium, *viz.*, designated as isotropic or random, in-plane, and out-of-plane according to the direction of alignment of the oscillator axis. The results of the calculations for the three structures presented above are summarized in Figs. 12, 13, and 14, respectively. Dramatic differences in the trajectories in the Δ–Ψ planes as functions of film thicknesses are clearly evident. These results demonstrate the high sensitivity of ellipsometric measurements in monitoring evolving structural characteristics in thin-film growth when the films exhibit anisotropic character. It also is clear that the appearance of anisotropic character in an evolving film may confuse the interpretation of the ellipsometric data if anisotropic effects are not taken into account. In such cases, the application of rigorous calculational methods can be of immense value in clarifying structural interpretations.

EFFECTS OF OPTICAL ANISOTROPY ON SPECTRO-ELLIPSOMETRIC DATA 311

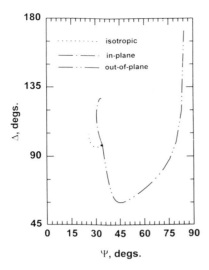

FIG. 12. The trajectories of calculated thickness-dependent ellipsometric angles in the $\Delta-\Psi$ plane for a three-phase system composed of a variable thickness film of the single oscillator medium (at three distinct orientations) supported on metallic Cu substrates. The incident angle is arbitrarily chosen at 70°, and the wavelength of incident light at the oscillator resonance wavelength of 500 nm. The film thickness is varied from 0 to 600 Å (see text for details).

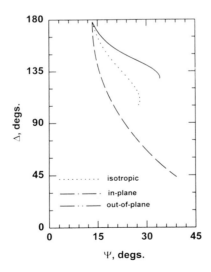

FIG. 13. The same plot as shown in Fig. 12, except that the substrate is a silicon (no oxide) surface (see text for details).

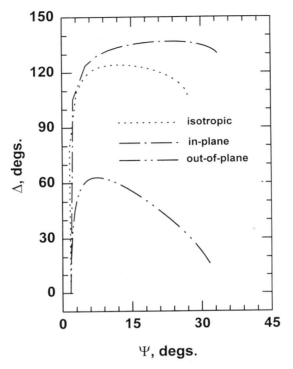

FIG. 14. The same plot as shown in Fig. 12, except that the substrate is an SbSI biaxial crystal surface oriented with the principal optic axes along the external laboratory coordinate axes. The Eulerian angles α, β, and γ for the SbSI substrate orientation chosen are exactly equal to zero. (see text for details.)

Acknowledgment

The authors acknowledge K. Vedam for continued encouragement during the course of this work and for many fruitful discussions. One of us (DLA) also acknowledges early discussions with D. Aspnes from which the original ideas for this work arose. Finally, support from the National Science Foundation (grant #DM900-1270) is acknowledged.

References

1. R. M. A. Azzam and N. M. Bashara, "Ellipsometry and Polarized Light," North-Holland, Amsterdam, 1976.
2. J. B. Theetan and D. E. Aspnes, *Annu. Rev. Mater. Sci.* **11**, 97 (1981), and selected references cited therein.
3. W. E. J. Neal, *Appl. Surf. Sci.* **2**, 445 (1979).
4. J. Barth, R. L. Johnson, and M. Cardona, in "Handbook of Optical Constants of Solids II" (E. D. Palik, ed.), Academic Press, Orlando, Florida, 1985, Ch. 10.
5. I. An and R. W. Collins, *Rev. Sci. Instrum.* **62**, 1904 (1991), and selected references therein.
6. E. Aspnes, in "Optical Properties of Solids: New Developments" (B. O. Seraphin, ed.), North-Holland, Amsterdam, 1976, Chapter 15.
7. B. Drevillon, *Thin Solid Films* **165**, 157 (1988).
8. K. Vedam, P. J. McMarr, and J. Narayan, *Appl. Phys. Lett.* **47**, 339 (1985).
9. See, for the most recent developments; "Proceedings of the 1st International Conference on Spectroscopic Ellipsometry," *Thin Solid Films* **233/234** (1993).
10. P. G. Snyder, M. C. Rost, H. Bu-Abbud, J. A. Woollam, and S. A. Alterovitz, *J. Appl. Phys.* **60**, 3293 (1985); J. A. Woollam, P. G. Snyder, A. W. McCormick, A. K. Rai, D. Ingram, and P. P. Pronko, *J. Appl. Phys.* **62**, 4867 (1987).
11. M. M. Ibrahim and N. M. Bashara, *J. Opt. Soc. Am.* **61**, 1622 (1971), and selected references cited therein.
12. F. H. P. M. Habraken, O. L. J. Gizeman, and G. A. Bootsma, *Surface Sci.* **96**, 482 (1980), and selected references cited therein.
13. See, for example: Y.-T. Kim, K. Vedam, R. W. Collins, and D. L. Allara, *Thin Solid Films* **193/194**, 353 (1991).
14. D. E. Aspnes, *J. de Physique* **44**, C10-3 (1983).
15. F. Hottier and J. B. Theetan, *J. Cryst. Growth* **48**, 466 (1980).
1. R. H. Muller and C. G. Smith, *Surf. Sci.* **96**, 375 (1980).
17. D. Drevillon, *Thin Solid Films* **130**, 165 (1985).
18. I. An, H. V. Nguyen, N. V. Nguyen, and R. W. Collins, *Phys. Rev. Lett.* **65**, 274 (1990).
19. A. Canillas, J. Campmany, J. L. Andújar, and E. Bertran, *Thin Solid Films* **228**, 109 (1993).
20. D. E. Aspnes, *Phys. Rev. B* **25**, 1358 (1982).
21. D. E. Aspnes, *Thin Solid Films* **89**, 249 (1982), and references cited therein.
22. J. P. Borgogno, B. Lazarides, and E. Pelletier, *Appl. Opt.* **21**, 22 (1982).
23. S. Y. Kim and K. Vedam, *Thin Solid Films* **166**, 325 (1988).
24. B. Law and D. Beaglehole, *J. Phys. D* **14**, 115 (1981).
25. J. C. Charmet and P. G. de Gennes, *J. Opt. Soc. Am.* **73**, 1777 (1983).
26. For example, see: A. Yariv and P. Yeh, "Optical Waves in Crystals," Wiley-Interscience, New York, 1984.
27. V. Narayanmurthy, *Science* **235**, 1023 (1987); M. B. Panish, *Science* **208**, 916 (1980).
28. See, for example, H. Angus Macleod, *J. Vac. Sci. Tech. A* **4**, 418 (1986), and selected references cited therein; I. Hodgkinson and Q.-H. Wu, *J. Opt. Soc. Am.* **A10**, 2065 (1993).
29. For example, K. Naegele and W. J. Plieth, *Surf. Sci.* **61**, 504 (1976).
30. For example, see: A. Ulman, "An Introduction to Ultrathin Organic Films," Academic Press, San Diego, 1991; S. Henon and J. Meunier, *Thin Solid Films* **234**, 471 (1993).
31. R. H. W. Graves, *J. Opt. Soc. Am.* **59**, 1225 (1969).
32. M. J. Dignam and M. Moskovits, *Appl. Opt.* **9**, 1968 (1970); M. J. Dignam, M. Moskovits, and R. W. Stobie, *Trans. Faraday Soc.* **67**, 3306 (1971).

33. D. den Engelsen, *J. Opt. Soc. Am.* **61**, 1460 (1971).
34. D. J. DeSmet, *J. Opt. Soc. Am.* **65**, 542 (1975).
35. M. Elshazly-Zaghloul, R. M. A. Azzam, and N. M. Bashara. *Surf. Sci.* **56**, 293 (1976).
36. A. F. Antippa, R. M. Leblanc, and D. Ducharme, *J. Opt. Soc. Am.* **3A**, 1794 (1986).
37. A. Y. Tronin and A. F. Konstantinova, *Thin Solid Films* **177**, 305 (1989).
38. R. Zhu, C. Lin, and Y. Wei, *Thin Solid Films* **203**, 213 (1991).
39. F. Meyer, E. E. de Kluizenaar, and D. den Engelsen, *J. Opt. Soc. Am.* **63**, 529 (1973).
40. M. L. Jones, H. H. Soonpaa, and B. S. Rao, *J. Opt. Soc. Am.* **64**, 1591 (1974).
41. A. R. Elsharkawi and K. C. Kao, *J. Opt. Soc. Am.* **65**, 1269 (1975).
42. D. E. Aspnes, *J. Opt. Soc. Am.* **70**, 1275 (1980).
43. M. K. Debe and D. R. Field, *J. Vac. Sci. Tech. A* **9**, 1265 (1991).
44. M. Born and E. Wolf, "Principles of Optics," Pergamon, Oxford, 1964.
45. P. Yeh, "Optical Waves in Layered Media," Wiley-Interscience, New York, 1980.
46. D. W. Berreman, *J. Opt. Soc. Am.* **62**, 502 (1972).
47. P. Yeh, *J. Opt. Soc. Am.* **69**, 742 (1979).
48. P. J. Lin-Chung and S. Teitler, *J. Opt. Soc. Am.* **A1**, 703 (1984); R. S. Weis and T. K. Gaylord, *J. Opt. Soc. Am.* **A4**, 1720 (1987).
49. A. N. Parikh and D. L. Allara, *J. Chem. Phys.* **96**, 927 (1992).
50. P. E. Laibinis, G. M. Whitesides, D. L. Allara, Y.-T. Tao, A. N. Parikh, and R. G. Nuzzo, *J. Am. Chem. Soc.* **113**, 7152 (1992).
51. In isotropic sample systems, the quantity $\tilde{r}_{pp}/\tilde{r}_{ss}(=\tilde{r}_p/\tilde{r}_s)$ is identically equal to ρ, whereas in cases of anisotropic sample systems, this is one of the measurable ratios; other ratios such as $\hat{r}_{sp}/\hat{r}_{ss}$ and $\hat{r}_{pp}/\hat{r}_{ps}$ can also be derived from ellipsometric experiment. (J. Lekner, *J. Opt. Soc. Am. A* **10**, 1579 (1993).).
52. R. H. Muller, *Surf. Sci.* **16**, 14 (1969).
53. See, for example: A. R. M. Zaghloul, R. M. A. Azzam, and N. M. Bashara, *Surf. Sci.* **56**, 87 (1976); M. Malin, K. Vedam, *ibid.*, **56**, 49 (1976); S. So, *ibid.*, **56**, 97 (1976).
54. G. T. Ayoub and N. M. Bashara, *J. Opt. Soc. Am.* **68**, 978 (1978).
55. "Handbook of Optical Constants of Solids II" (E. D. Palik, ed.), Academic Press, Orlando, Florida, 1985.
56. See, for example: M. R. Philpott, *Annu. Rev. Phys. Chem.* **31**, 97 (1980).
57. F. Stern, *in* "Solid State Physics" (F. Seitz and D. Turnbull, eds.), Academic Press, New York, 1963, Vol. 15, p. 299.
58. D. Bohm, "Quantum Theory," Prentice-Hall, Englewood Cliffs, New Jersey, 1963.
59. E. B. Wilson, J. C. Decius, and P. C. Cross, "Molecular Vibrations," McGraw-Hill, New York, 1950, p. 285.
60. M. Elshazly-Zaghloul and R. M. A. Azzam, *J. Opt. Soc. Am.* **72**, 657 (1982).
61. J. Lekner, *J. Opt. Soc. Am.* **A10**, 2059 (1993).
62. For example, see: J. Bennett, H. Bennett, *in* "Handbook of Optics," (W. G. Driscoll and W. Vaughan, eds.), McGraw-Hill, New York, 1978, Ch. 10.

Author Index

A

Abe, H., 35
Abeles, F., 50
Abelson, J. R., 52, 103
Acher, O., 2, 4, 7–9, 29–31, 38, 39, 41
Acosta-Ortiz, S. E., 9
Adachi, S., 97
Agranovich, V. M., 10
Aleksandrov, V. I., 239–241
Allara, D. L., 258, 282, 288, 290
Allen, J. T., 32
An, I., 53–58, 64, 70, 83, 84, 129, 148, 152, 154, 157, 159, 160, 163, 165–167, 171, 175, 181, 182, 269, 280
Anderson, D. A., 72
Angus, J. C., 61, 105
Antippa, A. F., 281
Aoki, T., 97
Armstrong, S. R., 2, 10, 15
Arndt, D. P., 192, 239, 241
Arwin, H., 94, 158, 169, 182
Ashcroft, N. W., 129, 131, 134, 135, 155
Aspnes, D. E., 2, 4, 6, 9–19, 23, 26, 27, 32–37, 41–43, 50–53, 59, 94, 129, 139, 148, 150, 153, 155, 158, 169, 182, 51, 196–199, 201–203, 205, 206, 254, 262, 280, 287, 289, 303
Ayoub, G. T., 291
Azzam, R. M. A., 2, 5, 10, 42, 43, 50, 51, 55, 148, 150, 158, 169, 196, 203, 280, 281, 288, 305

B

Bacon, D. D., 129, 139, 155
Badzian, A. R., 105
Badzian, T., 105
Bagley, B. G., 103
Banse, B. A., 35
Barrera, R. G., 11
Barth, J., 280
Bashara, N. M., 2, 5, 42, 43, 50, 51, 55, 148, 150, 158, 169, 195, 203, 280, 281, 288, 291
Batra, I. P., 253
Baumeister, P., 237, 238–240
Beaglehole, D., 280
Bedeaux, D., 128, 129, 139, 140, 173
Benferhat, R., 8
Bennett, H. E., 199, 305
Bennett, J. M., 199, 305
Berkovits, V. L., 2
Bernland, L. G., 168
Bernstein, A. I., 195
Berreman, D. W., 282, 288
Besselov, V. N., 2
Beyler, C. A., 12, 32, 35
Bezuidenhout, D. F., 237
Bhat, R., 2, 16, 17, 27, 32–37
Bhattacharya, S., 161
Bieganski, P., 180
Bimberg, D., 40, 41
Binder, K., 253
Biryukov, S. V., 252
Black, J. R., 149
Blanco, J. R., 139, 150
Blattner, R. J., 149
Blayo, N., 45
Bloomer, I., 61
Bohm, D., 298
Bootsma, G. A., 280, 281
Borel, J.-P., 168
Borgogno, J. P., 192, 239, 240, 280
Born, M., 281, 287, 288
Bornstein, 237
Bos, L. W., 131
Bouilov, L. L., 104
Bouree, J. E., 38
Braundmeier, A. J., 149
Brendel, R., 149
Breuer, H. D., 142
Briones, F., 10, 20–22
Bruggeman, D. A. G., 41, 43, 59, 60, 72, 82, 97, 99, 107, 110, 139, 155–157, 206, 210
Buchan, N. I., 35
Buckel, W., 168
Buckman, A. B., 50

Buffat, P., 168
Buhaenko, D. S., 32
Buhrman, R. A., 113
Bullot, J., 61, 85, 87

C

Cadoret, R., 52
Canillas, A., 45, 280
Cardona, M., 10, 73, 280
Cavese, J. M., 72
Chadi, D. J., 21
Chang, Y. C., 2, 9, 11, 16–18, 19, 23, 32
Charmet, J. C., 280
Chelikowsky, J. R., 102
Chen, J., 269
Chen, Q., 12, 32, 35
Chin, F. K., 148
Chindaudom, P., 199, 206, 213, 254, 263
Cinader, G., 41
Clarke, K. D., 237
Clausen, E. M., Jr., 241
Cody, G. D., 73, 75
Cohen, M. H., 139
Cohen, M. L., 102
Colas, E., 2, 27, 32–34, 36, 37
Collins, R. W., 50–52, 54–56, 72, 94, 103, 104, 110–115, 129, 135, 148, 152, 154, 157, 159, 160, 163, 165–167, 171, 175, 176, 181, 182, 196, 203, 205, 206, 280, 310
Cong, Y., 106, 108, 116
Cook, W. R., Jr., 250
Craighead, H. G., 129
Creighton, J. R., 35
Cross, L. E., 253
Cross, P. C., 299
Croteau, A., 250

D

Dalby, J. L., 6
Dapkus, P. D., 2, 12, 24, 25, 28, 32, 35
Dawson, R. M. A., 83
Debe, M. K., 281, 293
de Boeij, 2
Decius, J. C., 299
Defour, M., 2, 9, 29–31, 38, 41

de Gennes, P. G., 280
de Kluizenaar, E. E., 281
Del Sole, R., 2
Demchishin, A. V., 194, 216
Den Baars, S. P., 12, 32, 35
den Engelsen, D., 281, 293
de Nijs, J. M. M., 193, 203
Deppert, K., 2, 23, 26–29
Derjaguin, B. V., 104
De Smet, D. J., 281, 282
Deutscher, G., 161
Dey, S. K., 250
Dignam, M. J., 281
Dobierzewska-Mozrzymas, E., 180
Dobson, P. J., 168, 173, 179
Doremus, R. H., 139
Downs, M. J., 199, 238, 254
Drévillon, B., 2, 4, 6, 8, 9, 29–31, 38, 41, 45, 52, 280, 287
Druilhe, R., 38
Ducharme, D., 281
Dudkevich, V. P., 251, 253
Duncan, W. M., 45
Dzurko, K. M., 12, 32, 35

E

Edlinger, J., 195
Egan, W. G., 262
Elsharkawi, A. R., 281, 293, 294
Elshazly-Zaghloul, M., 281, 305
Emiliani, G., 192
Ennos, A. E., 264

F

Faraday, M., 127
Farmer, J. C., 52
Farrell, H. H., 2, 9, 11, 16–19, 21, 23
Feng, G., 52, 103
Feng, G. F., 102, 135
Field, D. R., 281, 293
Fischer-Colbrie, A., 40, 41
Florez, L. T., 2, 6, 9, 11–19, 23, 26, 27, 32–37
Flory, F., 195
Forouhi, A. R., 61
Forsbergh, P. W., Jr., 253

Fox, G. R., 251, 265, 267, 272
Foxon, C. T., 15
Fragstein, C. v., 134
Francis, S. M., 32
Fuchs, B. A., 210

G

Gallagher, A., 72, 76, 80
Gaylord, T. K., 282, 288
Giannelis, P., 263
Ginzburg, V. L., 10
Giri, A. P., 194, 195, 230
Gizeman, O. L. J., 280, 281
Gleason, K. K., 93
Gmitter, T. J., 2
Gottesfeld, S., 50
Goulding, P. A., 32
Granqvist, C. G., 60, 113, 129, 139, 180
Graves, R. H. W., 281
Gregory, S., 51
Grigorovici, R., 73
Grundmann, M., 40, 41
Guenther, K. H., 194, 195, 230, 243, 251
Guha, S., 61
Gururaja, R., 272

H

Habraken, F. H. P. M., 280, 281
Haertling, G. H., 259
Halford, J. H., 148
Hall, A. C., 50
Hallais, J., 31, 56
Hamakawa, Y., 250
Hanbucken, M., 70
Harbison, J. P., 2, 6, 9, 11–19, 21, 23, 26, 27, 32–37
Hariz, A., 12, 32, 35
Harris, E. P., 264
Hauge, P. S., 264
Hayashi, Y., 116
Hayman, C. C., 61, 105
Henck, S. A., 45
Henon, S., 281
Herrmann, R. I., 195

Hezel, R., 149
Hodgkinson, I., 281
Horikoshi, Y., 10, 12, 15, 19–22
Hottier, F., 31, 52, 56, 59, 196, 203, 280
Hren, J. J., 241
Hu, H., 265
Hunderi, O., 60, 129, 139, 168, 180

I

Ibrahim, M. M., 280
Iijima, K., 250
Inokuti, M., 131, 133, 153, 155

J

Jacobs, J. T., 253
Jaffe, B., 250
Jaffe, H., 250
Jarrett, D. N., 141
Jasperson, S. N., 6
Jellison, Jr. G. E., 199, 201, 212
Jeong, W. G., 12, 32, 35
Jeppesen, M. A., 257
Jeppesen, S., 2, 26, 29
Johnson, R. L., 280
Jones, M. L., 169, 180, 281, 291
Jönsson, J., 2, 23, 26–29
Josserand, Y., 8
Joyce, B. A., 10, 15
Junno, B., 23, 27, 28
Jushida, K., 253

K

Kamiya, I., 2, 15–17, 27, 32–36
Kamo, M., 104
Kanicki, J., 61
Kao, K. C., 281, 293, 294
Kar, S., 161
Kawashima, M., 12, 15
Keddie, J. L., 263
Kelly, M. K., 2, 11, 15, 16, 26
Keramidas, V. G., 2, 36
Keuch, T. F., 35
Khasieva, R. V., 2

Kikuchi, M., 93
Kim, H. G., 250
Kim, S. Y., 206, 216, 280
Kim, Y.-T., 52, 54, 55, 129, 148, 280
Kinbara, A., 128, 139, 140
King, R. J., 199, 238, 254
Kinsbron, E., 129, 139, 155
Kircher, C. J., 264
Kisilev, V. A., 2
Kizel, V. A., 199
Klazes, R. H., 73
Kleeman, W., 259
Klug, D. A., 10
Kobayashi, N., 12, 19
Koch, S. M., 2, 9, 29–31, 38, 41
Kodami, K., 35
Konstantinova, F., 281
Koza, M. A., 2, 36
Krasilov, Y. I., 199
Kreibig, U., 102, 128, 129, 134, 142, 178
Krishnan, S., 168
Kruangam, D., 61
Krupanidhi, S. B., 251, 265, 272
Kuang, A. X., 250
Kubo, H., 51
Kumar, V., 265
Kurabayashi, T., 35
Kurita, S., 61, 105
Kushida, K., 260
Kuwano, Y., 61

L

Laibinis, P. E., 282, 288, 310
Land, C. E., 259
Landolt, 237
Lastras-Martinez, A., 9
Laurence, G., 56
Lautenschlager, P., 99, 100, 107, 109
Law, B., 280
Lazarides, B., 192, 239, 240, 280
Leblanc, R. M., 281
Lee, H. Y., 250
Lekner, J., 286, 305
Li, Y. M., 54, 66, 68, 69, 75, 78–81
Lienert, 40, 41
Lin, C., 281
Lin-Chung, P. J., 282, 288

Lohner, T., 233
Loposzko, M., 250
Lu, Y., 93
Lucovsky, G., 77
L'vova, T. N., 2
Lynch, D. W., 131

M

Maa, B. Y., 2, 12, 24, 25, 28, 32, 35
Macleod, H. A., 251, 281
Mahan, A. H., 93
Majumder, K. S., 149
Makimoto, T., 12, 19
Maley, N., 52, 103
Malin, M., 288
Malitson, I. H., 59, 148, 206, 207–209, 212, 214, 217, 236, 238, 241, 257
Manghi, F., 2
Marbot, R., 6
Martin, P. J., 195
Masetti, E., 238, 240
Mathewson, A. G., 153, 155
Matsuda, A., 76, 80
Matsumoto, S., 104
Maxwell-Garnett, J. C., 43, 127, 139–141, 160
Mayadas, A. F., 135, 164
Mazor, A., 76, 82
McMarr, P. J., 196, 205, 232, 280
Mell, H., 75
Meessen, A., 128, 139
Meservey, R., 165
Messier, R., 61, 105, 194, 195, 230
Meunier, J., 281
Meyer, F., 281
Miller, J. C., 167, 168
Miller, J. N., 40, 41
Mochan, W. L., 11
Mochizuki, K., 35
Modine, F. A., 212
Molenbroek, A. M., 2
Molinary, E., 2
Moritani, A., 51
Moskovits, M., 281
Movchan, B. A., 194, 216
Mui, K., 73
Muller, A. B., 41
Muller, R. H., 50, 52, 280, 284, 287
Myers, H. P., 153, 155, 168

N

Naegele, K., 281
Nagata, H., 250
Nagatomo, T., 250
Naka, S., 252
Nakai, J., 51
Nakanisi, T., 27
Narayan, J., 196, 205, 232, 233, 280
Narayanamurthy, V., 281
Neal, W. E. J., 280
Neave, J. H., 10, 15
Neugebauer, C. A., 139, 149
Newnham, R. E., 249, 272
Nguyen, H. V., 54, 94–96, 100, 101, 103, 129, 135, 152, 154, 157, 159, 160, 163, 165–167, 171, 175, 176, 181, 183, 280, 310
Nguyen, N. V., 56, 148, 280, 310
Nigara, Y., 240
Niklasson, G. A., 60, 129, 139
Nishizawa, J., 35
Nordine, P. C., 168
Norman, J. E., 148
Norrman, S., 128, 173
Novikov, E. B., 2

O

Ogura, S., 238
Ohtsuka, N., 35
Okamura, T., 250
Okuyama, M., 250
Oldham, W. G., 58, 59
Omnes, F., 2, 9, 29, 30, 38, 41
Omoto, O., 250
Oron, M., 41
Osborn, J. A., 140
Ossikovski, R., 45
Oxley, A. E., 199
Ozeki, M., 35

P

Paget, D., 2
Panish, M. B., 281
Pappas, D. L., 105
Parey, J. Y., 8

Parikh, A. N., 258, 282, 288, 290
Parsons, R. R., 139
Pascual, E., 45
Pastrnak, J., 10
Paul, W., 72
Paul Hed, P., 210
Paulsson, G., 2, 23, 26–29
Payne, D. A., 250
Pelletier, E., 192, 239, 240, 280
Pemble, M. E., 2, 10, 15, 32
Perrin, J., 6
Peterson, L. D., 21
Petrich, M. A., 93
Petroff, P. M., 15, 32
Phillips, J. C., 73
Philpott, M. R., 296
Pickrell, D., 105
Piegari, A., 192, 238, 240
Plieth, W. J., 281
Poppe, G. P. M., 2
Pulker, H. K., 195

Q

Quinn, W. E., 51

R

Radosz, A., 180
Raizman, A., 41
Rao, B. S., 169, 180, 281, 291, 293, 294
Razegi, M., 2, 9, 29–31, 38, 41
Reimer, J. A., 93
Richter, W., 41
Roberts, S., 168, 173, 179
Robertson, R., 76, 80
Rosetti, G. A., Jr., 253
Roy, R. A., 194, 195, 230
Russev, S. H., 203
Ryan, R. E., 9, 15, 19
Rytz, D., 259

S

Safarov, V. I., 2
Sakurai, N., 35

Sales, B. C., 199, 201, 212
Sallet, V., 38
Samuelson, L., 2, 23, 26-29
Sayer, M., 250
Scaglione, 240
Scala, C., 139
Schantterly, S. E., 6
Schatzkes, M., 135, 164
Schlayen, B., 8
Schmidt, M. P., 61, 85, 87, 93
Schmidt, P., 2, 26, 29
Schmitz, B., 142
Schoijet, M., 252
Scholz, S. M., 41
Scott, G. D., 128
Scott, J., 72
Seddon, R. I., 195
Segall, B., 131, 133, 153
Selloni, A., 2
Sen, P. N., 139
Sennett, R. S., 128
Shafer, F. J., 259
Shamraev, V. N., 199
Sherrill, F. A., 252
Shiles, E., 131, 133, 153, 155, 156
Shirata, T., 15
Shtrikman, H., 41
Silverman, B. D., 253
Simondet, F., 31
Smith, C. G., 280
Smith, D., 237-240
Smith, D. Y., 131, 133, 153, 155
Smith, F. W., 73
Snyder, P. G., 280
So, S., 288
Solomon, I., 85, 93
Soonpaa, H. H., 169, 180, 281, 291
Souda, R., 35
Spiller, G. D. T., 70
Spitsyn, B. V., 104
Sreenivas, K., 250
Starke, A., 195, 238, 239
Stchakovsky, M., 8
Stephans, R. E., 217
Stern, F., 296
Stobie, R. W., 281
Stoner, E. C., 140
Strausser, Y. E., 149
Studna, A. A., 2, 4, 6, 9-19, 23, 26, 36, 59, 148, 150, 196-199, 201-203
Sturm, K., 129, 131, 134, 155
Sudoh, A., 128, 129, 139, 140
Sugiura, H., 15
Sullivan, B. T., 139
Surowiak, Z., 250

T

Takagi, M., 168
Takahashi, H., 128, 129, 139, 140
Tamargo, M. C., 2
Tanaka, H., 2, 16-18, 27, 32-36
Tanaka, K., 76, 80
Tauc, J., 73
Taylor, A. G., 2, 10, 15
Tedrow, P. M., 165
Teitler, S., 282, 288
Tesar, A. A., 210
Thacher, P. D., 259
Theeten, J. B., 50, 51, 59, 196, 203, 280
Thompson, M. J., 61
Thornton, J. A., 195
Tirabassi, A., 238, 240
Tran-Quoc, H., 85, 93
Triboulet, R., 38
Trolier-McKinstry, S., 249, 254, 263, 267, 272, 273, 276
Tronin, A. Y., 281
Tsai, C. C., 93
Tsarenkov, B. V., 2

U

Udayakumar, K. R., 249, 250, 253
Ulman, A., 281
Urban, F. K., 195
Usui, S., 93

V

Vancu, A., 73
van Silfhout, A., 2, 193, 203
Van Vechten, J. A., 73
Varghese R., 161
Vedam, K., 10, 52, 54, 55, 129, 148, 196, 199,

205, 206, 213, 216, 232, 254, 263, 280, 288
Venables, J. A., 70
Veprek, S., 80, 93
Violet, A., 6
Vlieger, J., 128, 129, 139, 140, 173
Vorlicek, V., 73

W

Ward, L., 141
Wassermeier, M., 15, 32
Weaver, A. L., 2
Wei, Y., 281
Weis, R. S., 282, 288
Weiser, G., 75
Wentik, D. J., 2
Westwood, D. I., 41
Wijers, C., 2
Wilkens, B. J., 9, 15, 19
Williams, R. H., 41
Wilson, E. B., 299
Wirick, M. P., 238
Wittels, M. C., 252
Wolfe, E., 281, 287, 288
Wood, D. M., 135
Woolf, D. A., 41
Woollam, J. A., 280
Wooten, F., 132
Wormeester, H., 2
Wu, L. S., 250
Wu, Q.-H., 281
Wu, Z., 250

Wurfel, P., 253

Y

Yagil, Y., 161
Yakovlev, V., 38
Yamaguchi, H., 12
Yamaguchi, S., 128, 139
Yamaguchi, T., 128, 129, 139, 140
Yamaka, E., 250
Yamauchi, Y., 12, 19
Yang, B., 195, 196, 216
Yang, B. Y., 52, 103
Yarbrough, W. A., 61, 105, 106
Yariv, A., 281, 310
Yeh, P., 258, 281, 282, 287, 288, 290, 310
Yi, G., 250
Yoon, S.-G., 250
Yoshida, S., 128, 139, 140
Yu, M. L., 35
Yu, Y. P., 10

Z

Zaghloul, A. R. M., 288
Zahn, D. R. T., 41
Zallen, R., 102, 135
Zhang, J., 10, 15
Zhou, Q. F., 250
Zhu, R., 281
Zuleeg, R., 250

Subject Index

A

Achromatic quarter-wave plate, 199, 254
Aluminum
  optical functions of bulk aluminum, 131–33
  optical functions of thin films, 150–83
Aluminum fluoride AlF_3
  microstructure and optical functions, 222, 227, 229
Aluminum oxide $\alpha\text{-}Al_2O_3$, 223, 235, 237, 241
  aluminum oxide $\gamma\text{-}Al_2O_3$
  microstructure and optical functions, 223, 227, 235, 238, 241
Aluminum thin films, 127–87
  nucleation and growth, 150–62
  optical functions of continuous films, 162–68
  optical functions of particle films, 168–83
  particles as nanocrystallite clusters, 179
Amorphous silicon, hydrogenated a-Si:H, 61–84
  optical functions of (a-Si:H), 70–73
  optical functions of thin films, 70–73
  during nucleation and growth, 70–73
Anisotropic materials, 258, 279–312
  anisotropic surface layer, 2–41
  reflectance anisotropy studies, 2–41
  anisotropic thin films, 309–312
  Brewster's angle for anisotropic substrates, 304
  spectro-ellipsometry of, 279–312
  simulation of SE spectra of, 295–312
  simulation of optical function tensors, 283–312
Anisotropic samples
  anthracene on glass, 292
  analysis methods, 281
  approximate ellipsometric models, 281
  bismuth tellurium sulfide crystal, 292
  biaxial, 281
  examples, 281
  surfaces and thin films, 281, 296
  uniaxial, 281, 304

B

Band gap of
  hydrogenated amorphous silicon a-Si:H, 73–75
  nucleating clusters of a-Si:H, 73–75
$BaTiO_3$ films on $SrTiO_3$
  microstructure and optical functions, 263
Bismuth tellurium sulfide crystal, 291
Brewster's angle, 301–302, 304
Bruggeman effective medium theory, 41–42, 59–61, 72, 107, 110, 153, 156, 162, 177, 206, 257

C

Cerium fluoride CeF_3
  microstructure and optical function, 221–222, 237
Compensator. See Achromatic quarter-wave plate
Cross-section transmission electron microscopy (XTEM), 232

D

Diamond film growth studies by RTSE, 105
  analysis of microstructural evolution, 108
  contamination by tungsten and its minimization, 108
  nucleation and growth using heater-filament method, 104
  Raman spectroscopy, 105, 111
  substrate treatment and annealing, 105
  true substrate temperature evaluated by RTSE, 108–09
Dispersion of refractive index. See Optimal Function

E

Effective medium theory (EMT), 59, 99, 153, 162, 177, 206

Effective medium theory —*cont.*
  Bruggeman effective medium theory, 41–43, 59, 107, 110, 153, 156, 162, 177, 206, 257
  generalized Maxwell-Garnett effective medium theory, 139–40, 177
  Lorentz-Lorenz theory, 156
  Maxwell-Garnett theory, 43, 127–28
Electric transition dipole moment vector, 295
  azimuthal angle, 299
  tilt angle, 299
Electromagnetic theory, of multilayered samples, 282
  4×4 transfer matrix methods
    Berreman's treatment, 282, 289
    Yeh's treatment, 282, 290
  general reflection ellipsometry experiment, 283
Electron mean free path, 133–39, 162–68, 176–78, 185–86
Ellipsometric parameters, 286, 289
Ellipsometry of Langmuir film of palmitic acid on water surface 6(18)
  of anthracene film on borosilicate glass substrate, 292
  of bismuth tellurium sulfide crystals, 291–92
  of metal-free pthalocyanine film on copper substrate, 293
Ellipticity, 285

F

Ferroelectric films, 249–76
  inhomogeneities in films, 250–75
  microstructure-electrical properties correlation, 272–75
  SE studies on ferroelectric films
    $BaTiO_3$ film on $SrTiO_3$, 263
    $(Pb, La)TiO_3$ film on sapphire, 256–61
    $PbTiO_3$ film on $SrTiO_3$, 260–62
    $(Pb, Zr)TiO_3$ film on sapphire, 264–69
    $(Pb, Zr)TiO_3$ film on Pt coated Si, 269–70
  size effect mechanisms, 250–53, 259
Films prepared by
  electron-beam evaporation (EB), 215
  evaporation of aluminum in high vacuum, using heated tungsten wire, 148
  ion-assisted deposition (IAD), 195, 239–44
  ion-beam sputtering (IBS), 240
  ionized cluster beam (ICB), 195
  magnetron sputtering, 83, 148
  metal organic chemical vapor deposition, 263
  molecular beam epitaxy (MBE), 50
  multi-ion-beam reactive sputtering (MIBERS), 254, 264
  plasma enhanced chemical vapor deposition (PECVD), 50, 62–63, 77–79
  reactive low-voltage ion plating (RLVIP), 195
  sol-gel technique 254, 260–61
Fresnel
  coefficient, 285
  reflectance matrix, 286

G

Gallium Arsenide
  atomic layer epitaxy (ALE) as studied by reflectance anisotropy, 35–38
  RA study of growth under atmospheric pressure AP MOCVD, 32–38
  RA study of growth under chemical beam epitaxy (CBE), 22–29
    growth oscillations under CBE conditions, 26–27
    in situ control of V/III ratio, 27–29
  RA study of growth under MBE conditions, 12–19
  RA study of growth under migration-enhanced atomic layer epitaxy (MEALE), 19–22
  RA study of growth under vacuum chemical epitaxy (VCE), 22–29
    surface transformation between As- and Ga-stabilized surfaces, 22–24
    in situ control of V/III ratio, 27–29
  surface reconstructions as studied by reflectance anisotropy, 32–34

Growth of lattice-matched multilayered structures by LP-MOCVD, 29–31
 RA study of GaInAs/GaAs/GaAs wells, 29–31
 RA study of GaInP/GaAs strures, 29–31
Growth of lattice-mismatched III–V and II–VI structures, 38–43
 RA study of CdTe/GaAs, 38–43
 RA study of InAs/InP, 38–43
 RA study of InP/GaAs, 38–43
 RA study of ZnTe/GaAs, 38–43

H

Hafnium fluoride HfF_4
 microstructural and optical function, 222, 229
Hafnium oxide HfO_2
 microstructure and optical function, 223, 225, 239
Hydrogenated amorphous silicon, (a-Si:H), 61–84
 optical function of (a-Si:H), 70–73
 optical function of thin films, 70–73
 during nucleation and growth, 70–73
Hydrogenated amorphous silicon–carbon alloys a- $Si_{1-x}C_x$:H, 85
 optical gaps from real time observations, 85–88
 role of H_2 dilution of reactive gases, 88–93

I

Inhomogeneity in thin films, 193–95, 250–51
Inhomogeneity transparent films, 193–95, 250–51
 optical characterization, 191, 249
 optical functions of ferroelectric films, 261, 268
 optical functions of fluoride films, 236
 optical functions of oxide films, 236
In situ SE studies on annealing of ferroelectric films, 264–70

K

Kramers-Kronig transformation, 296, 297

L

Langmuir film of palmitic acid on water surface, 291–92
Lanthanum Fluoride LaF_3 film
 microstructure and optical function, 222, 224, 238
(Lead, lanthanum) Titanate film (PLT)
 microstructure and optical functions, 258–59
(Lead, lanthanum) zirconate titanate film (PLZT)
 microstructure and optical functions, 259–60
(Pb, La)TiO_3 film on sapphire
 microstructure and optical function, 258–59
Lead titanate film $PbTiO_3$ (PT)
 microstructure and optical functions, 260–62
$PbTiO_3$ film on $SrTiO_3$
 microstructure and optical function, 260–62
(Pb, Zr)TiO_3 film on sapphire
 microstructure and optical function, 264–68
(Pb, Zr)TiO_3 film on Pt-coated Si
 microstructure and optical function, 269–70
Linear regression analysis (LRA), 58, 62, 64, 97, 146, 153–56, 169, 205

M

Magnesium oxide MgO
 microstructure and optical function, 217
Microcrystalline silicon μc-Si:H, 93
 optical and microstructural analysis, 94
 optical function, 94–104
 size effect in microcrystallites, 99

Model anisotropic medium, 295
  Gaussian oscillators, 295
Monitoring nucleation and growth of films by RTSE, 50–121, 127–87
  Aluminum film, 147
  hydrogenated amorphous silicon, a-Si:H, 61

O

Optical anisotropy, 279, 280
  in films, 168–69, 173–74, 178, 257
Optical functions of
  aluminum fluoride AlF_3, 236
  aluminum in bulk, 131–33
  aluminum oxide $\alpha\text{-}Al_2O_3$, 238, 241
  aluminum oxide $\gamma\text{-}Al_2O_3$, 236, 238, 241
  aluminum thin films, 150–83
    during nucleation and growth, 150–62
    liquid aluminum film, 167–68
    particle films, 168–83
  amorphous silicon, hydrogenated a-Si:H, 70–73
  anisotropic material-tensors-simulation, 298–300
  cerium fluoride CeF_3, 236, 237
  hydrogenated fluoride HfF_4, 236
  hafnium oxide HfO_2, 236, 239
  hafnium amorphous silicon a-Si:H, 70–73
  lanthanum fluoride LaF_3, 236, 238
  (Pb, La)TiO_3 film, 261
  $PbTiO_3$ film, 260
  liquid aluminum film, 167–68
  magnesium oxide MgO, 217
  microcrystalline silicon μc-Si:H, 94–104
  scandium fluoride ScF_3, 236, 243
  scandium oxide Sc_2O_3, 236, 239
  thorium oxide ThO_2, 236
  transparent fluoride films, 236–42
  transparent oxide films, 236–42
  vitreous silica, 214
  yttrium oxide Y_2O_3, 236, 240
  zirconium oxide ZrO_2, 236, 240
Optical gap of
  hydrogenated amorphous silicon a-Si:H, 73–75
  hydrogenated amorphous silicon–carbon alloys a-$Si_{1-x}C_x$:H, 85–88

Optical response functions, 287
  tensors, 287, 298

P

Percolation threshold in thin aluminum films, 161, 186
Polarized light
  linear, 284
  elliptical, 285
Pseudo-Brewster's angle, 301
Pseudo-dielectric function, 115

R

RA. See Reflectance anisotropy
RBS. See Rutherford Back Scattering
Real Time Ellipsometry at single wavelength, 50–51
Real Time Spectroscopic Ellipsometry (RTSE), 54–109, 148–87
  Aluminum films, 147–87
    optical functions of aluminum thin films in the nucleation and growth stages, 150–62
    optical functions of continuous films from real time observations, 162–68
    optical functions of particle films from real time observations, 168–83
  data collection and interpretation, 56
  diamond, 104
    analysis of microstructural evolution, 108
    substrate treatment and annealing, 105
    true substrate temperature evaluated by RTSE, 109
  Hydrogenated amorphous silicon a-Si:H, 61
    analysis of microstructural evolution, 62
    optical functions from real time observations, 70
    process–property relationship in a-Si:H, 75
  Hydrogenated amorphous silicon–carbon alloys a-$Si_{1-x}C_x$, 85
    optical gaps from real time observations, 85–88

SUBJECT INDEX

role of H_2 dilution of reactive gases, 88–93
instrumentation, 55, 69, 96, 108, 147
Microcrystalline silicon μc-Si:H, 93
  optical and microstructural analysis, 94
  size effects in microcrystallites, 99
Reflectance anisotropy (RA), 2–43
  applications, 11–43
    study growth of III–V semiconductors, 11–43
    study growth of GaAs and AlAs under MBE conditions, 12–19
    to migration-enhanced atomic layer epitaxy (MEALE), 19–22
    GaAs under chemical beam epitaxy (CBE) and vacuum chemical epitaxy (VCE) growth conditions, 22–29
  experimental, 8–10
  growth of GaAs under atmospheric pressure AP MOCVD, 32–38
    surface reconstructions, 32–34
    atomic layer epitaxy, 35–38
  growth of lattice-matched multilayered structures by LP MOCVD, 29–31
  growth of lattice-mismatched III–V and II–VI structures, 38–43
  theory, 3–8
Reflectance anisotropy spectroscopy (RAS). *See* Reflectance anisotropy
Reflectance difference spectroscopy (RDS). *See* Reflectance anisotropy
Refractive index and its dispersion with photon energy. *See* Optical function
Rotation matrix, 300
RTSE. *See* Real time spectroscopic ellipsometry
Rutherford back scattering (RBS), 232

S

Scandium fluoride film ScF_3
  microstructure and optical functions, 222, 230
Scandium oxide film Sc_2O_3
  microstructure and optical functions, 223, 226, 239, 241
Scanning electron microscopy (SEM), 232
SE. *See* Spectroscopic ellipsometer

Simulation of anisotropic effects in ellipsometry, 295
  variable angle experiments, 301
  variable ambient medium experiments, 308
  variable thickness measurements, 310
  spectroscopic experiments, 305–08
  orientation of crystal principle axis, 305
Simulation of SE spectra of anisotropic materials, 295–312
  for single-interface ambient–substrate system, 300–308
    single-wavelength, variable angle measurements, 301
    effects of oblique orientation of crystal principal axes
    with plane of incidence, 305
    simulations of general angle dependence, 301
    test of a common approximation, 302
    spectroscopic measurements, 305
    variation of the ambient medium, 308
  multiple-interface, ambient–film–substrate systems, 309
  of optical function tensors of anisotropic materials, 283–312
Spectro-ellipsometry, 279
Spectroscopic Ellipsometer (SE)
  achromatic quarter-wave plate, 198, 254
  analysis of SE data, 205–07
  film characterization of, 191, 249
  nonlinearity of detector, 202–05
  optical functions of
    aluminum films in the disordered state, 168–83
    hydrogenated a-Si, 70–75
    inhomogeneous transparent films, 192
    transparent fluoride films, 234
    transparent oxide films, 234
    aluminum in bulk, 131–33
    aluminum thin films, 150–83
      during nucleation and growth, 150–62
  optical multichannel analyzer detector. *See* Photodiode array detector
  photodiode array detector, 54
  rotating analyzer system, 196, 205
  rotating polarizer system, 55
Structure zone model, 194, 195
Surface roughness, 150, 153–56, 210–13, 221–27, 233

T

Tauc's method of determining the band gap of amorphous films, 73, 86
Thorium oxide film ThO_2
  microstructure and optical function, 223, 226, 236

V

Variable angle spectroscopic ellipsometry (VASE), 301
Vitreous silica, 207–15
  optical function, 213, 214
  surface roughness, 210–13

X

XTEM. *See* Cross-section transmission electron microscopy

Y

Yttrium oxide Y_2O_3 film
  microstructure and optical functions, 223, 231, 236, 240

Z

Zirconium oxide ZrO_2 film
  microstructure and optical functions, 223, 227, 236, 240